SAT-Based Scalable Formal Verification Solutions

Series on Integrated Circuits and Systems

Series Editor: Anantha Chandrakasan
Massachusetts Institute of Technology
Cambridge, Massachusetts

SAT-Based Scalable Formal Verification Solutions
Malay Ganai and Aarti Gupta
ISBN 978-0-387-69166-4, 2007

Ultra-Low Voltage Nano-Scale Memories
Kiyoo Itoh, Masashi Horiguchi and Hitoshi Tanaka
ISBN 978-0-387-33398-4, 2007

Routing Congestion in VLSI Circuits: Estimation and Optimization
Prashant Saxena, Rupesh S. Shelar, Sachin Sapatnekar
ISBN 978-0-387-30037-5, 2007

Ultra-Low Power Wireless Technologies for Sensor Networks
Brian Otis and Jan Rabaey
ISBN 978-0-387-30930-9, 2007

Sub-Threshold Design for Ultra Low-Power Systems
Alice Wang, Benton H. Calhoun and Anantha Chandrakasan
ISBN 978-0-387-33515-5, 2006

High Performance Energy Efficient Microprocessor Design
Vojin Oklibdzija and Ram Krishnamurthy (Eds.)
ISBN 978-0-387-28594-8, 2006

Abstraction Refinement for Large Scale Model Checking
Chao Wang, Gary D. Hachtel, and Fabio Somenzi
ISBN 978-0-387-28594-2, 2006

A Practical Introduction to PSL
Cindy Eisner and Dana Fisman
ISBN 978-0-387-35313-5, 2006

Thermal and Power Management of Integrated Systems
Arman Vassighi and Manoj Sachdev
ISBN 978-0-387-25762-4, 2006

Leakage in Nanometer CMOS Technologies
Siva G. Narendra and Anantha Chandrakasan
ISBN 978-0-387-25737-2, 2005

Statistical Analysis and Optimization for VLSI: Timing and Power
Ashish Srivastava, Dennis Sylvester, and David Blaauw
ISBN 978-0-387-26049-9, 2005

Malay Ganai
Aarti Gupta

SAT-Based Scalable Formal Verification Solutions

Malay Ganai
NEC Labs America
4 Independence Way
Princeton, NJ 08540
USA

Aarti Gupta
NEC Labs America
4 Independence Way
Princeton, NJ 08540
USA

Series Editor:
Anantha Chandrakasan
Department of Electrical Engineering and Computer Science
Massachusetts Institute of Technology
Cambridge, MA 02139
USA

ISBN 978-1-4419-4341-5 e-ISBN 978-0-387-69167-1

Printed on acid-free paper.

9 8 7 6 5 4 3 2 1

springer.com

Dedication

This book is dedicated to all those who continuously strive to produce better algorithmic and engineering solutions to complex verification problems.

Preface

"Engineering is the profession in which a knowledge of the mathematical and natural sciences gained by study, experience, and practice is applied with judgment to develop ways to utilize, economically, the materials and forces of nature for the benefit of mankind" —Engineers Council for Professional Development (1961/1979)

Functional verification has become an important aspect of the IC (Integrated Chip) design process. Significant resources, both in industry and academia, are devoted to bridge the gap between design complexity and verification efforts. SAT-based verification techniques have attracted both industry and academia equally. This book discusses in detail several latest and interesting SAT-based techniques that have been shown to be scalable in an industry context. Unlike other books on formal methods that emphasize theoretical aspects with dense mathematical notation, this book provides algorithmic and engineering insights into devising scalable approaches for an effective and robust realization of verification solution. We also describe specific strengths of the various approaches in regards to their applicability. This books nicely complements other excellent books on introductory or advanced formal verification primarily in two aspects:

First, with growing interest in SAT-based approaches for formal verification, this book attempts to bring various emerging SAT-based scalable verification techniques and trends under one hood. In the last few years, several new SAT-based techniques have emerged. Not all of these are covered by other books: Hybrid SAT Solver, Efficient Problem Representation, Customized SAT-based Bounded Model Checking, Verification using Efficient Memory Modeling, Distributed SAT and SAT-Bounded

Model Checking, Proof-based Iterative Abstraction, High-level Bounded Model Checking, SAT-based Unbounded Model Checking, and Synthesis for Verification Paradigm.

Second, and more importantly, due to the practical significance of these techniques, they are appropriate for direct implementation in industry settings. In this book, we describe how these techniques have been architected into a verification platform called *VeriSol* (formerly *DiVer*) which has been used successfully in the industry for the last four years. We also share our practical experiences and insights in verifying large industry designs using this platform.

We strongly believe that the techniques described in this book will continue to gain importance in the verification area, given that the verification complexity is growing at an alarming rate with the design complexity. We also believe that that this book will provide useful information about foundation work for future verification technologies.

The book expects the reader to have a basic understanding of formal verification, model checking and issues inherent in model checking. The book primarily targets researchers, scientists and verification engineers who would like to get an in-depth understanding of scalable SAT-based verification techniques that can be further improved. The book also targets CAD tool developers who would like to incorporate various SAT-based advanced techniques in their products. Currently, colleges do not emphasize adequately the algorithmic and engineering aspects of designing a verification tool. Such practices should be encouraged, as a good infra-structure is required to produce quality research. We strongly believe that this book will motivate such activities in the future.

Here is the outline of the book: With an introduction and background on current design verification challenges for model checking techniques in Chapters 1 and 2 respectively, we divide the rest of the book into five parts, each with 1-4 chapters. Part I describes the underlying infrastructure — efficient problem representation and SAT-solvers — to realize scalable verification algorithms. Parts II-IV describe SAT-based model checking algorithms for various verification tasks such as accelerated falsification, robust proof methods, and iterative abstraction/refinement, respectively. Part V gives detail of an industry tool *VeriSol* and several industry cases studies. It also covers future trends in SAT-based model checking such as, synthesis for verification paradigm, and high-level model checkers, to further improve the scalability.

We would like to express our deep gratitude to NEC Laboratories America, Princeton, NJ for providing the opportunities and the infrastructure to carry out the research, and Central Research Laboratories, Tokyo Japan for packaging and deploying our technology to the end-users. Individually we would like to thank Dr. Pranav Ashar for his numerous and valuable insights; Dr. Kazutoshi Wakabayashi and Akira Mukaiyama for stimulating us with the verification challenges that are encountered in an industry setting. We also acknowledge the support of other team members in building the tool: Dr. James Yang, Dr. Lintao Zhang, Dr. Chao Wang, and Dr. Pankaj Chauhan.

We also thank all those who are involved in the publication of the book, especially Carl Harris and his colleagues of Springer Publishing Company. Last, but not the least, we thank our families for their patience during the project.

Dr. MALAY K. GANAI
Dr. AARTI GUPTA

Princeton, New Jersey 2006

Contents

List of Figures

List of Tables

1 DESIGN VERIFICATION CHALLENGES

1.1 Introduction

Verification ensures that the design meets the specification and has become an indispensable part of a product development cycle of a digital hardware design. Cost of chip failure is enormous due to high cost of respins and delayed tape-out, resulting in loss of opportunity to launch product on time in a highly competitive market. With the increasing design complexity of digital hardware, functional verification has become increasingly on the critical path of the cycle [1], requiring expensive and time-consuming efforts, as much as 70% of the product development cycle. As per the 2002/2004 functional verification study conducted by *Collett International Research* (as reported by *EETimes.com* [2]), functional/logic flaws account for 75% causes for respins of more than two-thirds of IC/ASIC designs to reach volume. Of these 75% flaws, more than 80% are due to design errors and remaining are due to incorrect/incomplete specification, internal and external IPs. Market forces mandate scalable verification solutions and radical shifts in design methodology to overcome the difficulty in verifying complex designs. Not surprisingly, traditional "black-box" verification methodology is giving way to "white-box" verification methodology, where more than half of the engineers in the design team are verification engineers who are getting involved in the early phase of design and specification.

1.2 Simulation-based Verification

Conventionally, designs are verified using extensive simulation. A model of the design is built in software, to which small monitors are added. These monitors check for failures of the design assertions. Large numbers of input sequences, called *tests*, are applied to this model; these tests are generated by (possibly biased) random test pattern generators, or by hand. If for a given test the assertion is violated, the corresponding monitor enters a "violation" state, flagging the failure. The effectiveness of simulation sequences, i.e., the *test-bench,* is assessed using several coverage metrics: *code coverage* [3, 4], *tag coverage* [5], *event coverage* [6], and *state machine coverage* [7-10]. *Code-based coverage* includes statement, branch, sub-expression, and path coverage. *Tag coverage* evaluates the observability of possible incorrect evaluation represented as tags at the circuit outputs. *Event coverage* is measured by activating the coverage models on the event trace. *State machine coverage* is based on the number of distinct states visited and transitions occurred in an abstracted design.

A simulation-based verification approach is simple, scales well with design size and has traditionally been the *de facto* workhorse for functional verification. However, it cannot guarantee completeness of coverage and hence, design correctness. More disturbingly, for practical designs, the fraction of design space which can be covered by simulation is vanishing small; resulting in a significant probability of *respin severity bugs* undetected in the design even after substantial simulation efforts [11]. Besides diminished coverage, development and debugging of test-benches is a non-trivial time-consuming process, often mandating the verification team members to understand the design behaviors, features to be tested, and interpreting the simulation results.

1.3 Formal Verification

Formal Verification (FV) refers to mathematical analysis of proving or disproving the correctness of a hardware or software system with respect to a certain unambiguous specification or property. The methods for analysis are known as *formal verification methods*, and unambiguous specifications are referred as *formal specifications*. Formal verification complements simulation but with higher complexity of analysis, where mathematical analysis is done on an abstract model of the system, modeled using finite state machines or labeled transition systems[1]. Formal verification can provide complete coverage and can therefore, ensure design correctness with respect to the properties specified, without requiring any test-bench. However, one should not

[1] Other mathematical models also used are Petri nets, timed automata, hybrid automata, process algebra, operational semantics, denotation semantics, and Hoare's Logic.

construe the formal verification to produce a "defect-free" design as it is impossible to formally specify what defect-free really means. To reiterate, *formal verification can ensure the correctness of a design only with respect to certain properties that it is able to prove.*

Formal verification can be broadly classified into two methods: *Model Checking* and *Theorem Proving.* Model Checking [12] consists of a systematic exhaustive exploration of all states and transitions in a model. It is implemented using explicit or implicit state enumeration techniques on a suitably abstracted model, and proving or disproving the existence of "defect" states in the model. Theorem Provers [13, 14], on the other hand, use mathematical reasoning and logical inference to prove the correctness of the systems and often, require a "theorem prover guru" with substantial understanding of the system-under-verification.

1.3.1 Model Checking

Model checking is an automated technique and hence, more popular in the industry as an alternate verification strategy to simulation. The applications of model checking are typically classified as *equivalence checking* and *property checking.* In *equivalence checking*, a "golden model" is used as a reference model to check if the given implementation model has a "defect". In *property checking*, the correctness properties describing the desirable/undesirable features of the design are specified using some formal logic (e.g. temporal logic) and verification is performed by proving or disproving that the property is satisfied by the model. In this book, we will focus our discussion on efficient property checking techniques that make formal verification practical and realizable.

Model checking techniques, in practice, are inherently limited by the state-explosion problem, i.e., the fact that the number of states is exponential in the number of state elements (e.g. registers, latches) in the design. Model checking approaches are broadly classified into two, based on state enumerations techniques employed: *explicit* and *implicit* (or symbolic). Explicit model checking techniques [15, 16] store the explored states in a large hash table, where each entry corresponds to a single system state. A system with as few as a hundred state elements amounts to a state space with $\sim 10^{11}$ states. Understandably, model checkers need to pay special attention to scalability of the techniques used. Symbolic model checking techniques [17] store sets of explored states symbolically by using characteristic functions represented by canonical/semi-canonical structures, and traverse the state space symbolically by exploring a set of states in a single operation. Canonical structures such as Binary Decision Diagrams (BDDs) [18, 19] allow constant time satisfiability checks of Boolean expressions, and are used to perform symbolic Boolean function manipulations [17, 20-22]. Though these BDD-based methods have greatly improved scalability in comparison to explicit state enumeration techniques, by and large, they are limited to designs with a few hundred state holding elements, which is not even at the level of an individual designer subsystem. This is mainly due to frequent space-outs and severe performance degradation [23] as BDDs constructed in the course of symbolic traversal grow extremely large, and BDD size is critically dependent on variable ordering. Though several variable ordering heuristics to reduce BDD sizes have been proposed [24-27], in many cases BDDs are hard to optimize [28, 29]. Several variations of BDDs such as Free BDDs [30], zBDDs [31], partitioned-BDDs [32] and subset-BDDs [33] have also been proposed to target domain-specific application; however, in practice, they have not scaled adequately for industry applications.

In a quest for robust and scalable approaches, research has been heavily directed toward separating Boolean reasoning and representation. Boolean Satisfiability (SAT), which has been studied over several decades, has emerged [34] as a workhorse for Boolean reasoning primarily due to many recent advances in DPLL-style [35] SAT-solvers [35-41]. Efficient Boolean representation [42, 43] such as semi-canonical representations, that are simple and reduced, are also emerging [44] as a *de facto* structure due to their less sensitivity to variable ordering and compact representation compared to BDDs. SAT, together with efficient representation, have become a viable alternative to BDDs for model checking applications. This helped to make SAT-based symbolic model checking techniques both realizable and practical.

With emerging power of Boolean reasoning, various robust and scalable SAT-based techniques are simultaneously developed to

- *target* specific verification tasks such as falsification, proofs, and abstraction/refinements;
- *address* current design features such as embedded memories and multiple clocks domains;
- *address* complex specifications due to the presence of nested clocks;
- *overcome* limitation of computation resources of a single workstation.

We see a clear preference in verification communities: *BDD-based Methodists are becoming SAT-based Methodists.*

1.4 Overview

In this book, we discuss various SAT-based formal verification methods that we have developed and applied in an industry setting, especially to address the scalability and performance issues that have been major limitations in BDD-based methods. These techniques comprise robust Boolean reasoning [41], efficient problem representation [42, 45], accelerated bug finding techniques such as bounded model checking (BMC) [45, 46], proof techniques such as induction and unbounded model checking (UMC) [47], and improved abstraction and refinement approaches [48, 49]. These methods efficiently handle designs with complex features such as embedded memories [50, 51], and multiple clock-domains [52], with complex clocked specifications. These methods are implemented in our verification platform *VeriSol* (formerly *DiVer*) [53], and have been applied successfully in industry for the last four years (as of 2006) to verify large hardware designs. Specifically, *VeriSo*l drives the *Property Checker* in NEC's *CyberWorkBench* (CWB) [54, 55], a high level design and synthesis environment that automatically generates RTL (Register Transfer Level) designs and properties from high-level behavioral descriptions. Based on this verification platform, we share our practical experiences and insights in verifying large industry designs. We also use *VeriSol* as the primary model checking workhorse in our software verification platform, *F-Soft* [56].

Scalability of formal verification tools will always remain an open research problem, as design complexity continues to grow. In this book, we discuss the following two trends that have shown some potential in mitigating this problem: Synthesis-For-Verification, and High-level Model Checkers.

Synthesis-For-Verification (SFV) paradigm [57, 58] addresses generation of "verification aware" models to improve the effectiveness of verification techniques. In particular, we discuss how a high-level synthesis (HLS) tool can be guided within its existing infrastructure to obtain "verification-friendly" models that are relatively easy to model check. Such an approach also leverages off the various advancements in verification techniques as discussed in this book.

High-level model checkers [59-62] are applied at word-level models to cope with inherent limitations of formal techniques at the bit level — due to requirement of finite datapaths, inefficient translations into SAT, and loss of high-level design information. Thus, high-level model checkers have potential to scale up to industry designs. We discuss later how the performance of high-level model checkers can be improved using high-level information extracted from the high-level models, on-the-fly simplification and model transformations.

Rest of the chapter is organized as follows: In Section 1.5, we describe various verification tasks; in Section 1.6, we describe the design verification challenges; and in Section 1.7, we present the organization of the rest of the book.

1.5 Verification Tasks

In order to address scalability and handle large designs, model checking techniques are often divided into *Falsification*, *Proof methods* and *Abstraction/Refinement techniques*, based on three main tasks, i.e., finding counter-example or bugs, proving the correctness of the specification, and obtaining smaller models that facilitate verification, respectively. Each task is geared towards resolving the correctness of the given specifications. This is based on the observation that complete state space traversal, forward or backward, is unnecessary for resolving many specifications. Most of the earlier works on symbolic model checking did not scale well partly due to the "over-ambitious goal" to combine the proof and falsification methods, thereby, explore, and store the reachable state space to verify a given set of properties for a design. In the recent years, several specialized techniques have been developed for each of these tasks; each addressing the capacity and performance issues in verifying large designs. Moreover, the results obtained from one task can either be used by other tasks to speed up their performance, or can be mathematically reasoned about together to deduce the outcome. Moreover, as these tasks have complementary goals, one can easily instrument them to execute concurrently in a distributed environment.

Falsification

Falsification refers to verification methods where the goal is to show that the given property *does not hold* for the model and generate a counter-example trace that exposes the bug corresponding to the violation of the property. Falsification methods are also used for generating witnesses for desirable (reachable) properties arising from various coverage metrics (also used in simulation-based verification). Clearly, it is not necessary to explore all the reachable states to guarantee falsification. Furthermore, using efficient state space exploration techniques, storing of the reached states can be avoided to overcome the state explosion problem. Incorporation of such strategies in our SAT-based BMC framework [53], we made the falsification approach scalable for designs with as many as 10K state holding elements and one million gates. In discussion with actual hardware designers, it has been commonly found that they are less concerned with formal proofs of correctness of the designs than they are with finding bugs as early as possible; thereby making falsification one of the most popular verification tasks. Evidently, SAT-based BMC for falsification is now synonymous with model checking efforts in the industry.

Proof methods

Proof methods refer to verification techniques where the goal is to show that the specified property holds for the model. Proofs, as noted above, are not so useful to the designers directly. However, as falsification approaches like SAT-based BMC are incomplete — i.e., they require additional stopping criteria before the correctness of a property can be proved [66, 67] — proof techniques such as induction [67] and unbounded analysis [47, 68-71] can be used in determining the completeness threshold [46, 67, 72, 73] and thereby, guaranteeing the correctness of the property. One can use inductive invariants, externally provided or automatically generated [73], to obtain shorter completeness bound. Proof methods, based on unbounded analysis, typically store some representation of the explored state space. To address the scalability, it is desirable to have the following: compact and efficient state representation, efficient state exploration techniques, and good abstraction techniques. Semi-canonical representations such as reduced circuit graphs [42, 43] can be used for storing states [46, 47], and have shown better scalability than canonical structures such as BDDs and its variants. Further, SAT-based quantification techniques [47, 68, 71] are found to be promising for efficient exploration of state space. Proof-based abstraction methods have been shown to generate small abstract models that do not have spurious violations up to a bounded length. With several advancements, as discussed in this book, proof techniques are routinely applied to industry designs.

Abstraction/Refinement

Abstraction/Refinement refers to modeling technique [74-77], where the goal is to reduce the model by removing unnecessary details but preserve the validity of the property. Typically, an abstract model has additional behavior in comparison to a *concrete* (actual) model. Such an abstraction is called a *sound abstraction.* If the property holds on a sound abstract model, then it also holds on the *concrete* model. On the other hand, if the property does not hold on the abstract model, then either it could be a true violation in the concrete model or a spurious violation (also called a *false negative*). In the latter case, some of the behaviors (states, or state transitions) need to be removed from the abstract model so that the spurious violation is eliminated at the least. Such a process is referred to as *refinement.* Refinement tends to increase the model size as additional design details are added, and make it harder for the proof-methods in general. Proof-based abstraction techniques [48, 49, 77, 78] generate abstract models automatically with sufficient design details such that spurious violations are eliminated for a bounded depth. Many a times, proofs for the correctness of the properties are readily obtained on such abstract models without the requirement for expensive refinements.

Today's model checking algorithms comprise of the specialized techniques developed for each of the above tasks, and a tightly integrated environment where interplay of these tasks are exercised. Developing and integrating them have been quite a challenge to the verification community. We will discuss efforts in this direction that have made it possible to verify large scale industry designs.

1.6 Verification Challenges

As the model checking problem has high theoretical complexity, P-SPACE complete to be precise [12], a complete state space traversal may be unavoidable in the worst case. We, as verification community, should not be deterred by this complexity barrier, but rather look out for "intelligent" algorithms to solve the most common cases efficiently to make formal verification viable and alive. Challenges encountered during verification arise from design features, verification techniques and methodology used. In the following, we list specific challenges that are addressed to some extent in this book, and are in need of more research by the community.

1.6.1 Design Features

Here we discuss two design features – embedded memories and multiple clocks – that are quite common in current SoC designs. These have posed serious challenges to the hardware verification community.

Design with Embedded Memories

Embedded memories with multiple read and write ports are quite common and widely used in current designs. According to the Semiconductor Industry Association (SIA) Roadmap, as reported by *EETimes* [2], embedded memories comprise more than 70% of the SoC. These embedded memories on SoC support diverse code and data requirements arising from an ever increasing demand for data throughput in applications ranging from cellular phones, smart cards and digital cameras.

However, these embedded memories add further complexity to formal verification tasks, due to an exponential increase in state space with each additional memory bit. With explicit modeling of large embedded memories, the search frequently becomes prohibitively large to analyze. In this book, we discuss efficient memory modeling techniques that capture the memory semantics without explicitly modeling each memory bit. These techniques have significantly improved the scalability of falsification and proof methods.

Multiple Clocks

A continuing push for high performance and low power designs has led to several features that have become norms of System-on-Chip (SoC) design — the use of multiple clocks, clocks with arbitrary frequency ratios, multiple-phased clocks, gated clocks, and latches (opposed to flip-flops). An integrated solution to verify such multi-clock systems, combined with clocked specifications that exploits recent advancements in SAT-based BMC, has been quite challenging. We discuss such an effort involving efficient clock modeling and dynamic simplification schemes, and customized translations of clocked properties within our BMC framework.

1.6.2 Verification Techniques

Some verification techniques have shown better potential than others. We, as a research community, continuously strive to improvise such winning approaches. Here we discuss some of those approaches and challenges that require special attention.

Customization for Common Properties

Customizing verification tasks using dedicated algorithms for the most commonly specified properties is an effective strategy to scale verification. In various discussions with designers, it has been found that most of the specifications used are simple *safety* (also known as *assertion*) or *liveness* properties, with safety properties accounting for more than 90% of the the properties, and liveness accounting for most among the remaining 10% properties. The designers, in general, avoid complicated nesting of properties. In this book, we discuss property-specific customization of model checking algorithms for simple safety and liveness properties with limited nesting. We strongly believe that the scalability of a *generalized* model checking algorithms, i.e., to handle all type of nesting of properties, will always be an issue. Further, such an approach typically carries extra overhead when checking most commonly used properties. Thus, we focus on devising dedicated algorithms for specific properties that are in demand.

Efficient State Space Representation

Verification tasks such as proof methods require efficient representations for storing explored states, and efficient symbolic traversal schemes based on these representations. BDDs and its variants are sensitive to variable ordering and have a tendency to blow-up in memory. Unlike BDDs that are canonical, semi-canonical circuit graphs have shown great potential in making the various verification tasks scalable. In this book, we will discuss one such simplified and reduced circuit graph that has played a central role in improving scalability of various verification algorithms.

Distributed Environment

The computing power and memory available in a single computing resource is always a limitation for the verification tasks. Since the verification tasks have complementary goals, distributing these tasks and executing them concurrently over several computing resources can be done easily. However, distributing a single verification task over multiple distributed resources can be quite challenging. In this book, we discuss the distribution of one such task — falsification using SAT-based BMC — over a network of workstations in order to overcome the memory limitation of a single server.

High-level Verification

To cope with increasing design complexity and demand to reduce design cycle time, the focus has shifted towards supporting languages and specifications at a level higher than the RTL (Register Transfer Level), and methodologies for high-level design, synthesis and verification. At the Boolean-level of design representation, the formal verification techniques are quite mature and advanced. Unfortunately, we do not see a similar level of maturity and advancements in verification efforts at higher levels of abstraction. This is mainly due to higher theoretical complexity, and a wide engineering gap between theoretical and practical solutions at the higher levels. To reduce this gap, we discuss a framework for a high-level model checker and several novel techniques that extract high-level design information to make model checking relatively easier. As we shall see in the book, these techniques not only overcome the inherent limitations of Boolean-level model checkers, but also allow adaptation of state-of-the-art techniques that have made Boolean-level model checkers successful.

System-level Verification

Verifying system-level designs remains a challenge to the verification community. Though SAT-based formal methods have improved scalability over BDD-based methods, they lack the scalability required for system-level verification. Faced with twin dilemmas of diminished coverage through simulation and the computational complexity of complete system verification, semi-formal approaches have been proposed [63-65] to augment simulation with symbolic verification techniques to increase the coverage of the state space and thereby, improve the chances of finding design bugs. Although simulation will remain the primary workhorse for functional verification in the foreseeable future, yet formal verification techniques are being increasingly adopted by the design industry to overcome the limitations of simulation.

1.6.3 Verification Methodology

Design-For-Verification (DFV) methodology [80], i.e., exporting designer's intent to verification tools, has been quite effective in improving verification efforts. The underlying principle is to leverage designers' intent and knowledge to strengthen verification tasks by passing some of the burden to the designers. Such a methodology, however, requires reliable insight from the designers and cannot be automated in general. Here, we describe various emerging methodologies that can be automated, and their challenges.

Synthesis for Verification Paradigm

High-level synthesis (HLS) [54, 81], also known as *Behavioral Synthesis*, is a process of generating a structural RTL model from a system-level model (SLM) — algorithmic description written in high-level languages such as C/C++ or SystemC [82]. HLS is usually targeted at meeting three main design constraints: performance, area, and power (PAP). Functional verification, on the other hand, ensures that the designs at various levels of implementation such as SLM, structural RTL, and gate level netlist, meet the functional specifications. Most of the current research emphasis is on improving the scalability of model checking techniques and reducing the model size. However, not much attention is paid to the effects of various design-modeling entities that can severely limit the capabilities of such verification techniques. Observe that the design optimized for PAP, is often not optimized for verification, and therefore, adversely affects the performance and scalability of the verification tools. Thus, by separating the design optimized for PAP from the one optimized for correctness check, one can hope to reduce the verification burden. We discuss a new paradigm Synthesis-For-Verification (SFV) [57, 58] that involves synthesizing "verification-aware" designs that are most suitable for functional verification. Furthermore, we recommend that SFV methodology should be applied along with DFV methodology to obtain the full benefit of verification efforts. In other words, the design input to high-level synthesis (HLS) techniques should have verifiability features as proposed in the DFV methodology. Note, SFV paradigm requires support only from automated synthesis approaches, such as HLS, and can be easily automated, in contrast to DFV methodology which requires designers' reliable insights. We believe such a paradigm of generating "verification aware" models will improve the overall verification efforts, and we demonstrate its effectiveness with our experiments using our tools on industry designs.

Verification Procedure

Combining various verification tasks has been a formidable challenge for the verification community. There is no clear procedure that has been shown to work in all cases. Based on our practical experience, we describe various verification flows that attempt to combine several tasks, and that have been successfully applied to verify industry designs.

Specifications

Specification refers to some "correct" behaviors of the system that are intended. Obtaining complete specifications — i.e., expressing everything "correct" about the system — has been another challenge for ensuring a "defect-free" design; often requiring design and verification teams to collaborate as early as possible in the design phase. Standards such as PSL [83] and System Verilog Assertions [84] are promoting assertion-based methodology [85] to allow proliferation of assertion-ready designs and verification IP. The benefits of assertion-based verification methodology are well-recognized by designers due to the simplicity of specifying assertions to capture "worry" cases, hot spots and undesired behaviors of complex designs, as well as by the formal verification community due to its relative ease in analysis. As famously quoted by Einstein, "*Everything should be made as simple as possible, but not simpler*", assertion-based verification methodology is definitely one-step closer to the challenge, but definitely not an end. In this book, we discuss some of the efforts towards capturing the designers "intent" while generating the properties automatically.

Property Decomposition and Grouping

Design details in the cone-of-influence (COI) [86] of a single property can be significantly smaller compared to the entire design. However, verifying a single property at a time could be quite expensive due to repetitive search and exploration of overlapping state spaces. On the other hand, although verifying all properties together can take advantage of learning from previous overlapping search space, the model in the COI of all the properties can be quite large. Property decomposition — i.e., partitioning the property group into several small but overlapping property sub-groups, provides a trade-off between scalability and performance.

1.7 Organization of Book

In Chapter 2, we provide some basic background on bounded/unbounded model checking, unclocked/clocked specifications, and describe briefly the state-of-the-art DPLL style SAT solvers and various algorithmic advancements that are the key to the success of SAT-based formal methods. The structure of the rest of the book is described by the following diagram.

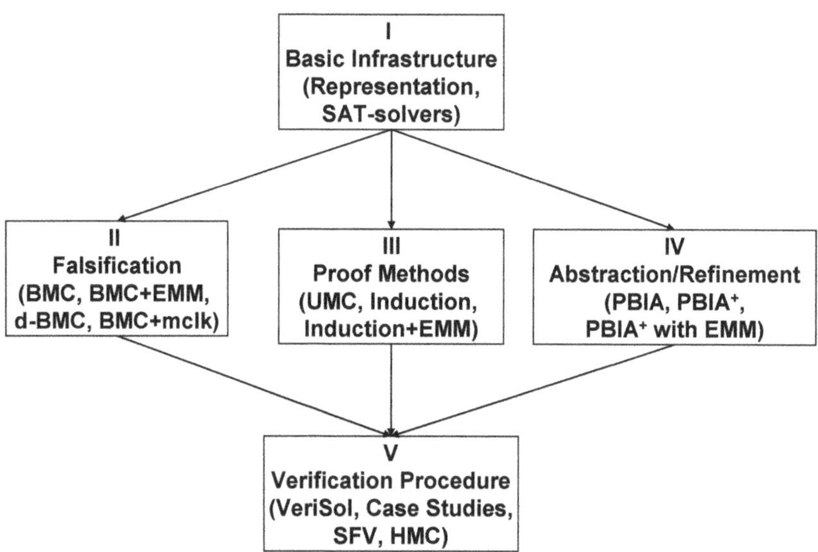

In Part I (Chapters 3—4), BASIC INFRASTRUCTURE, we discuss the elements of a basic infrastructure that are required to build scalable verification algorithms.

In Chapter 3, EFFICIENT BOOLEAN REPRESENTATION (Representation), we present a semi-canonical circuit graph representation and efficient on-the-fly compression algorithms to obtain a simple and reduced graph. This circuit graph is used to symbolically represent transition systems, model checking problem instances, explored states, and Boolean problems for efficient Boolean reasoning.

In Chapter 4, HYBRID DPLL-STYLE SAT SOLVER (SAT-Solvers), we discuss a hybrid SAT solver that combines the best of the SAT solvers of various application domains, and works directly on the semi-canonical circuit graph representation. When such a solver is augmented with incremental and BDD-based learning capabilities, it becomes ideally suited for SAT-based model checking applications.

We describe SAT-based techniques that utilize the efficient circuit graph representation and hybrid SAT solver for various verification tasks in Parts II-IV.

In Part II (Chapters 5—8), FALSIFICATION, we discuss various accelerated techniques to find bugs and witnesses.

In Chapter 5, SAT-BASED BOUNDED MODEL CHECKING (BMC), we present customized SAT-based Bounded Model Checking approach for solving simple safety and nested liveness properties. Customization involves property-specific problem partitioning, incremental formulation, and incremental learning. We also discuss various bug-finding accelerated techniques such as dynamic circuit simplification, SAT-based incremental learning, lightweight and effective BDD-based learning integrated with hybrid SAT solver.

In Chapter 6, DISTRIBUTED SAT-BASED BMC (d-BMC), we describe techniques to distribute hybrid SAT and SAT-based BMC over a network of workstations in order to overcome the memory limitation of a single server, for performing deeper search for very large industry designs. For the sake of scalability, at no point in the BMC computation does a single workstation have all the information. Such an approach scales well in practice, with acceptable communication cost.

In Chapter 7, EFFICIENT MEMORY MODELING IN BMC (BMC+EMM), we describe an abstract interpretation of embedded memory called Efficient Memory Modeling used in BMC. It provides a sound and complete abstraction that preserves the memory semantics, thereby, augmenting the capability of SAT-based BMC to handle designs with large embedded memory without explicitly modeling each memory bit.

In Chapter 8, BMC FOR MULTI-CLOCK SYSTEMS (BMC+mclk), we describe an integrated solution within the SAT-based BMC framework for verification of synchronous multi-clock systems with nested clocked specifications.

In Part III (Chapters 9—10), PROOF METHODS, we discuss various SAT-based proof techniques.

In Chapter 9, PROOF BY INDUCTION, (Induction, Induction+EMM), we describe induction-based proof techniques augmented with automatically generated inductive invariants such as reachability invariants. For embedded memory systems, we also describe techniques to model arbitrary initial memory state precisely and thereby, provide inductive proofs using SAT-based BMC for such systems.

In Chapter 10, UNBOUNDED MODEL CHECKING (UMC), we describe an efficient circuit-based cofactoring approach for SAT-based quantifier elimination that significantly improves the performance of pre-image and fixed-point computation in SAT-based UMC. We also describe customized formulations for determining completeness bounds for safety and liveness using SAT-based UMC, rather than using loop-free path analysis. These formulations, comprising greatest fixed-point and least fixed-point computations, handle nested properties efficiently using SAT-based quantification approaches.

In Part IV (Chapter 11), ABSTRACTION/REFINEMENT, we discuss various techniques for generating an abstract model that preserve the property correctness, including techniques specialized for design with embedded memories.

In Chapter 11, PROOF-BASED ITERATIVE ABSTRACTION (PBIA, PBIA$^+$, PBIA$^+$ with EMM)), we describe an iterative resolution-based proof technique using SAT-based BMC to generate an abstract model that preserves the property up to a bounded depth. We also describe use of lazy constraints and other SAT heuristics (PBIA$^+$) for further reducing the size of the abstract model and for improving verification of the abstract model. This technique when applied iteratively on the abstract model obtained in the previous step further reduces the abstract model size. Later, we describe techniques (PBIA$^+$ with EMM) to combine proof-based iterative abstraction with EMM techniques to identify irrelevant memory and ports.

In Part V (Chapters 12—13), VERIFICATION PROCEDURE, we describe our SAT-based verification framework, combination of strategies for various verification tasks, and a new paradigm to improve further the scalability of current verification framework.

In Chapter 12, SAT-BASED VERIFICATION FRAMEWORK (VeriSol, Case Studies), we discuss the verification platform *VeriSol* that has matured over 4 years (as of 2006) and is being used extensively in several industry settings. Due to an efficient and flexible infrastructure, it provides a very productive verification environment for research and development. We also analyze the inherent strengths/weaknesses of various verification tasks, and describe their interplay as applied on several industrial case studies, highlighting their contribution at each step of verification.

In Chapter 13, SYNTHESIS FOR VERIFICATION (SFV, HMC) we describe a Synthesis-for-Verification paradigm (SFV) to generate "verification friendly" models and properties from the given high-level design and specification. We explore the impact of various high-level synthesis (HLS) parameters on the verification efforts, and discuss how HLS can be guided to synthesize "verification aware" models. We also describe a framework for high-level BMC (HMC) and several techniques to accelerate BMC. Such techniques overcome the inherent limitations of Boolean-level BMC, while allowing integration of state-of-the-art techniques that have been very useful for Boolean-level BMC.

2 BACKGROUND

2.1 Model Checking

Model checking is a method to verify whether a model obtained from hardware or software system satisfies a formal specification, written as temporal logic formulas such as *Linear Time Logic* (LTL) [87] or *Computational Tree Logic* (CTL) [88, 89]. The model, capturing the behavior of the system, is expressed as a Kripke Structure as defined below.

Definition 2.1 A Kripke Structure is a quadruple $K=<S,I,T,L>$ where
S, is the finite set of states
$I \subseteq S$, is the set of initial states
$T \subseteq S \times S$, is the set of transitions
$L: S \rightarrow 2^V$, is the labeling function with variables set V such that $L(s)$ denotes the set of variables that hold true in state $s \in S$.

Kripke structures do not make any distinction between inputs, outputs or state variables, and transitions are not guarded. Such structures determine the set of computations as an unrolled tree structure from a root state.

Definition 2.2 An LTL formula φ in non-negated form (NNF) is expressed as

$\varphi = true \mid false \mid p \mid \neg p \mid \varphi_1 \wedge \varphi_2 \mid \boldsymbol{X}\, \varphi \mid \boldsymbol{F}\, \varphi \mid \boldsymbol{G}\, \varphi \mid [\varphi_1\, \boldsymbol{U}\, \varphi_2] \mid [\varphi_1\, \boldsymbol{R} \varphi_2] \mid [\varphi_1\, \boldsymbol{W}\, \varphi_2]$, where

p:	Atomic Proposition (AP)		
$\boldsymbol{X}$:	*Next* operator	$\boldsymbol{F}$:	*Eventually* operator
$\boldsymbol{G}$:	*Always* operator	$\boldsymbol{R}$:	*Release* operator
$\boldsymbol{U}$:	*Until* operator	$\boldsymbol{W}$:	*Weak Until* operator

The semantics of an LTL formula is interpreted over a computation path, i.e., sequence of states, $\pi = s_0, s_1, s_2,\ldots$ where for every $j \geq 0$, $s_j \in 2^{AP}$ is the subset of atomic propositions that hold in the jth position of π. Symbolically, we write $\pi |= f$ to denote that f is valid along the path π and is defined recursively as follows (*iff* is shorthand for *if and only if*, and $\pi^i = s_i, s_{i+1}, s_{i+2},\ldots$)

$\pi \models true$	$\pi \not\models false$	
$\pi \models p$	*iff*	$p \in s_0$,
$\pi \models \neg p$	*iff*	$p \notin s_0$,
$\pi \models \varphi_1 \wedge \varphi_2$	*iff*	$\pi \models \varphi_1$ and $\pi \models \varphi_2$
$\pi \models \boldsymbol{X}\varphi$	*iff*	$\pi^1 \models \varphi$
$\pi \models \boldsymbol{F}\varphi$	*iff*	$\exists i\ \pi^i \models \varphi$
$\pi \models \boldsymbol{G}\varphi$	*iff*	$\forall i\ \pi^i \models \varphi$
$\pi \models \varphi_1 \boldsymbol{U} \varphi_2$	*iff*	$\exists i\ \pi^i \models \varphi_2\ and\ \forall j, j< i\ \pi^j \models \varphi_1$
$\pi \models \varphi_1 \boldsymbol{R} \varphi_2$	*iff*	$\forall i\ \pi^i \models \varphi_2\ or\ \exists j, j< i\ \pi^j \models \varphi_1$
$\pi \models \varphi_1 \boldsymbol{W} \varphi_2$	*iff*	$\pi \models \varphi_1 \boldsymbol{U} \varphi_2\ or\ \pi \models \boldsymbol{G}\varphi_1$

The rewrite rules among ***F***, ***G***, ***R***, and ***U*** operators are as follows:

$\boldsymbol{F}\varphi \equiv true\ \boldsymbol{U}\ \varphi \equiv \neg \boldsymbol{G} \neg \varphi$

$\boldsymbol{G}\varphi \equiv false\ \boldsymbol{R}\ \varphi \equiv \neg \boldsymbol{F} \neg \varphi$

$\varphi_1 \boldsymbol{R}\ \varphi_2 \equiv \neg(\neg\varphi_1\ \boldsymbol{U} \neg\varphi_2)$

An LTL formula φ is *universally valid* in a Kripke Structure M, (symbolically, $M \models A\varphi$) *iff* for all paths π in M such that $s_0 \in I$, $\pi \models \varphi$. Similarly, an LTL formula φ is *existentially valid* in a Kripke Structure M, (symbolically, $M \models E\varphi$) *iff* there exists a path π in M, such that $s_0 \in I$, $\pi \models \varphi$. Checking universal validity of an LTL formula φ is same as checking $\neg\varphi$ is *not* existentially valid.

2.1.1 Correctness Properties

We will focus mainly on simple reachable, safety and liveness properties with limited nesting.

- A *reachable property*, denoted as ***EFp***, specifies that a particular state is reachable from present state.
- A simple *safety property*, denoted as ***AGp***, specifies that on all paths (***A***) of a system, globally (***G***) in each state of the path, the property *p* holds. Safety properties are used to express "nothing bad will take place in the system, ever".
- A *liveness property*, denoted as ***AFp***, specifies that on all paths (***A***), eventually *p* will hold in some state in the path. Liveness properties

express behaviors such as "something good will take place eventually, always" [90].

A counter-example to a safety property is a finite length execution trace, while that to a liveness property is an infinite length trace. In the presence of *fairness constraints* — i.e., constraints that a set of states should occur infinitely often along a path — a counter-example trace to a liveness property should also satisfy such fairness constraints. For a finite-state system, such an infinite trace counter-example consists of a loop on states where the "good" never happens and all fairness constraints are satisfied at least once. Safety properties include most common properties like deadlock detection, mutual exclusion, invariants, coverage (state/block/line) checking, and promptness requirements. Liveness properties, on the other hand, include subtle behaviors like starvation freedom, making progress, termination, guaranteed service and receiving service.

2.1.2 Explicit Model Checking

Model checking techniques based on explicit state enumeration such as *Murφ* [15, 91] and *SPIN* [16, 92] use an explicit representation of states and transitions in the system, and enumerate all reachable states explicitly. To overcome the state explosion, several state reduction techniques are employed such as symmetry detection, reversible rules, repetition constructors, partial order methods and incomplete techniques such as probabilistic verification. Other techniques such as parallelization, caching and using storage disks are also used to make the approach scalable. Explicit model checkers have serious limitations due to explicit enumeration of large state space and are largely unviable for verifying hardware designs.

2.1.3 Symbolic Model Checking

Symbolic model checkers such as *SMV* [17], use symbolic representation of the states using BDDs [18, 19] and traverse the state space symbolically using efficient symbolic algorithms [17, 20-22]. Core steps in symbolic model checking are the *image* and *pre-image* computations, which compute the set of states reachable in single step from/to a given set of states via the transition relations. These operations are shown below:

$$\begin{aligned} Img(\delta, X) &= \exists X, U.\ F(X) \wedge \delta(X, Y, U) \\ PreImg(\delta, Y) &= \exists Y, U.\ T(Y) \wedge \delta(X, Y, U) \end{aligned} \tag{2.1}$$

where the variable sets X, Y and U denote the present state, next state and primary input variables, respectively; and F, T, and δ denote the next states, the current states and the transition relation, respectively. The state sets are represented as BDDs, BEDs [43], or combination [68] of CNF clauses and BDDs, and symbolic computation such as conjoining and existential quantification are carried out on these representations. For large designs, these operations can cause a blow up in a memory size. Use of partitioned transition relation [32] and partitioned symbolic traversal [93] has been shown to improve scalability of these operations to some extent.

A basic algorithm *Model_check* for symbolic model checking simple safety properties can be formulated as shown in Figure 2.1. Let B be the bad states, in which the property p does not hold, and I the set of initial states. The forward reachability analysis starts from the initial states I and computes the image operations (*Img*) in successive states. If the bad states intersect with explored states so far, the algorithm returns *CEX* indicating "violation found" else it continues till it explores all reachable states without any violation, indicating a proof of correctness. Backward reachability analysis is similar to the forward analysis, except that the exploration starts backwards from the bad states and uses symbolic *preimage* computations.

```
Synopsis:  Model check simple safety property
Input:     Initial states I, Transition relation δ,
           Bad States B
Output:    CEX/Proof
Procedure: Model_check

R=F=N=I;
while (N ≠ ∅ ) {// fixed point?
  if (B ∩ N ≠ ∅ )
    return CEX; // Violation found
  F = Img(δ, N);
  N = R \ F;
  R = R ∪ N; //Forward reachable states
}
return Proof; // No Violation Possible
```

Fig. 2.1: Symbolic forward traversal algorithm for safety property

2.2 Notations

We describe the basic notations used in BDD-based and SAT-based model checking applications, described later.

Let V be a finite set of n variables over the set of Boolean values $B \in \{0,1\}$. A *literal* is a variable $v \in V$ or its negation $\neg v$. A *cube* c is a product of one or more distinct literals. A *minterm* over a set of m variables is a product of m distinct literals in the set. A minterm m is contained in a cube c, $m \in c$ iff $c \wedge m = m$. A *clause* is a disjunction of one or more distinct literals and is identified as a set of literals. In the sequel, clause and cube will be used interchangeably to represent the same set of literals. A cube, minterm, or clause are considered trivial if they have both v and $\neg v$. In the sequel, we consider only non-trivial cubes, minterms, and clauses. A *CNF formula* is a conjunction of clauses and is identified as a set of clauses.

A *Boolean function* f is a mapping between Boolean spaces f: $B^n \rightarrow B$. Let $f(v_1 \ldots v_n)$ be a Boolean function of n variables. We use $supp(f)$ to denote the support set $\{v_1 \ldots v_n\}$ of f. In the sequel, we will use function to denote a Boolean function. The positive and negative cofactors of $f(v_1 \ldots v \ldots v_n)$ with respect to a variable v are $f_v = f(v_1 \ldots 1 \ldots v_n)$ and $f_{\neg v} = f(v_1 \ldots 0 \ldots v_n)$, respectively. Existential (Universal) quantification of $f(v_1 \ldots v \ldots, v_n)$ with respect to a variable v is $\exists v\, f = f_v \vee f_{\neg v}$ ($\forall v\, f = f_v \wedge f_{\neg v}$). The cofactor f_c of f by a cube c is obtained by applying a sequence of positive (negative) cofactoring on f with respect to variable v ($\neg v$) in c. A function is a sum of minterms over its support variables and is identified with a set of minterms. The *onset* of a function f is a set of minterms m_i s.t. $f(m_i)=1$. A sum and product of functions over the same domain can be viewed as the union ($\cup$) and intersection ($\cap$) of their onsets, respectively. A function f is said to imply ($\Rightarrow$) another function g, if the onset of f is a subset ($\subseteq$) of the onset of g. In the sequel, we use $f \subseteq g$ to denote that f implies g.

For a finite state machine, let $U=\{u_1 \ldots u_k\}$ denote the set of k input variables and $X=\{x_1 \ldots x_m\}$ denote the set of m state variables, where $U \cup X = V$ and $U \cap X = \varnothing$. In the sequel, we denote a latch (state holding element) with a state variable. A state $s \in S$ (S is a finite set of states) is a minterm over X. A state cube (and a state clause) consists of only state variables. A characteristic function Ω_Q: $X^m \rightarrow B$ represents a set of states Q ($\subseteq S$), i.e., $Q = \{s \mid \Omega_Q(s)=1 \wedge s \in S\}$. An assignment is a function $\alpha: V_\alpha \rightarrow B$ where V_α ($\equiv dom(\alpha)$) $\subseteq V$. An assignment α_T is called *total* when $V_\alpha = V$; otherwise, it is called *partial*. An assignment α is a *complete satisfying assignment* for a formula f when f evaluates to 1 after applying $v=\alpha(v)$ for all $v \in V_\alpha$. We sometimes refer to a complete satisfying assignment as a *satisfying solution*. A satisfying solution is called *total* if the satisfying assignment is *total*, else called *partial*. We equate an assignment α with a cube that contains all v such that $\alpha(v)=1$ and all $\neg v$ such that $\alpha(v)=0$. We extend the domain of α to literals, i.e., by $\alpha(\neg v)=1$ we mean $\alpha(v)=0$. A *satisfying state cube* is $\alpha \downarrow X$ (projection of α onto state variables X) and a *satisfying input cube* is $\alpha \downarrow U$ (projection of α onto input variables U).

2.3 Binary Decision Diagrams

We discuss a popular graph-based representation called Binary Decision Diagrams (BDDs) [18, 19] used for representing Boolean functions. A BDD is a directed acyclic graph in which each vertex is labeled by a Boolean variable and has outgoing arcs labeled as "then" and "else" branches. Each arc leads either to another vertex or to a terminal labeled 0 or 1. The value of the function for a given assignment to the inputs is determined by traversing from the root down to a terminal label, each time following the then (else) branch corresponding to the value 1 (0) assigned to the variable specified by the vertex label. The value of the function then equals the terminal value.

In an ordered BDD, all vertex labels must occur according to a total ordering of the variables. In a reduced ordered BDD (ROBDD), besides the ordering criteria, two more conditions are imposed:

a) two nodes with isomorphic BDDs are merged, and
b) any node with identical children is removed.

These conditions make a reduced ordered BDD a canonical representation for Boolean functions. In the sequel, we will use the term BDD to refer to a ROBDD.

Researchers [24-27] have developed several heuristics to obtain a good variable ordering that produce compact BDD representation. Even though finding a good BDD variable ordering is not easy [28, 29], for some functions such as integer multiplication, there does not exist an ordering that gives a sub-exponential size representation [28].

We show the effect of two different variable orderings on the size of a BDD for a given function $f = a_1b_1 + a_2b_2 + a_3b_3$; (example from [19]). As shown in Figure 2.2(a), the BDD requires 8 vertices for an ordering $a_1 < a_2 < a_3 < b_1 < b_2 < b_3$; whereas for an ordering $a_1 < a_2 < a_3 < b_1 < b_2 < b_3$, 16 vertices are required to represent the same function as shown in Figure 2.2(b).

BDDs have many desirable properties for symbolic Boolean manipulation [17, 20-22]. Many binary operations such as AND and OR on Boolean functions can be implemented efficiently as graph algorithms applied to BDDs. Given two BDDs of size m and n, the complexity of a binary operation is $O(m.n)$. Thus, symbolic manipulations involving quantifications $\exists$ and $\forall$ have polynomial complexity in the sizes of BDDs. As BDDs are canonical, tautology checking and complementation can be achieved in constant time. Due to these nice features, BDDs have become primary workhorse for symbolic model checking [17, 20-22]. BDDs are used to represent transition system and state sets as characteristic functions. The primary limitation of the BDD-based approaches, however, is the scalability, i.e., BDDs constructed in course of verification grow extremely large, resulting in space-outs or severe performance degradation due to paging [23]. Moreover, BDDs are not very good representations of state sets, especially when the sharing of the nodes is limited. Choosing a right variable ordering for obtaining compact BDDs is very important. Finding a good ordering is often time consuming and/or requires good design insight, which is not always feasible. Several variations of BDDs such as Free BDDs [30], zBDDs [31], partitioned-BDDs [32] and subset-BDDs [33] have also been proposed to target domain-specific application; however, in practice, they have not scaled adequately for industry applications. Therefore, BDD-based approaches are limited to designs containing of the order of a few hundred state holding elements; this is not even at the level of an individual designer subsystem.

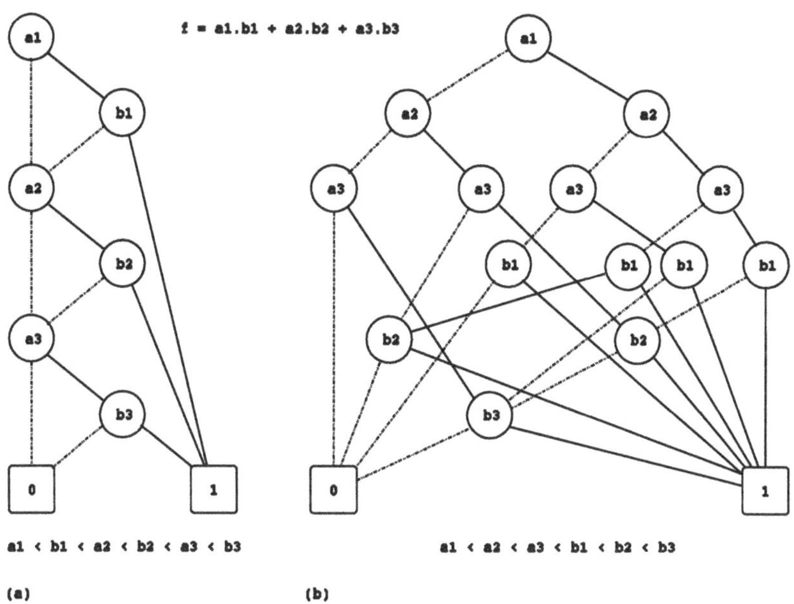

Fig. 2.2: Effect of variable ordering on BDD size: (a) with ordering $a_1 < b_1 < a_2 < b_2 < a_3 < b_3$, and (b) with ordering $a_1 < a_2 < a_3 < b_1 < b_2 < b_3$. A solid line denotes a then-branch and a dashed line denotes an else-branch

2.4 Boolean Satisfiability Problem

The Boolean Satisfiability (SAT) problem is a well-known constraint satisfaction problem, with many applications in the fields of VLSI Computer-Aided Design (CAD) and Artificial Intelligence. Given a propositional formula, the Boolean Satisfiability problem is to determine, whether there exists a variable assignment under which the formula evaluates to true, or to prove that no such assignment exists. The SAT problem is known to be NP-Complete [94]. In practice, there has been tremendous progress in SAT solver technology over the years, summarized in a survey [34].

Most SAT solvers use a Conjunctive Normal Form (CNF) representation of the Boolean formula. In CNF, the formula is represented as a conjunction of clauses, each clause is a disjunction of literals, and a literal is a variable or its negation. Note that in order for the CNF formula to be satisfied, each clause must also be satisfied, i.e., at least one literal in the clause has to be *true*. A clause with at least two unassigned literals is called *unsatisfied clause*. A *unit clause* is a clause where all literals are false, except one which is unassigned, referred to as the unit literal. A *conflicting clause* is a clause where all literals are false. A Boolean circuit can be encoded as a satisfiability equivalent CNF formula [95, 96]. Alternatively, for SAT applications arising from the circuit domain, it may be more efficient to modify the SAT solver to work directly on the Boolean circuit representation.

Most modern SAT solvers are based on a DPLL-style [35] procedure, shown as *SAT_solve* in Figure 2.3, with three main engines: *decision, deduction*, and *diagnosis*. Here, we focus on the state-of-the-art techniques involved that have allowed the basic SAT algorithm to scale to problem instances with as many as millions of variables. The basic SAT algorithm is a branch-and-search algorithm. A variable that is unassigned, is called a *free* variable. Initially, all variables are unassigned.

```
Synopsis:   Check Boolean Satisfiability
Input:      Boolean Problem
Output:     SAT/UNSAT
Procedure:  SAT_solve

if (Deduce()=CONFLICT) return UNSAT; //Pre-process
while(Decide()=SUCCESS) { //branch
  while(Deduce()=CONFLICT) { //constraint propagation
    blevel = Diagnose(); // conflict-driven learning
    if (blevel == 0) return UNSAT;
    else backtrack(blevel); //backjump to blevel
  }
}
return SAT;
```

Fig. 2.3: DPLL-style SAT Solver

Pre-processing of the formula is done by a *deduction* engine without making any decision. The *deduction* engine applies the *unit clause rule* [35] repeatedly on *unit* clauses until no such clause exists or a conflicting clause is detected. As per this rule, the unit literal is set to *true* in order to satisfy the corresponding unit clause. Consecutive application of this rule is also called *unit propagation* or *Boolean Constraint Propagation* (BCP). The deduction engine can also optionally apply *pure literal rule*, where a variable that appears as only positive (negative) literal in remaining unsatisfied clauses is set to *true* (*false*) [35].

The main loop (lines 7-13) in the procedure *SAT_solve* begins with invocation of *decision* engine, wherein a *free* variable is assigned a value (also called *branching literal*). The decision on the branching literal is critical to the performance of the SAT solver. After the branch, the *deduction* engine is called to apply BCP until no more *unit* clause exists or a *conflict* occurs, i.e., conflicting values are implied on a variable. In the former case, the *decision* engine is called again to make a decision at next level; otherwise, a *diagnosis* engine is called to resolve the conflict that involves learning reasons for conflict, and *backtracking.* Note that backtracking corresponds to undoing the assignments made during BCP up to a previous level on the decision stack derived from conflict analysis. Advanced SAT solvers use backjumping or non-chronological backtracking [36-38, 40, 97] during conflict analysis and derive backtrack level and conflict-driven learnt clauses based on implication graph. Diagnosis engine, in some sense, is a correction procedure to the decision strategy that provides guidance to the search procedure. Effectiveness of the guidance is measured by number of backtracks before the problem is resolved. A bad decision strategy will cause more frequent backtracks, affecting the SAT performance overall. Backtrack to decision level 0 indicates that the problem is unsatisfiable (UNSAT). On the other hand the problem is said to be satisfiable (SAT) if the decision engine is unsuccessful in finding any free variable to branch on.

We now discuss various techniques developed to improve each of the SAT engines.

2.4.1 Decision Engine

The choice of the branching literal is critical to the performance of a SAT solver. Many heuristics have been proposed in the past [98, 99] which seem to work on certain classes of problems More recently, a decision strategy called *Variable State Independent Decaying Sum* (VSIDS) [38] has been shown to be both effective and efficient for many classes of problem instances. This strategy improves upon a previously proposed strategy on the use of literal count [98], wherein the current branching literal has the highest occurrence in currently unresolved clauses. In VSIDS strategy, the positive and negative scores of variables are computed initially based on the occurrence of positive and negative literals in the clauses. Whenever a clause is learned and added, the score of its variable is increased dynamically by a constant. Further, scores of variables are periodically divided by a constant, i.e., decreased, to give more weight to the recently added learned clauses. A branching literal is the highest (positive or negative) scoring variable with that phase. As opposed to [98], VSIDS score does not exactly reflect the occurrence of literal in currently unresolved clauses, but has been found to be very effective and efficient. In another variation [40] of VSIDS, one can also increase the score of all the literals in the clauses that cause a conflict, instead of only those that appear in a conflict-driven learned clause. Further, one can decide to satisfy those conflict-driven clauses added, starting with the most recent ones [40]. This scheme has been shown to be quite effective and robust.

2.4.2 Deduction Engine

BCP constitutes the core of SAT algorithm, taking about 80% of the total time in SAT. Any improvement in BCP directly translates into overall performance gain. We discuss a successful and highly optimized BCP scheme called *lazy update*, first introduced in Chaff [38], which is an improvement over BCP scheme based on head/tail pointer proposed in SATO [36]. In lazy update scheme,

1. For each clause, only two *non-zero literals* are watched.
2. The clause state is updated lazily when the watched literals coincide.
3. Re-positioning of watch pointers after backtrack is not required as they are guaranteed to watch non-zero literals after backtrack.

Note, a literal assignment does not necessarily imply re-positioning of watch pointers in a clause as that literal may not be watched. Also, backtrack cost, modulo variable unassignment, is constant time.

The implications generated during BCP are stored in a FIFO queue or assignment queue. In other words, implications are processed in the order in which they are generated. Since the actual assignment is postponed until it is de-queued, such a scheme is also called *lazy assignment* [100]. In contrast, one can also process the implications as they are generated, essentially in a LIFO order. Such processing of implications is also referred to as *eager assignment* [100]. In lazy assignment, the watch pointers may point to a literal that is implied but not yet assigned value *false* in the queue. Also, watch pointers can get updated even if the clause is satisfied. Though detection of unit and conflict clauses is easy, detection of satisfiable clauses requires additional book-keeping and invariant maintenance which increase the backtracking cost. One can avoid such overhead in head/tail pointer scheme [36] at the cost of backtrack. Recently, Biere *et al.* have also proposed [100] an efficient way to access the other watched literal to determine if the clause is satisfied. Experimental results, however, have shown that gain due to zero-cost backtrack using simple structure is significantly higher than the overhead of repositioning of the watch pointers using strict invariants [36] or use of additional pointer de-referencing [100].

In the following, we illustrate the lazy update with an example as shown in Figure 2.4. The clause ($-V_1 + V_4 + -V_7 + V_{11} + V_{15}$) has *watch pointers* (shown as arrow)on literals V_4 and V_{15}. With variable assignment V_{15}=0 at decision level 4 (denoted as @4), the right watch pointer moves left to point at V_{12}. With assignments V_7=1@5 and V_{11}=0@5, watch pointers are not repositioned. With V_4=0@5, the clause becomes unit clause with unit literal V_{12}=1. Later, with backtrack to level 4, all variables get unassigned in the clause except V_1 (assuming V_1 got assigned before level 4). Note, the watch pointers are not repositioned during backtrack and thus, backtrack is effectively zero-cost process. Further when V_{12}=1@5, the right watch pointer is not moved. However, with V_4=0@5, the left pointer moves even though the clause is satisfied with V_{12}=1.

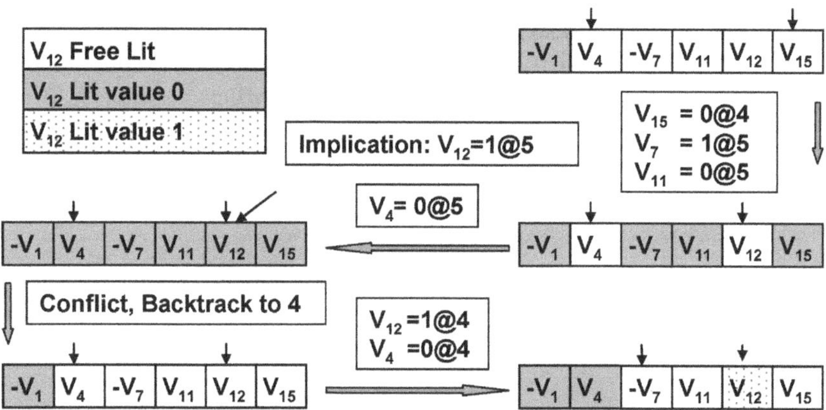

Fig. 2.4: Lazy update of a CNF clause during BCP

For clauses with many literals, a lazy update scheme performs significantly better than a head/tail pointer scheme. This also works well in practice, as conflict-driven learning tends to add very long clauses. Other forms of deduction such as *pure literal rule* [35] and equivalence reasoning [101] can also be applied but are not so robust and efficient in practice.

2.4.3 Diagnosis Engine

A conflict occurs when opposite values are implied during BCP on a variable. When such a conflict is detected, the conflict analysis routine is called to resolve the conflict. The analysis could be as simple as flipping the last decision variable that has not been previously flipped, and sometimes referred to as *chronological backtracking*. It is not very effective in pruning the search space. In contrast, conflict analysis could be more involved, wherein resolution process is applied on the implication graph [36, 37, 97] to learn conflict clauses for *conflict-driven learning* (for learning conflict clauses) and *non-chronological backtracking* [37, 102] (i.e., back-jumping to an earlier decision level).

The conflict-driven learned clauses are generated by applying the following *resolution rule*: $(x+ y)\wedge(-x + z) \Rightarrow (y + z)$. As per unit clause rule, when a clause becomes a unit clause, it *implies* a true value on the unit literal. Such a clause becomes the *antecedent* of the corresponding variable. Thus, every implied variable will have an antecedent clause. The implications are typically stored in the form of an *implication graph*, a digraph with the vertices as variables and edges between the variables corresponding to the antecedent clause of the *to-vertex*. Vertices, with no incoming edges, denote decision variables. A conflict occurs whenever the graph has two vertices for the same variable with opposite values. A small example of an implication graph leading up to a conflict is shown in Figure 2.5.

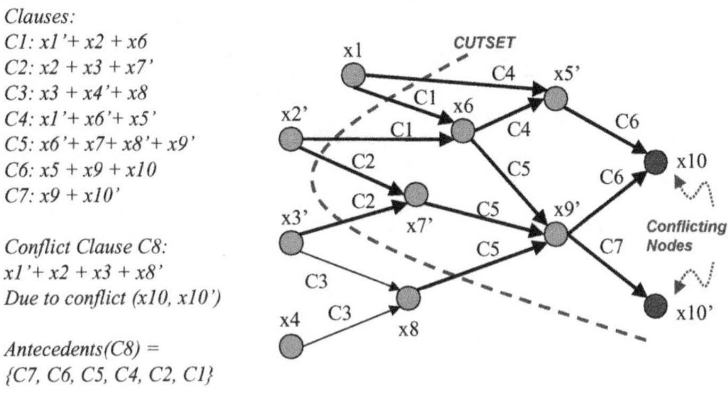

Fig. 2.5: Example for Conflict Analysis

Conflict analysis works on the implication graph by first bipartitioning the graph into a reason and a conflict side, as shown in Figure 2.5. All the vertices on the reason side that have at least one edge to the conflict side constitute the reasons for the conflict. Such a bipartition is called a *cut*. Each cut corresponds to various learning schemes [103]. Conflict analysis then traverses back the edges leading to the conflicting nodes up to any cutset in this graph. A conflict clause is then derived from the variables feeding into the chosen *cutset*. A key feature of a learned conflict clause is that it is also the result of resolution on all the antecedent clauses, which are traversed up to the chosen *cutset*. In order to make the conflict clause an *asserting clause* (equivalently, *unit clause)*, a cut is chosen such that the unit literal corresponds to the variable with the current decision level, and backtrack level chosen is the maximum decision level (except current decision level) of the variables in the conflict clause.

An *unique implification point* (UIP) [37] is a vertex at the current decision level that dominates both the vertices corresponding to the conflicting variable. (A variable x is said to dominate y *iff* any path from the decision variable of the decision level of x to y needs to go through x.) In *firstUIP* learning scheme, only the UIP at the current decision level is computed, which becomes the unit literal in the conflict clause after backtrack. Other learning schemes such *lastUIP*, *allUIP* and min-cuts requires additional resources for computation and have not been found to be so effective [103] as *firstUIP*.

Conflict-driven learning results in addition of conflict clauses to the SAT problem in order to prevent the same conflict from occurring again during the search. Additionally, information recorded during conflict analysis has been used very effectively to provide a *proof of unsatisfiability*, as described next.

2.4.4 Proof of Unsatisfiability

There have been several independent efforts aimed at extracting resolution-based proofs of unsatisfiability from a CNF-based SAT solver [48, 77, 78, 104, 105]. They are based on recording additional information during conflict analysis and conflict-driven learning such as antecedents (reasons) and their association with learned clauses. As part of the check, these techniques also identify a *subset* of clauses from the original problem, called the *unsatisfiable core*, such that the clauses are sufficient for implying the unsatisfiability. Similar SAT-based proof analysis techniques have also been proposed independently in the context of refinement, and abstraction-based verification methods [48, 77, 78]. We describe the generation of a proof of unsatisfiability in the following.

When a SAT solver determines that a given problem is unsatisfiable, it does so because there is a conflict without any decisions being taken. A conflict analysis can again be used to record the antecedents for this *final conflict.* This time, the learned conflict clause, i.e., a resolution of all its antecedents is an empty clause. Therefore, this *final resolution proof tree* (also called *final conflict graph*) constitutes a proof of unsatifiability [35], except that it may include some learned clauses. Recall that a learned clause is itself a resolution of the antecedents associated with it. Therefore, by recursively substituting the resolution trees corresponding to the learned clauses into the final resolution tree, a resolution proof only on the original clauses can be obtained. These original clauses constitute an *unsatisfiable core,* i.e., they are sufficient for implying the unsatisfiability. In practice, the resolution tree is created only if a check is needed for the unsatisfiability result [104, 105]. For the purpose of identifying an *unsatisfiable core*, a marking procedure is used, which starts from the antecedents of the final conflict graph, and recursively marks the antecedents associated with any marked conflict clause. At the end, the set of marked original clauses constitutes an *unsatisfiable core*.

The existing resolution-based proof analysis techniques have been described for SAT solvers that use a CNF representation of the Boolean problem. In Chapter 4, we discuss how these techniques can be adapted for hybrid SAT solvers that work directly on circuit structures.

2.4.5 Further Improvements

There has been ongoing research [106] on further improving the SAT solvers. We briefly mention some of the most relevant efforts that have shown promising results on certain problem instances and involve low overhead.

Frequent Restarts

The state-of-the-art SAT solvers also employ a technique called *random restart* [38] for greater robustness. The first few decisions are very important in the SAT solver. A bad choice could make it very hard for the solver to exit a local non-useful search space. Since it is very hard to determine *a priori* what a good choice might be for decisions, the restart mechanism periodically undoes all decisions and starts afresh. The learned clauses are preserved between restarts; therefore, the search conducted in previous rounds is not lost. By utilizing such randomization, a SAT solver can minimize local fruitless search.

Non-conflict-driven Back-jumping

This refers to a backjump to an earlier decision level (not necessarily to level 0), without detecting a conflict [107]. It is a variation of frequent restarts strategy, but guided by the number of conflict-driven backtracks seen so far. The goal is to quickly get out of a "local conflict zone" when the number of backtracks occurring between two decision levels exceeds a certain threshold. For hard problem instances, such strategy has shown promising results.

Frequent Clause Deletion

Conflict-driven learned clauses are redundant, and therefore, deleting them does not affect satisfiability of the problem. Conflict clauses, though useful can become an overhead especially due to increased BCP time and due to large memory usage. Such clauses can be deleted based on their relevance metric [38] which is computed based on number of unassigned literals. One can also compute relevance of a clause based on its frequent involvement in conflict [40].

Clause Shrinking

Effectiveness of conflict-driven learning scheme is very hard to determine *a priori.* In general, a shorter conflict-clause is useful, as it prunes a larger search space. One scheme is to shrink the conflict clause by identifying a sufficient subset of the literals required to generate the conflict [108]. Using the conflict literals as decision variables, one applies BCP and stops as soon as a conflict is detected. In many cases, fewer literals in the conflict clause are involved.

Early Conflict Detection in Implication Queue

The implication queue (in lazy assignment) stores the newly implied variables during BCP. If a newly implied variable is already in the queue with an opposite implied value, one can detect conflict early without doing BCP further [100].

Shorter Reasons First

Several unit clauses can imply the same value on a variable at a given decision level. With an intuition that shorter unit clauses decrease the size of implication graph, and hence, the size of conflict clauses, the implications due shorter clauses are given preference [100].

2.5 SAT-based Bounded Model Checking (BMC)

Bounded Model Checking refers to a model checking approach where the number of steps in forward traversal of the state space are bounded. The approach reports either "violation found" or "no violation possible within the bounded depth". Symbolic simulation can be considered as bounded model checking, restricted to invariants checking. Such an approach does not mandate storing state space and hence, is found to be more scalable and useful.

In SAT-based BMC [66, 67], the specifications are expressed in LTL (Linear Temporal Logic). To keep the discussion simple, we consider formulas of the existential type ***E****f* where f is a temporal formula with no path quantifiers. In the sequel, we sometimes drop the quantifier ***E*** and implicitly imply an existential LTL formula, if not obvious from the context. Given a Kripke structure M, an LTL formula f, and a bound k, the translation task in BMC, i.e., henceforth denoted as $BMC(M,f,k)$, is to construct a propositional formula $[M, f]_k$, such that the formula is satisfiable if and only if there exists a witness of length k. $[M, f]_k$ is defined as follows:

$$[M, f]_k = [M]_k \wedge [f]_k, \text{ where} \tag{2.2}$$

$$[M]_k = I(s_0) \wedge \wedge_{0\le i<k} T(s_i, s_{i+1}) :: \textit{initial and circuit constraints}$$

$$[f]_k = ((\neg L_k \wedge [f]_k^0) \vee (\vee_{0\le l\le k} ({}_lL_k \wedge {}_l[f]_k^0))) :: \textit{property translation}$$

$${}_lL_k = T(s_k, s_l) :: \textit{loop transition from state } s_k \textit{ to } s_l \ (l \le k)$$
$$L_k = \vee_{0\le l\le k} \, {}_lL_k :: \textit{disjunction of loops of length up to } k$$

Since the set of states is finite, a symbolic encoding in terms of Boolean variables denoted as s is used to represent a state, and $s_0 \ldots s_k$ is used to represent a finite (k+1)-length sequence of states.

The first set of constraints, denoted $[M]_k$, ensures that this sequence is a valid path in M. The first part of this formula imposes the constraint that the first state in the sequence should be an initial state I. The second part of this formula imposes the constraint that every successive pair of states should be related according to the transition relation T. Note that this second part corresponds to an unrolling of the sequential design for k steps, and results in an increasing SAT problem size with increasing k.

The next set of constraints $[f]_k$ ensures that this valid path in M satisfies the LTL formula f. This involves a case split, depending upon whether or not the path is a *(k, l)* -loop, as shown in Figure 2.6 (taken from [66]). The case of a loop from state *k* to state *l* can be translated as $_lL_k = T(s_k, s_l)$. The case where there is no loop can be translated as $\neg L_k$ where $L_k = \vee_{0\leq l\leq k}\, {}_lL_k$. Let $[f]_k^{\,0}$ denote the translation for formula *f* in the no loop case, and $_l[f]_k^{\,0}$ denote the translation for *f* in the *(k, l)*-loop case (detailed definitions are in [66]).

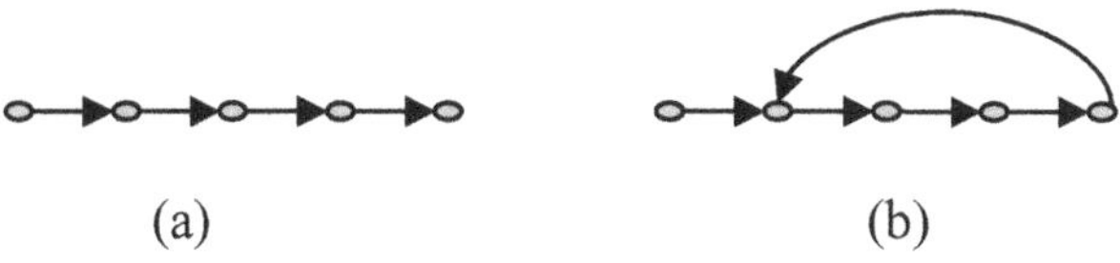

(a) (b)

Fig. 2.6: BMC check (a) No loop (b) Loop

For our discussion later, we call the first conjunct $[M]_k$ as the *circuit constraints*, and the second conjunct $[f]_k$ as *property translation*. Note that the general translation is monolithic, i.e., the entire propositional formula is generated for a given *k*. This formula is then checked for satisfiability using standard SAT solvers [35].

In practice, the search for longer witnesses is conducted by incrementing the bound *k*. This works well when a witness does exist. However, in case there is no witness, an additional proof technique is needed to conclude that the property is indeed false. In particular, it is sufficient to examine all *k* up to the diameter of the finite state machine [66]. Use of additional constraints such as loop-free paths, shortest paths, and inductive invariants [73], have also been proposed in a similar setting for proving safety properties [67, 109].

2.5.1 BMC formulation: Safety and Liveness Properties

A counter-example to a safety property is a finite trace. Most used safety properties are simple, i.e., $\mathbf{G}(\neg p)$ where *p* is a propositional atom. A standard translation for the negation of the property can omit the loop condition since a finite trace is sufficient for falsifying the safety property, as shown below:

$$[\neg\mathbf{G}(\neg p)]_k = [\mathbf{F}(p)]_k = [M]_k \wedge [\vee_{0\leq t\leq k}\; p^t]_k \;;;\; p^t \equiv \textit{node } p \textit{ at depth } t \tag{2.3}$$

Recall, a liveness property expresses that eventually “something good will take place” along an execution path. Presence of fairness constraints $B = \{f_1,\ldots,f_b\}$ puts additional requirement that each $f_r \in B$ should hold infinitely often along the execution path. Liveness properties $\mathbf{F}(\neg q)$, $\mathbf{G}(p \rightarrow \mathbf{F}(\neg q))$ in the presence of fairness constraints *B* expressed as $\mathbf{G}(\wedge_{1\leq r\leq b}\mathbf{F}(f_r))$ are:

$$\boldsymbol{G}(\wedge_{1\le r\le b}\boldsymbol{F}(f_r)) \rightarrow \boldsymbol{F}(\neg q)$$
$$\boldsymbol{G}(\wedge_{1\le r\le b}\boldsymbol{F}(f_r)) \rightarrow \boldsymbol{G}(p\rightarrow \boldsymbol{F}(\neg q)) \tag{2.4}$$

The standard LTL translations [66, 110, 111] for the negation of the above properties where p, q are propositional atoms are as follows:

$$[\neg(\boldsymbol{G}(\wedge_{1\le r\le b}\boldsymbol{F}(f_r)) \rightarrow \boldsymbol{F}(\neg q))]_k = [\boldsymbol{G}(q \wedge (\wedge_{1\le r\le b}\boldsymbol{F}(f_r)))]_k$$
$$= [M]_k \wedge [\vee_{0\le l\le k}({}_lL_k \wedge (\wedge_{1\le r\le b}(\vee_{l\le j\le k} f_r^j)) \wedge (\wedge_{0\le i\le k} q^i))]_k \tag{2.5}$$

$$[\neg(\boldsymbol{G}(\wedge_{1\le r\le b}\boldsymbol{F}(f_r)) \rightarrow \boldsymbol{G}(p\rightarrow \boldsymbol{F}(\neg q)))]_k$$
$$= [\boldsymbol{F}(p \wedge \boldsymbol{G}(q \wedge (\wedge_{1\le r\le b} F(f_r))))]_k$$
$$= [M]_k \wedge [\vee_{0\le l\le k}({}_lL_k \wedge (\vee_{0\le t\le k}(p^t \wedge (\wedge_{min(t,l)\le i\le k} q^i) \wedge (\wedge_{1\le r\le b}(\vee_{min(t,l)\le j\le k} f_r^j)))))]_k \tag{2.6}$$

A counter-example to a liveness property $\boldsymbol{G}(\boldsymbol{F}(f)) \rightarrow \boldsymbol{G}(p\rightarrow \boldsymbol{F}(\neg q))$ in the presence of a fairness constraint f is shown in Figure 2.7. Note, the fairness requirement is that f should hold at least once between the looping states i.e., s_l and s_k.

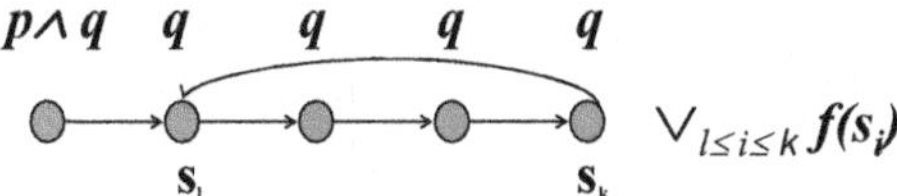

Fig. 2.7: Debug trace for $\boldsymbol{G}(p\rightarrow \boldsymbol{F}(\neg q))$ with fairness f

Another form of liveness property that is commonly used is *bounded liveness*, which is typically expressed as $(q\wedge_{1\le n\le M}\boldsymbol{X}{:}n(q)$ and $\boldsymbol{G}(p\rightarrow q\wedge_{1\le n\le M} \boldsymbol{X}{:}n(q))$ where M is the liveness bound and $\boldsymbol{X}{:}n$ is shorthand hand for $\boldsymbol{X}\wedge\ldots\wedge\boldsymbol{X}$, repeated n times.

Safety properties have generated a lot of interest among researchers due to their relative importance in practice. Recently, there have been several effective SAT-based techniques – BMC [66, 110], induction [67, 73], unbounded model checking (UMC) [47, 68, 71, 112], and proof-based abstraction [48, 78] – that can handle safety properties very efficiently compared to BDD-based approaches like symbolic model checking [17]. Though many of these techniques can be applied to checking all properties in principle, there has been a general lack of efforts dedicated to liveness, barring a few described later in Chapters 5 and 10.

As reported in a recent workshop [113], one school of thought believes that rigorous reasoning about liveness is tedious and therefore, it is not worth the effort given finite resource for verification. Another school of thought, along similar lines, believes that liveness without a bound is not useful and bounded liveness can be reduced to a safety property; therefore, we should use safety checking only. We believe, like an alternative school of thought, that liveness is important and that it too deserves dedicated model checking techniques. We list some of the main reasons cited:

- If we restrict verification to only safety checking, it would be impossible to detect subtle design errors that prevent some behavior from happening eventually.
- Translation of bounded liveness to safety is expensive especially when the bound is large. Moreover, designers may not always provide a safe bound.
- Fairness constraints are used to capture the interaction of the design with the environment. Since formal verification is usually applied at the block level, fairness is needed to model the environment correctly. Therefore, liveness with fairness is important to capture subtle system behaviors.
- Model checking for all LTL properties can be reduced to checking of a fairness constraint.

In general, liveness is handled as a standard LTL property, by converting the negated formula to a Buchi automaton and checking for language emptiness of its product with the design model [114]. Such a general translation may not be very effective, and many improved translations [115-117] have been proposed. SAT-based BMC [66] handles liveness as part of a standard translation of an LTL formula to a propositional satisfiability problem where the challenge is to find a state loop. BMC is incomplete without a *completeness bound* [66, 67, 72], and generally it is difficult to obtain such a bound. However for safety checking, the completeness has been addressed by several other methods: induction [67, 73], SAT-based UMC [47, 68, 71, 112], and proof-based abstraction [48, 78]. This inspired some researchers [118] to reduce the liveness property to a safety property and use the complete safety checking algorithms. However, this translation is based on a state recording approach that doubles the number of state bits in the design. Such a translation adds significant overhead to safety checking algorithms and is therefore, not very practical. In another recent work [119], authors reason about a model composed with Buchi automata obtained from translation of the negated LTL formula, and provide completeness bound for a fairness constraint based on loop-free path analysis. Such bounds may be quite large in practice, and are therefore not very useful.

Recently we have proposed [46] SAT-based techniques geared towards verifying liveness properties of the form $\boldsymbol{F}(q)$ and $\boldsymbol{G}(p \rightarrow \boldsymbol{F}(q))$ with fairness B. These techniques can be extended to handle commonly used LTL properties. We use dedicated SAT-based translations for bounded model checking of non-safety properties, and determine the completeness bounds for liveness using SAT-based unbounded analysis. This is discussed in more detail in Chapters 5 and 10.

2.5.2 Clocked LTL Specifications

Property variables that involve gates with support from state elements in multiple clock domains require the use of clocks in the formula to avoid ambiguities. The Property Specification Language (PSL) standardized by Accellera [83] has formal semantics for specifying clocked properties using the clock operator @, based largely on the work of Eisner *et al.* [120]. A clocked LTL specification *(f)@clk*, expressed under the context *clk* using clock operator @, can be equivalently translated [120] into an un-clocked LTL specification (with an implicit global clock tick) $T^{clk}(f)(\equiv (f)@clk)$ where $T^{clk}(f)$ is defined recursively using the following rules *R1-6*:

$$
\begin{aligned}
&R1\text{: } T^{clk}(p) = \neg clk\ \mathbf{U}\ (clk \wedge p)\ ;;f \text{ is propositional atom } p \\
&R2\text{: } T^{clk}(\neg f) = \neg T^{clk}(f) \\
&R3\text{: } T^{clk}(f_1 \wedge f_2) = T^{clk}(f_1) \wedge T^{clk}(f_2) \\
&R4\text{: } T^{clk}(\mathbf{X} f) = \neg clk\ \mathbf{U}\ (clk \wedge \mathbf{X}(\neg clk\ \mathbf{U}\ (clk \wedge T^{clk}(f)))) \\
&R5\text{: } T^{clk}(f_1 \mathbf{U} f_2) = (clk \rightarrow T^{clk}(f_1))\ \mathbf{U}\ (clk \wedge T^{clk}(f_2)) \\
&R6\text{: } T^{clk}((f)@clk1) = T^{clk1}(f)
\end{aligned}
\quad (2.7)
$$

Rules for other LTL operators ***F***, ***G*** and ***W*** can be derived from the above rules. Note, that in [120], the authors have differentiated temporal operators and propositional atoms as weak or strong using the strength operator **!**. A strong atomic proposition *p***!** does not hold on an *empty path*, compared to a weak atomic proposition *p* which does. For traditional temporal semantics, there is always a current state and hence the problem of an empty path does not arise. However, for clocked semantics, there may not be a current state when a clock in the specification stops ticking. For most practical scenario we will assume, henceforth, that clocks in the specifications are always ticking. Under that assumption, the strength operator **!** can be dropped. We first make some crucial observations regarding the rules *R1*, *R4* and *R6*.

As per rule *R1*, *p@clk* holds at the current state, if

a) *p* holds and *clk* ticks at the current state, or
b) *clk* ticks next at a state where *p* also holds.

Similarly, as per rule *R4*, ***(Xf)****@clk* takes us two *clk* ticks into the future if the *clk* does not hold in the current state. The rule *R6* disallows accumulation of clocks in presence of nesting, allowing only the innermost specified clock to supersede the outer ones.

For example, a clocked LTL formula

$$\boldsymbol{F}(p \wedge (\boldsymbol{X}q@clk1)@clk)$$

gets translated into an equivalent unclocked LTL formula

$$\boldsymbol{F}(p \wedge (\neg clk \boldsymbol{U}(clk \wedge \boldsymbol{X}(\neg clk \boldsymbol{U}(clk \wedge (\neg clk1 \boldsymbol{U}(clk1 \wedge q)))))))$$

The general translation scheme for clocked properties tends to generate large nested LTL formulas that can limit the effectiveness of a standard BMC solver.

Any property specific translation, as discussed later, can get complicated due to subtleties in the clocked specifications, as illustrated in the following example. Consider two clocked specifications, P1 and P2.

P1: ***F****(ctr2[0] * (****X****(ctr2[0]))@clk2_r_d)*
P2: ***F****((ctr2[0] * ****X****(ctr2[0]))@clk2_r_d)*

In P1, the *ctr2*[0] (bit 0 of *ctr2*) is clocked by the global clock, *gclk*, while in P2, it is clocked by *clk2_r_d* as shown in Figure 2.8. One can verify that the witness state for P1 is at *gclk*=6 where *ctr2*[0]=1 and also two *clk2_r_d* ticks later *ctr2[0]*=1 at *gclk*=13. On the other hand, P2 does not have a witness on the path shown. These subtleties in the clocked specifications add further complexity to the property-specific BMC customization method [46].

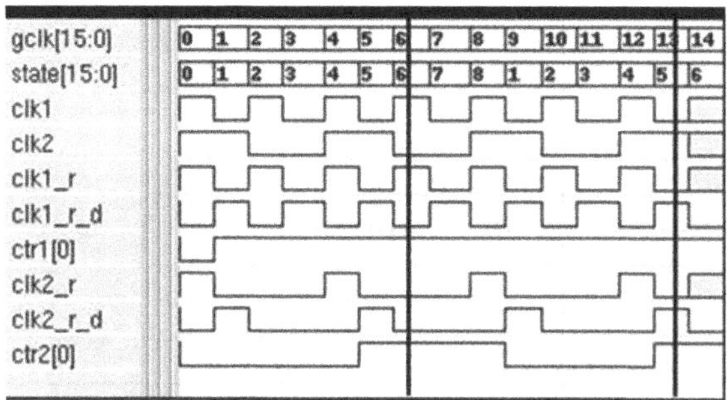

Fig. 2.8: Example: Timing diagram for clocked specification

2.6 SAT-based Unbounded Model Checking

Recall, a standard pre-image computation is given by

$$Pre\text{-}Image(\delta, Y) = \exists_{U,Y}\, \delta(X,Y,U) \wedge T(Y) \tag{2.8}$$

A pre-image computation involves existential quantification of next state variables Y and input variables U. Such an existential quantification can be accomplished using SAT by enumerating all possible solution cubes over X variables in the conjunct $\delta(X,Y,U) \wedge T(Y)$ and adding blocking constraints that exclude previously enumerated solution cubes. A basic algorithm for SAT-based existential quantification based on cube enumeration [68, 71] is shown in Figure 2.9. The procedure *All_sat* takes a function $f(A,B)$ with support variables from sets A and B and returns $C=\exists_B f(A,B)$ after quantifying the variables from the set B. The procedure *SAT_solve* is called repeatedly in line 7 on the constrained problem $f=1 \wedge C=0$ where C represents the set of solutions enumerated so far. Typically, the next call to *SAT_solve* is made from the previous SAT state [47, 69, 71]. The enumerated cube c, in line 9, is obtained by keeping only the A set literals from the assignment cube α in line 8. Successive enumeration of cube c stops in line 7 when C represents $\exists_B f(A,B)$ exactly. The blocking constraints are represented by BDDs [68], zBDDs [71], or CNF [69]. Enlarging the cube c is done by redrawing the implication graph [71] or by applying a justification procedure [121] as a post-processing step. Note that when the procedure *All_sat* is called with substitutions $f \leftarrow \delta(X,Y,U) \wedge T(Y)$, $A \leftarrow X$, and $B \leftarrow U \cup Y$, it returns the *pre-image* of the set $T(Y)$.

```
1.  Synopsis:   SAT-based Quantification (cube enumeration)
2.  Input:      function f, input keep set A,
                input quantifying set B
3.  Output:     ∃B f(A,B)
4.  Procedure:  All_sat
5.
6.  C=∅; //initialize constraint
7.  while (SAT_solve(f=1∧C=0)=SAT) {
8.      α=Get_assignment_cube();
9.      c=Get_enumerated_cube(α,B); //obtain, ∃B α
10.     C= C ∨ c;
11. }
12. return C;   // return when no more solution
```

Fig. 2.9: SAT-based existential quantification using cube enumeration

We show a standard least fixed-point computation for temporal logic operator *EF* using the procedure *Fixed_point_EF* in Figure 2.10. $R(X)$ is the set of states that satisfy ***EF***$(f(X))$ and is updated with pre-images in line 8. Note that the exclusion of $R(X)$ in the pre-image computation using *All_sat* as shown in line 10 is equivalent to a typical fixed-point check for $R(X)$.

Note that the number of steps required to reach a fixed-point gives the *longest shortest backward diameter*, a *completeness bound* for BMC [66, 67]. Moreover, one can use the *reachability constraint C(X)* (over-approximated reachable states from initial state) [73], as a *care set* in line 10 (i.e., *All_sat* is called with $f \leftarrow \delta \wedge T \wedge \neg R \wedge C$) to potentially get a fixed-point in less number of pre-image steps than the backward diameter [67].

```
1.  Synopsis:    Least fixed-point using All_sat
2.  Input:       property node f (function of state vars X)
3.  Output:      set of states satisfying EF(f(X))
4.  Procedure:   Fixed_point_EF
5.
6.  i=0; R(X)=∅; T(X)=f(X); //initialize
7.  while (T(X)≠∅) {// fixed-point reached?
8.   R(X)=R(X)∨ T(X); //add i^th pre-image of f
9.   //Compute pre-image state for T but not in R
10.  T(X)=All_sat(δ ∧T(<X←Y>)∧¬R(X), X, U∪Y) ;
11. }
12. return {R(X),i}; // returns states satisfying EF(f(X))
```

Fig. 2.10: Least fixed-point computation using SAT

2.7 SMT-based BMC

In Boolean-level BMC, the translated formula is expressed in propositional logic and a Boolean SAT solver is used for checking satisfiability of the problem. Several state-of-the-art techniques (a survey in [122]) exist for Boolean BMC that have led to its emergence as a mature technology, widely adopted by the industry. However, there are several limitations of a propositional translation and use of a Boolean SAT Solver. Some of these are as follows:

- A propositional translation in the presence of large data-paths leads to a large formula; which is normally detrimental to a SAT-solver due to increased search space.
- Data-path sizes need to be known explicitly *a priori,* before unrolling of the transition relation. For unbounded data-path, additional range-analysis of the program/design is required to obtain conservative but finite data-path sizes.
- High-level information is lost during Boolean translation and therefore, needs to be re-discovered by the Boolean SAT solver, often with a substantial performance penalty.

High-level BMC, as shown in Figure 2.11, overcomes the above limitations of a Boolean-level SAT-based BMC; wherein, a BMC problem is translated typically into a quantifier-free formula in a decidable subset of first order logic, instead of translating it into a propositional formula. A Satisfiability Modulo Theory (SMT) solver is then used to determine the satisfiability of the formula [123-131].

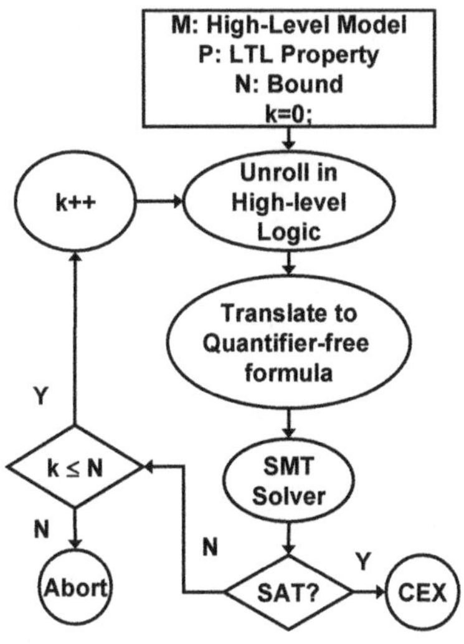

Fig. 2.11: SMT-based BMC

2.8 Notes

We have provided an overview of model checking, bounded model checking, and SAT solvers and their advancements. We require the reader to have a basic understanding of the material described in this chapter. We have also provided many references, which can be followed to get a complete in-depth understanding of the issues and challenges that will be addressed in the subsequent chapters. For ease of understanding, we have attempted to use uniform notation across chapters. Any deviations are either explicitly explained, or are implied from the context.

PART I: BASIC INFRASTRUCTURE

In Part I, we discuss the elements of a basic infrastructure that are required to build scalable verification algorithms in Chapters 3 and 4.

In Chapter 3, EFFICIENT BOOLEAN REPRESENTATION, we present a semi-canonical circuit graph representation and efficient on-the-fly compression algorithms to obtain a simple and reduced graph. This circuit graph is used to symbolically represent transition systems, model checking problem instances, explored states, and Boolean problems for efficient Boolean reasoning.

In Chapter 4, HYBRID DPLL-STYLE SAT SOLVER, we discuss a hybrid SAT solver that combines the best of the SAT solvers of various application domains, and works directly on the semi-canonical circuit graph representation. When such a solver is augmented with incremental and BDD-based learning capabilities, it becomes ideally suited for SAT-based model checking applications.

We describe SAT-based techniques that utilize the efficient circuit graph representation and hybrid SAT solver for various verification tasks in Parts II-IV.

3 EFFICIENT BOOLEAN REPRESENTATION

3.1 Introduction

Many traditional tasks in computer-aided design such as equivalence or property checking, logic synthesis, static timing analysis, and automatic test-pattern generation require an efficient representation of combinational circuits in terms of a network of Boolean primitives. Many practical problems derived from the above-mentioned applications have a high degree of structural redundancy. There are three main sources for this redundancy:

- First, the primary netlist derived from an RTL specification contains redundancies generated by language parsing and processing. In our experiments, we found 30-50% redundancy in generated netlist gates in the industrial designs.
- A second source of structural redundancy is inherent to the actual problem formulation. For example, a Miter structure [134] — built for the purpose of equivalence checking — is globally fully redundant when designs are equivalent. It also contains a large number of local redundancies in terms of identical substructures used in both designs to be compared.
- A third source of structural redundancy originates from repeated invocations of Boolean reasoning on similar problems derived from overlapping parts of the design. For example, the individual paths checked during false path timing analysis are composed of shared sub-paths which get repeatedly included in subsequent checks. Similarly, a combinational equivalence check of large designs is decomposed into a series of individual checks of output and next state functions which often share a large part of their structure. Further, BMC problem instances

arising from increasing unroll depths have large overlaps, and we shall see further details in subsequent chapters.

In general, identifying functionally redundant nets is computationally hard. As a special case, structurally equivalent nets in combinational loop-free circuits can be detected in linear time. As an example, in [135], a method is described that employs structural hashing to identify and merge isomorphic sub-circuits.

In this chapter, we discuss a generalized low overhead hashing scheme that identifies functionally identical subcircuits of bounded size, independent of their actual structural implementation. The method hashes each subcircuit onto a unique functional signature, which is then used to implement the function. As a result, two structurally different but functionally identical subcircuits get mapped onto the same canonical implementation, and hence are automatically merged during their construction. Moreover, such representation is significantly less sensitive to the order in which it is built. The described technique uses a constant-time table lookup scheme to obtained reduced graph with insignificant computational overhead. The further reduction of circuit graph can be achieved using sophisticated and computationally expensive techniques such as BDD sweeping [135, 136] and/or SAT sweeping [137]. Here we focus on low overhead multi-level structural hashing techniques to obtain on-the-fly reduction of circuit graphs. Such on-the-fly reduced graph has various applications such as, for representing transition relations, enumerated state sets, BMC unrolled problem instances, and efficient Boolean reasoning. We now describe a sketch of the rest of this chapter.

Overview

In Section 3.2, we give a brief survey of alternative representations that are not-canonical and are used in various applications. In Section 3.3, we describe a multi-level hashing technique that extends a structural hashing method by also identifying a bounded set of functionally equivalent sub-circuits that are not necessarily structurally isomorphic and, thereby, eliminates a large amount of network redundancy. We give experimental results in Section 3.4 to show that this technique directly results in a reduction of the memory requirements for storing the circuit network, and improves the efficiency of the algorithms using these representations. We also show that multi-level hashing technique outperforms previous hashing technique proposed in combinational verification. In Section 3.5, we describe further simplification using external constraints. Finally, in Section 3.6, we summarize our discussion.

3.2 Brief Survey of Boolean Representations

3.2.1 Extended Boolean Decision Diagrams (XBDDs)

Trading off compactness of Boolean function representations with canonicity for efficient reasoning in CAD applications has been the subject of many publications. BDDs [18, 19] represent one extreme of the spectrum which map Boolean functions onto a canonical graph representation. Deciding whether a function is a tautology can be done in constant time, at the possible expense of an exponential graph size. XBDDs [138] propose to divert from the strict functional canonicity by adding function nodes to the graph. The node function is controlled by an attribute at the referencing arc and can represent an AND or OR operation. Similar to BDDs, complementation is expressed by a second arc attribute and structural hashing identifies isomorphic subgraphs on the fly. The proposed tautology check is similar to a technique presented in [139] and is based on a recursive inspection of all cofactors. This scheme effectively explores the BDD branching structure sequentially, resulting in exponential runtime for problems where BDDs are excessively large.

3.2.2 Boolean Expression Diagrams (BEDs)

Another form of a non-canonical function graph representation is BEDs [43]. BEDs use 5 basic two-input Boolean operators: *NAND* ($\overline{\wedge}$), *OR*($\vee$), *RIMP*($\Rightarrow$), *LIMP*($\Leftarrow$), and *BIMP*($\Leftrightarrow$). Complementation is achieved by absorbing the inversion in the fanout operators. The innovative component of BEDs is the application of local functional hashing, which maps any four-input sub-structure onto a canonical representation. This mapping significantly reduces the graph size by eliminating local functional redundancies on the fly. However, it misses structural similarities internal to substructure. To illustrate this, consider a small design shown in Figure 3.1. The output g is represented as $(a \vee b) \wedge abc$ in Figure 3.1(a), which has actually a simpler equivalent representation given by abc. Figure 3.1(b) shows an equivalent BED representation where the internal nodes are represented as follows: l as $(a \vee b)$, $r1'$ as $\neg(a \wedge b)$, r' as $(r1' \Leftarrow c) \equiv (r1' \vee \neg c)$, finally g as $\neg(l \Rightarrow r') \equiv \neg(\neg l \vee r')$. Clearly, this representation misses the reduced form of g. Later when we describe our generalized hashing scheme, we discuss how we identify the reduced structure of g.

Tautology check proposed in [43] is based on converting a BED structure into a BDD by moving the variable from the bottom to the top of the structure. Similar to many pure cut-point based methods, this approach is extremely sensitive to the ordering in which the variables are pushed up.

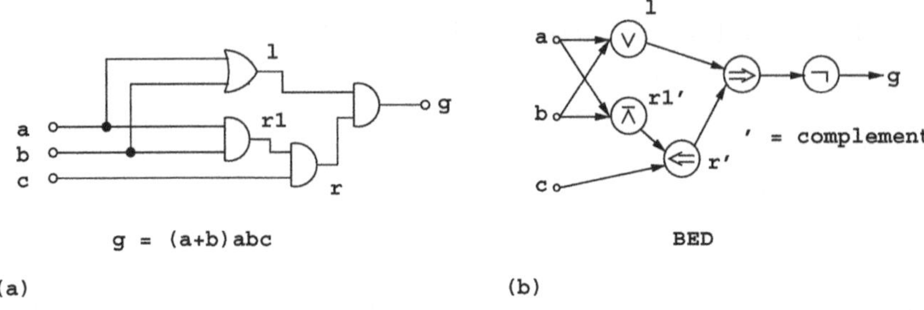

Fig. 3.1: Example for graph comparison: (a) a netlist with output g=$(a \vee b) \wedge abc$, (b) equivalent BED representation. The symbol $\neg$ denotes negation

3.2.3 AND/INVERTER Graph (AIG)

AND/INVERTER graph (AIG) representation [135, 136] refers to a circuit graph where a vertex denotes two-input AND gate (or a primary input) and an attribute on edge denotes *inversion*. During network construction, all Boolean operations are converted into *AIG* graph. Similar to efficient BDD implementations [140], each vertex is entered into a hash table using the identifier of the two predecessor vertices and their edge attributes as key. We refer to this hashing scheme as structural hashing (also referred to as *2-input hashing*) scheme [135]. This hash table is applied during graph construction to identify isomorphic subnetworks and to immediately map them onto the same subgraph. Without loss of generality, we will focus our description on the previously proposed uniform *AIG* network representation.

In the following, we first revisit the structural hashing scheme and then describe its generalization to include functional hashing (referred to as multi-level hashing) to obtain *Reduced AIG*. Figure 3.2 shows the pseudo-code *And_2* for the structural hashing algorithm. Constant folding is applied first (in lines 7-12) to reuse existing vertices without a hash lookup. In a non-trivial case, two input vertices of AND node are ranked before a hash lookup, to capture single-level structural isomorphism (in lines 16-17). Hash lookup is done (in line 19) to check the possibility of reuse of an existing vertex; otherwise, a new AND vertex is created and inserted in the hash table, using the procedure *New_and_vertex* (line 23).

```
Synopsis:   Boolean AND using structural hashing
Input:      Edges p1, p2
Output:     Boolean AND of p1 and p2
Procedure:  And_2
//constant folding
if (p1 == const_0) return const_0;
if (p2 == const_0) return const_0;
if (p1 == const_1) return p2;
if (p2 == const_1) return p1;
if (p1 == p2)      return p1;
if (p1 == ¬p2)     return const_0;
//Each vertex has unique rank. Rank of edge is
//rank of vertex.
if (Rank(p1) > Rank(p2))
   Swap(p1,p2); // Order edges for local canonicity
p = Hash_lookup(p1,p2);
if (p != NULL) return p;
//create And node and hash it
return New_and_vertex(p1,p2);
```

Fig. 3.2: Pseudo-code for 2-input structural hashing

The construction of the circuit graph for a simple example is illustrated in Figure 3.3. Part (a) represents a Miter circuit built for proving equivalence of nets x and y which are functionally identical but have different structural implementations. The applied labeling identifies functional equivalent nets by using identical numbers with one or more apostrophes. Figure 3.3(b) shows the result of the graph construction using the algorithm *And_2* (Figure 3.2). The vertices of the graphs represent AND functions and the filled dots on the edges symbolize the INVERTER attribute. Note that the functions $\neg(a \wedge b)$ (net 1) of the upper circuit and $(\neg a \vee \neg b)$ (net 1') of the lower circuit are identified as structurally equivalent (modulo inversion) and mapped onto the same vertex in the graph model. No other parts of the two circuits could be merged due to the limited scope of this technique.

A natural way to increase the scope of the structural hashing method is to divert from the two-input graph model and use vertices with higher fanin degree. The set of possible functions of a vertex with more than two inputs cannot be encoded with edge attributes. Instead, the vertex function is represented by an index that is hashed along with the input identifiers to find structurally identical circuit parts. Since the number of possible vertex functions grows exponentially, this method is only practical for vertices with up to four inputs. For the given circuit example, Figure 3.3(c) shows the graph model based on vertices with a maximum fanin degree of four. Note that this method can identify the equivalence of the net pair (5, 5') but still fails for pair (7, 7').

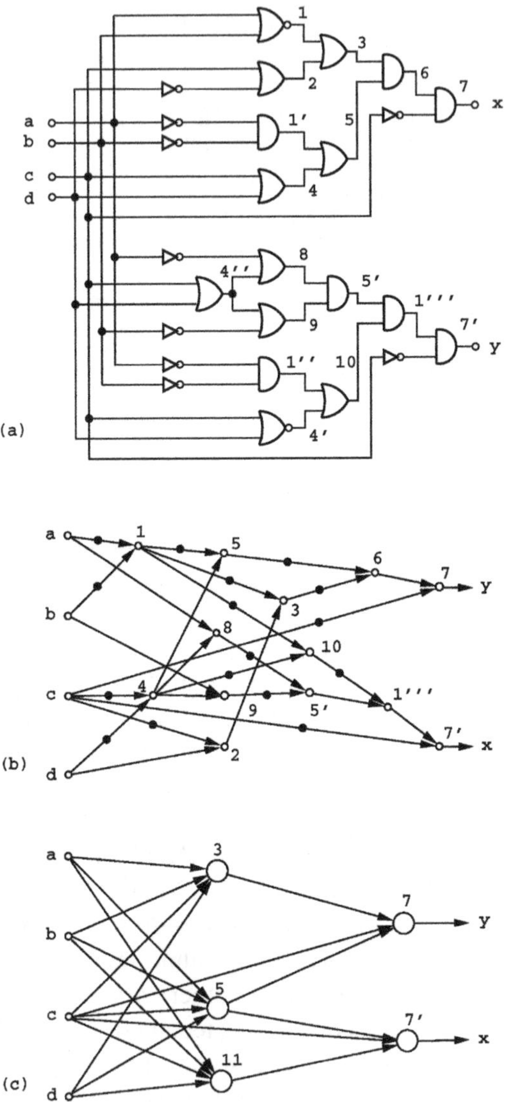

Fig. 3.3: Example for circuit graph construction: (a) a functionally redundant Miter structure generated to check functional equivalence of outputs x and y, (b) circuit graph using structural hashing constructed according to [135] with two-input vertices, (c) circuit graph with four-input vertices

In the following section, we describe our multi-level hashing method that combines the fine granularity of the original two-input graph model with the larger scope of detecting functional equivalent nodes. We demonstrate that this approach can identify all functional equivalent vertex pairs for the given example.

3.3 Functional Hashing (Reduced AIG)

The approach described is a generalization of the structural hashing method described in [135] to obtain *Reduced AIG.* Instead of considering only the two immediate inputs (also called *left* and *right* child) of a vertex i.e., one level backwards, the structural analysis is extended to two or three levels preceding a vertex for functional hashing. Thus, the resulting granularity to identify functionally identical vertices is comparable to the granularity of the hashing technique based on four-input vertices. However, by applying this method *recursively* on all intermediate vertices in an *overlapping manner*, this approach can also take advantage of structural similarities that otherwise remain internal to four-input vertices.

The algorithm *And*, in Figure 3.4, gives the pseudo-code for the recursive AND operation. It outlines the overall flow of the multi-level hashing scheme. The symbol $\neg$ is used to denote the complementation of the inverter edge attribute. Similar to typical BDD data structures, this attribute is implemented by a single bit of the reference pointer.

The first part performs constant folding and a table lookup for identical pre-existing vertices, similar to Figure 3.2. Next, a multi-level lookup scheme is used to convert the local function of the grandchildren (children of left and right vertices) into a canonical representation, by applying a suitable procedure *And_3* or *And_4*. During network construction from the primary inputs, the first level of vertices does not have four grandchildren and therefore must be treated specially. If both immediate children are primary inputs, the algorithm calls the function *New_and_vertex* to allocate the data structure (line 23). If only one of the children is a primary input, a canonical three-input substructure is created using the procedure *And_3* (line 27). If both children are AND vertices, then a canonical four-input substructure is created using the procedure *And_4* (line 33). Note that the procedures, *And_3* and *And_4*, call *And* recursively. To avoid infinite recursion, the procedure *New_and_vertex* gets called whenever a cycle condition is detected.

Multi-level table is pre-computed by considering all possible combinations of two or three levels of structure and mapping their unique signature to a canonical representation. This mapping then can be obtained in a constant time table lookup. In the following subsections 3.3.1 and 3.3.2, we discuss the algorithms to obtain unique signature and to build canonical substructure for three and four inputs, respectively.

```
Synopsis:  Boolean AND using multi-level hashing
Input:     Edges p1, p2
Output:    Boolean AND of p1 and p2
Procedure: And

if (p1 == const_0) return const_0;
if (p2 == const_0) return const_0;
if (p1 == const_1) return p2;
if (p2 == const_1) return p1;
if (p1 == p2)      return p1;
if (p1 == ¬p2)     return const_0;

//Each vertex has unique rank. Rank of edge is the
//rank of vertex. Order edges for local canonicity
if (Rank(p1) > Rank(p2))
  Swap(p1,p2);

p = Hash_lookup(p1,p2);
if (p != NULL) return p;

//p1, p2 are variables
if (Is_var(p1) && (Is_var(p2))
  return New_and_vertex(p1,p2);

//only p1 is variable
if (Is_var(p1))
return And_3(p1,p2);

//Only p2 is variable
if (Is_var(p2))
  return And_3(p2,p1);

return And_4(p1,p2);// p1, p2 are both AND nodes
```

Fig. 3.4: Pseudo-code for *And* using functional hashing

3.3.1 Three-Input Case

The construction of non-redundant three-input substructures, using the procedure *And_3*, is done in two steps as shown in the Figure 3.5. In the first step, a unique 8-bit signature is computed (line 32) using the procedure *3_input_signature* from the three-input structure. The signature is then used for table lookup (line 36) that returns an index pointing to a non-redundant implementation of the function. The idea is that the signatures of all substructures with equivalent functions will produce identical indices and therefore result in the same structural implementation.

```
Synopsis:   Compute signature for 3-input structure
Input:      Edges L (from VAR node), R (from AND node)
Output:     signature
Procedure:  3_input_signature
//get children of R
RL = Left_child(R);
RR = Right_child(R);
//Set i-th bit sig[i] if RHS predicate is true
sig[0] = (Non_inv(L) == Non_inv(RL));
sig[1] = (Non_inv(L) == Non_inv(RR));
sig[2] = (Rank(L) > Rank(RL));
sig[3] = (Rank(L) > Rank(RR));
sig[4] = Is_inv(RL);
sig[5] = Is_inv(RR);
sig[6] = Is_inv(L);
sig[7] = Is_inv(R);
return sig;
Synopsis:   Compute AND for 3-input structure
Input:      Edges L (from VAR node), R (from AND node)
Output:     Boolean AND of L and R
Procedure:  And_3
//get non-inverted edge for L and children of R
l = Non_inv(L);
rl = Non_inv(Left_child(R));
rr = Non_inv(Right_child(R));
Signature = 3_input_signature(L,R);
//Table lookup with signature to obtain
//pre-computed canonical sub-structure
index = LOOKUP3[signature];
switch(index) {
  ...
  Case 3: {p = And(l,rl);} //Example in Figure 3.6
  Case 4: {p = l;}
  ...
  Case 33: {p = ¬l;}
  ...
}
return p;
```

Fig. 3.5: Pseudo-code for *3_input_signature* and *And_3*

The second argument *R* passed to the procedures points to an AND node, i.e., non-leaf vertex. The signature is composed of three components that identify the substructure and input order in an unambiguous manner. The first two signature bits indicate whether any of the children *RL* (read as left child of R) and *RR* (read as right child of R) are identical to *L*. The next two bits encode the rank order of all three inputs. This information is essential to make the merging of the substructure independent of the actual input sequence. Note that the algorithm *And_2* (see Figure 3.2) ensures by construction that the children *RL* and *RR* are different and that *Rank(RL)* < *Rank(RR)*. The last four signature bits reflect the setting of the inverter attributes within the substructure using the procedure *Is_inv*.

Figure 3.6 provides a small example to illustrate the described mechanism. Both structures $f = b \wedge \neg(a \wedge \neg b)$ and $h = \neg b \wedge \neg(b \wedge \neg c)$ are first characterized by the signatures $sig_f = \langle 10100110 \rangle$ and $sig_h = \langle 11100001 \rangle$ (assuming $Rank(a) < Rank(b) < Rank(c)$) which then get mapped to the indices $index_f = 4$ and $index_h = 33$, respectively. As outlined in the pseudo-codes (line 39, Figure 3.5), both structures are then re-implemented as $f = b$ and $h = \neg b$ shown in the Figure 3.6(b).

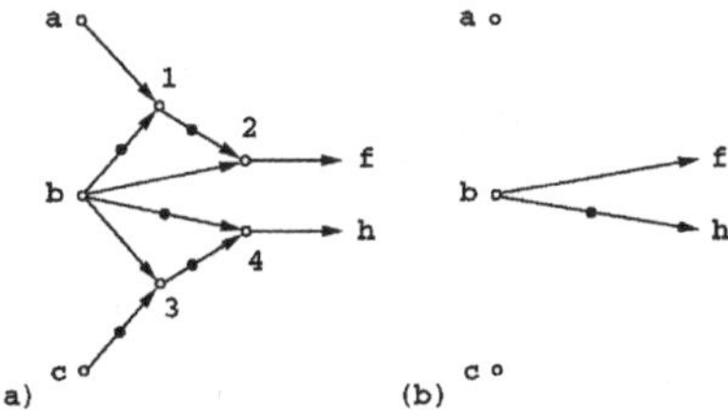

Fig. 3.6: A 3-input example: (a) pre-lookup structure $sig_f = \langle 10100110 \rangle$ and $sig_h = \langle 11100001 \rangle$ (b) post-lookup structure $f = b$ and $h = \neg b$

3.3.2 Four-Input Case

Similar to the three-input case, the construction of four-input substructures is based on a signature computation and table lookup to point to a non-redundant construction rule for the corresponding function. The outlines of the pseudo-code *4_input_signature* and *And_4* are shown in Figure 3.7. Using the design example shown in Figure 3.1, we describe how the algorithm *And_4* reduces the output *g* to an equivalent simpler representation. The algorithm first performs an analysis of the local structure and computes a 22-bit signature that uniquely reflects

1. the equality relationship of the four grandchildren (bits 0-4),
2. ordering among the grandchildren using the procedure *Rank* (bits 4-7)
3. the inverter attributes of the size edges using the procedure *Is_inv* (bits 8-13)
4. sharing among grandchildren, i.e., if any immediate predecessor are shared using the procedure *Share* (bits 14-21)

```
1.  Synopsis:   Compute signature for 4-input structure
2.  Input:      Edges L, R (Both from AND nodes)
3.  Output:     signature
4.  Procedure:  4_input_signature
5.
6.  //get children of L and R
7.  LL = Left_child(L); LR = Right_child(L);
8.  RL = Left_child(R); RR = Right_child(R);
9.
10. //Set i-th bit sig[i] if RHS predicate is true
11. sig[0]  = (Non_inv(LL) == Non_inv(RL));
12. sig[1]  = (Non_inv(LR) == Non_inv(RR));
13. sig[2]  = (Non_inv(LL) == Non_inv(RR));
14. sig[3]  = (Non_inv(LR) == Non_inv(RL));
15. sig[4]  = (Rank(LL) > Rank(RL));
16. sig[5]  = (Rank(LR) > Rank(RR));
17. sig[6]  = (Rank(LL) > Rank(RR));
18. sig[7]  = (Rank(LR) > Rank(RL));
19. sig[8]  = Is_inv(LL);    sig[9]  = Is_inv(LR);
20. sig[10] = Is_inv(RL);    sig[11] = Is_inv(RR);
21. sig[12] = Is_inv(L);     sig[13] = Is_inv(R);
22. sig[14] = Share(L,RL);   sig[15] = Share(L,RR);
23. sig[16] = Share(R,LL);   sig[17] = Share(R,LR);
24. sig[18] = Share(LL,RL);  sig[19] = Share(LR,RL);
25. sig[20] = Share(LL,RR);  sig[21] = Share(LR,RR);
26. return sig;
27.
28. Synopsis: Compute AND for 4-input structure
29. Input: Edges L,R (both from AND nodes)
30. Output: Boolean AND of L and R
31. Procedure: And_4
32.
33. l  = Non_inv(L); r = Non_inv(R);
34. ll = Non_inv(Left_child(L));
35. lr = Non_inv(Right_child(L));
36. rl = Non_inv(Left_child(R));
37. rr = Non_inv(Right_child(R));
38.
39. Signature = 4_input_signature(L,R);
40.
41. //Table lookup with signature to obtain
42. //pre-computed canonical sub-structure
43. index = LOOKUP4[signature];
44. switch(index) {
45.     ...
46.     Case 144: {p = And(rl,rr);}//Use in Figure 3.8
47.     ...
48.     Case 245: {p = And(And(¬l,rl),rr);} //Figure 3.8
49.     ...
50. }
51. return p;
```

Fig. 3.7: Pseudo-code for *4_input_signature* and *And_4*

The signature is then categorized into two sets of cases:

- The first set includes the local structures with shared grandchildren, i.e., $\exists i,\ 0 \leq i \leq 3\ sig[i]=1$. Here all substructures with identical or complemented Boolean functions get mapped to an isomorphic implementation, effectively removing local redundancy. Line 46 of *And_4* provides an example for case 144 illustrating this transformation for the function $(a \vee b) \wedge ab$. Figure 3.8(a) shows a particular instance of such a structure. In general, this mechanism is called recursively, which often results in a significant global reduction of the graph.

- The second set of cases does not share any grandchildren, i.e., $\forall i$, $0 \leq i \leq 3$ $sig[i]$=0. The signature distinguishes between the four cases of different polarity combinations of the children. For each case, if any sharing is found between grandchildren or great-grandchildren, specific rewriting rules are applied. Note, in Figure 3.8(b), $LL=\neg a$, $LR=\neg b$, $RL=rl$, $RR= c$, $L=\neg l$, and $R=r$. Clearly, all grandchildren are different, but *Share*(*L*,*RL*) is *true*. The AND expression *And*(*L, R*) are rearranged to *And*((*And*(*L*, *rl*), *rr*), as shown in case 245 (line 48 in Figure 3.7), so that case 144 (line 46 in Figure 3.7) can be applied on the nested AND. This gives the reduced structure as shown in the right of the Figure 3.8(b).

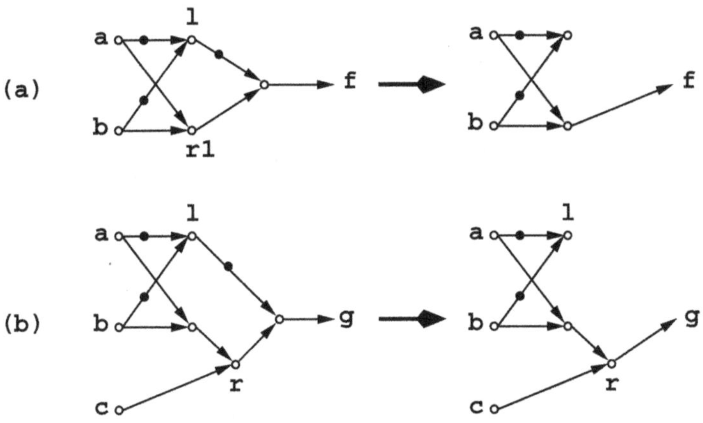

Fig. 3.8: Local reduction of output of the netlist in Figure 3.1 using procedure *And_4* of Figure 3.7, (a) reduction of using case 144, (b) reduction of using case 245 and then recursively using case 144

3.3.3 Example

In this section, we first demonstrate that the multi-level hashing scheme can find the functional identity of the two output nets of the example shown in Figure 3.3. Figure 3.9 illustrates how the network representation is built step-by-step.

As shown in part 3.9(a), the creation of the first set of vertices 1–5 is identical to Figure 3.3(b). Figure 3.9(b) illustrates that the construction of nets 6 and 7 yields an optimized structure for both vertices, removing the redundancy in the original description. The next part shows the creation of vertices 8–10, again without any achievable reduction by the multi-level hashing scheme. In contrast to the 2-input hashing method, the presented multi-level scheme can then identify the equivalence of vertices 5' and 5. This successively results in the merging of vertices 1''' and 7' with 1 and 7, respectively, which proves the equivalence of the two outputs.

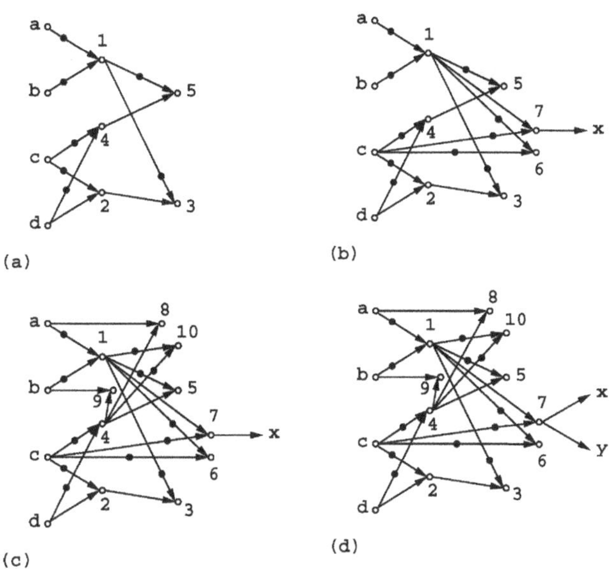

Fig. 3.9: Example of Figure 3.3 for circuit graph construction using multi-level hashing: (a) vertices 1–5 are identical to Figure 3.3(b), (b) vertices 6 and 7 yield more compact structures, (c) vertices 8–10 yield no compaction, (d) vertices 5', 1''', and 7' are merged with vertices 5, 1, and 7, respectively

In a second example, we show the effectiveness of the proposed multi-level hashing in combinational equivalence checking using a circuit example from a high-performance microprocessor design. This is illustrated graphically in Figure 3.10.

We selected an output pair, which had 97 inputs, 1322 gates for the specification and 2782 gates for the implementation as shown in Figure 3.10(a). When using the two-level hashing scheme, as shown in Figure 3.10(b), the number of nodes in specification and implementation reduces to 164 and 1473 respectively. While using the multi-level hashing scheme, as shown in Figure 3.10(c), there is a significantly larger reduction in the number of nodes; the number of nodes in specification and implementation reduces to 87 and 139 respectively. This clearly demonstrates the superiority of the multi-level hashing scheme over a two-input hashing scheme.

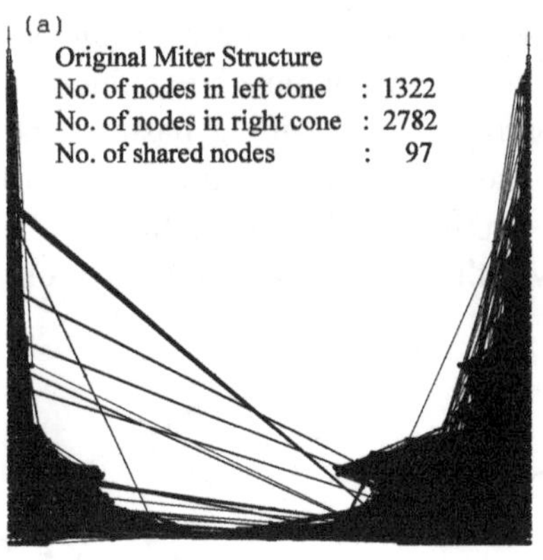

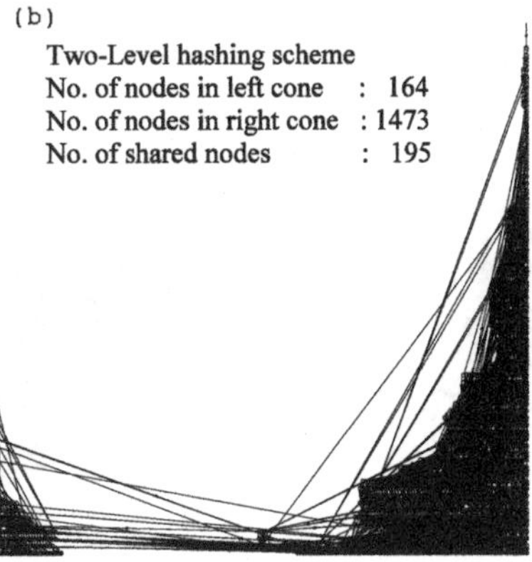

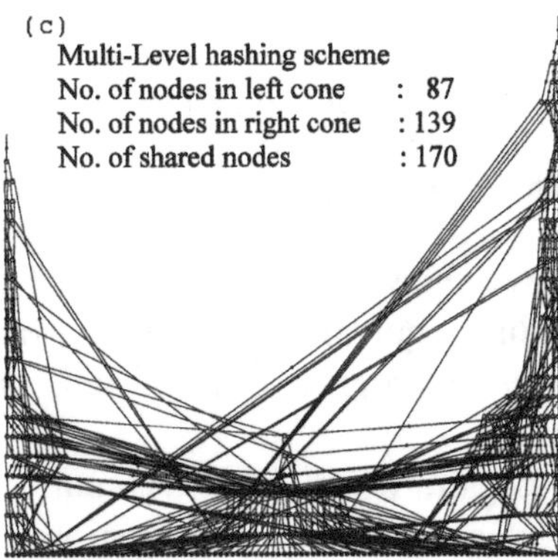

Fig. 3.10: Miter structure of Industry Example (a) without hashing (b) with two-input (or 2-level) hashing scheme (c) with multi-level hashing scheme

3.4 Experiments

We implemented the presented multi-level hashing algorithm in the equivalence checking tool [141]. In order to evaluate its effectiveness, we performed extensive experiments using 488 circuits randomly selected from a number of microprocessors designs. The circuits range in size from a few 100 to 100K gates, the size distribution is shown in Figure 3.11. The number of outputs and inputs per circuit range from a few 100 to more than 10,000. The experiments were performed on a RS/6000 model 270 with a 64-bit, two-way Power3 processor running at 375 MHz and 8 GBytes of main memory.

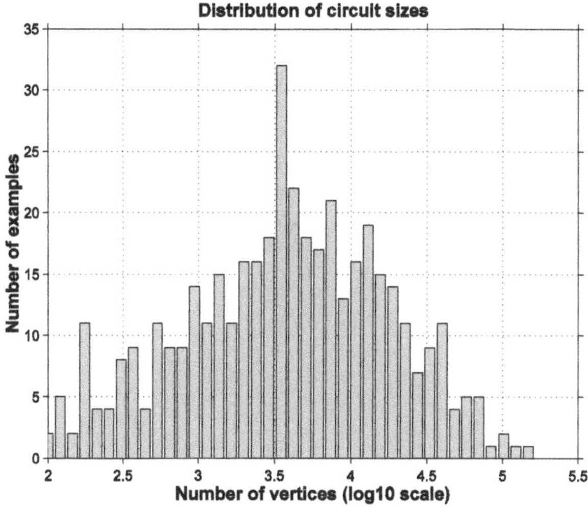

Fig. 3.11: Distribution of circuit sizes

In the first experiment, we constructed the circuit graph for the RTL specification only and compared the graph size using the newly presented multi-level hashing method with that using the two-input hashing method and BEDs. In Figure 3.12, we show the AIG size and runtime comparison between two-input hashing and multi-level hashing.

The histogram for the size reduction of the circuit graphs is plotted in Figure 3.12(a). Part (b) displays the computing overhead needed to achieve this reduction. As shown, on average the given sample of circuit representations can be reduced by 50% with minimal overhead in computing resources. For some design examples, we do find an increase in graph sizes. Similarly, in Figure 3.13, we show graph size and runtime comparison between multi-level hashing and BEDs. The histogram for the size comparison of the circuit graphs is plotted in Figure 3.13(a). Part (b) displays the runtime comparison between the two approaches. As shown, on average the multi-level hashing can achieve 40% reduced graph size as compared to BEDs for the given sample of circuit representations. Also, multi-level hashing has better runtime behavior as compared to BEDs.

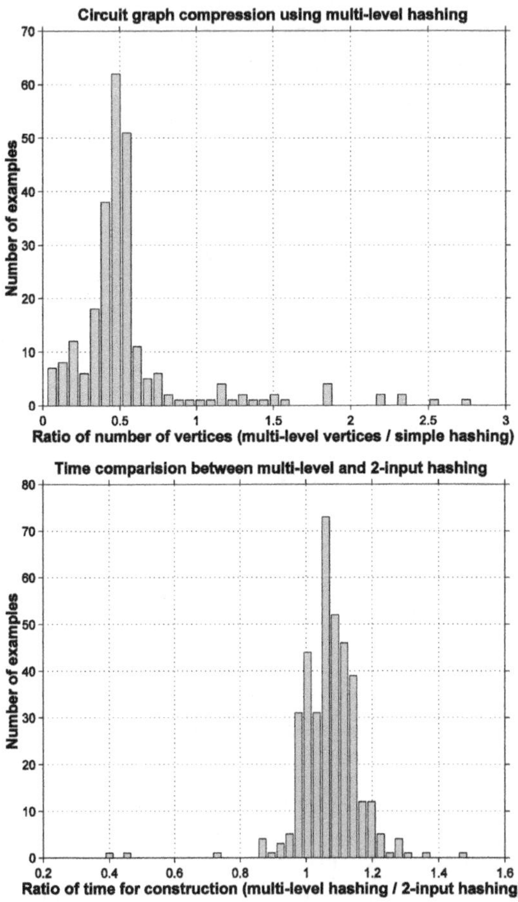

Fig. 3.12: AIG size comparison using multi-level vs 2-input hashing: (a) Compression ratio (ratio of the size of circuit built by multi-level to that by 2-input), (b) Time overhead (ratio of the time taken for circuit building by multi-level hashing to that by 2-input)

In the second experiment, we build the actual Miter circuit for performing the functional equivalence check between the RTL and transistor-level representation using the two schemes. We then compare our approach with two-input hashing method and BEDs in solving Miters as shown in Figure 3.14.

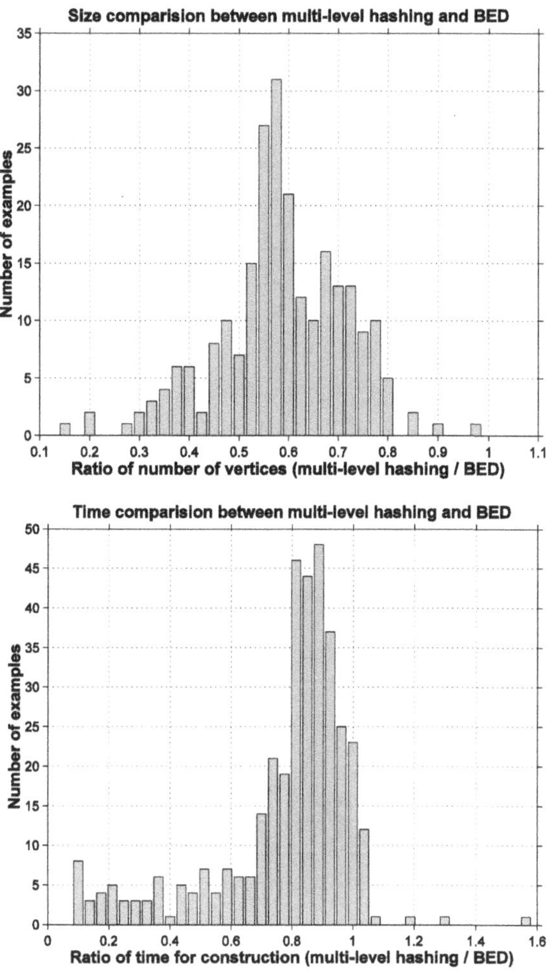

Fig. 3.13: Graph size comparison AIG with BEDs: (a) Reduction ratio (ratio of the size of circuit built by multi-level hashing to that by BEDs), (b) Time overhead (ratio of the time taken for circuit building by multi-level hashing to that by BEDs)

No additional equivalence checking algorithm (e.g. BDD sweeping [135]) has been applied, therefore only a subset of the outputs can be fully verified with the two structural hashing methods. Figure 3.14(a) shows for each circuit what fraction of the outputs can be solved with the two-input method and the multi-level method on the X-axis and Y-axis, respectively. For example, for the circuit indicated, the two-level hashing scheme could solve 31 of the Miter outputs whereas the multi-level hashing scheme was able to solve 86 of them. It can be seen that all crosses are in the upper triangle of the diagram, indicating that the multi-level hashing could always solve more problems than two-input hashing without engaging more powerful verification algorithms. Similar comparison with BEDs is shown in Figure 3.14(b). On average, we solve more number of Miters compared to that solved by BEDs.

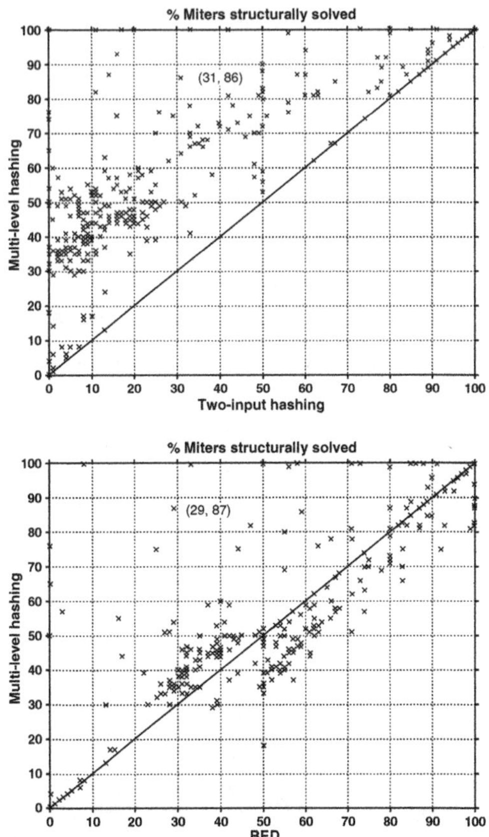

Fig. 3.14: Fraction of Miters solved: (a) comparison between multi-level and 2-input hashing (for the example indicated, two-input hashing solves 31% while multi-level hashing solves 86% Miters), (b) comparison between multi-level and BEDs (for the example indicated, BED solves 29% while multi-level hashing solves 87% Miters)

3.5 Simplification using External Constraints

The simplification algorithm described in the previous sections may not be able to identify and canonicalize the intermediate circuit nodes. In order to improve the simplification further, one can use external Boolean reasoning to identify the equivalent pairs and tautology nodes in the circuit.

Using such information, one can further simplify the circuit using the procedure *Merge* [135] as shown in Figure 3.15. As we shall see later, such information is typically obtained while solving BMC instances and has been exploited in the past [45] to simplify the BMC instance problem.

The function *Get_canon* returns the *canon* of the node and, *Set_canon* sets the canon for the node. Here, a *canon* represents the smallest structure functionally equivalent to the node. It is possible that the *canon* for a node that is functionally equivalent to tautology, may not be *const_1* or *const_0* node. The procedure *Merge* is called recursively for merging *p1* to *p2*, propagating the effect backward on its left and right children (lines 13-17) and forward on the fanout nodes (lines 18-23) .

```
1.  Synopsis:    Simplify using merging
2.  Input:       Edges p1 merged to p2
3.  Output:      Simplified structure
4.  Procedure:   Merge
5.
6.  //obtain canon
7.  c1 = Get_canon(p1);
8.  c2 = Get_canon(p2);
9.  if (c1 == c2) return c1;
10. if (Rank(c1) < Rank(c2)) Swap(c1,c2);
11. Set_canon(c1,c2); //c2 is new canon for c1
12. //Backward Merging
13. if ((c2 == const_1 && !Is_inv(c1)) ||
14.      (c2 == const_0 && Is_inv(c1))) {
15.    Merge(Left_child(c1),1);
16.    Merge(Right_child(c1),1);
17. }
18. //Forward simplification
19. forall fanouts fout of c1 {
20.   cout = AND(Get_canon(Left_child(fout),
21.              Get_canon(Right_child(fout));
22.   Merge(fout,cout);
23. }
```

Fig. 3.15: Pseudo-code for structural merging

3.6 Comparing Functional Hashing with BDD/SAT Sweeping

In [135-137], a two-step approach for identifying functionally identical nets is presented. First, during the actual network construction, structural hashing is applied to consolidate isomorphic subcircuits. Then, a BDD sweeping [135, 136] and/or SAT sweeping [137] technique are used to further identify and merge functionally equivalent nets. Although BDD sweeping is quite general and can identify all functional identical nets as long as the BDD representation does not grow too large, this method has a considerable overhead caused by the BDD manipulation. The functional hashing technique presented in this chapter is not as general as the BDD sweeping method. However, it is based on a constant-time table lookup scheme that, if applicable, avoids the significant overhead of the BDD processing.

Furthermore, BDD sweeping requires the identification of intermediate nets as cutpoints from which the sweeping process restarts. Finding meaningful cutpoints is not an easy task and there are no robust heuristics in general [142]. The presented functional hashing technique completely avoids cutpoint selection by working on all subcircuits of a fixed depth. Moreover, as exploited in [135], an upfront structural compression of the network increases the average fanout of intermediate nets and as a result enhances the effectiveness of all selected cutpoints. In essence, the presented functional hashing method does not replace the cutpoint-based BDD sweeping or SAT sweeping techniques; it rather increases its robustness through the additional static compression. Since our graph representation preserves the AND clustering, the hashing can take advantage of its commutative property which makes it significantly less sensitive to the order in which the structure is built.

3.7 Summary

This chapter describes an efficient algorithm to simplify logic network representations during construction. The method is based on a table lookup hashing scheme that maps each subcircuit onto a unique functional signature, which is then used to implement the function. We demonstrate that for industrial circuits we can compress the circuit representation by up to 90% with insignificant computational overhead. This compression directly results in a reduction of the memory requirements to store the network. Moreover, the application of the proposed method in functional equivalence checking results in a significantly larger number of outputs that can be verified without engaging algorithms that are more computationally expensive techniques such as BDD/SAT sweeping.

3.8 Notes

The subject material described in this chapter are based on the authors' previous works [42], [39] © ACM 2001, and [136] © 2002 IEEE.

4 HYBRID DPLL-STYLE SAT SOLVER

4.1 Introduction

The Boolean Satisfiability (SAT) problem has extensive applications in VLSI CAD. Recent advances in SAT solvers based on Conjunctive Normal Form (CNF) representation have resulted in significantly improved performance. In particular, innovative techniques for decision variable selection [38], Boolean constraint propagation (BCP) [36, 38], and backtracking with conflict analysis based learning [37, 38] have led to high-performance CNF-based SAT solvers like Chaff [38]. For circuit application domains such as Automatic Test Pattern Generation (ATPG) [143], equivalence checking [39], and Bounded Model Checking (BMC) [66], the Boolean reasoning problem is typically derived from the circuit structure. This has also led to interest in circuit-based SAT solvers [39, 143-145], which use circuit specific knowledge to guide the search.

A Hybrid SAT solver such as [41] uses a hybrid approach to take advantages of both circuit-based and CNF-based SAT techniques. In this hybrid approach, the original logic formula is processed in circuit form, and the learned clauses are processed separately in CNF. Such hybrid technique leads to a consistent performance improvement of up to a factor of 3 over state-of-the-art CNF-based SAT solvers [38] on representative problem applications. It can also be augmented with lightweight BDD-based deductive learning, as discussed later in Chapter 5, within a bounded region for further improving the Boolean search [146]. We have found a Hybrid SAT solver to be the most robust framework for circuit applications, and use it as the primary Boolean reasoning workhorse for SAT-based model checking algorithms described in later chapters. In this chapter, we describe the key insights for its wider acceptance as a *defacto* Boolean search engine for circuit applications.

We discuss the important differences between the CNF and circuit-based approaches, and highlight how one can reap benefits from both by employing state-of-art innovations in BCP, decision variable selection, and backtracking. In the following, we discuss issues in the past efforts of using one solver or the other in circuit domain applications, and the need for a hybrid SAT solver.

Circuit-based Information Adds Overhead to CNF-based Solver

In general, attempts to include circuit structure information into CNF-based solvers have been unsuccessful or partially successful [147] due to significant overhead. Furthermore, it is also difficult to integrate CNF-based solvers with other useful techniques like BDD sweeping and dynamic circuit transformation [39]. On the other hand, in the hybrid approach, the original formula is retained as a circuit structure similar to the circuit-based SAT approach.

Circuit to CNF Translation

Though translation from circuit to CNF is polynomial time [95, 96], there is an overhead of copying the entire circuit structure into a CNF data structure, especially, when such translations are invoked repeatedly. For large circuits, or for BMC application at higher depths, such translation consumes both memory and time. Though circuit clauses are generally short, in the hybrid approach it is unnecessary to incur such overhead as the problem is retained in circuit form.

Circuit-based Learning is Inefficient

Conflict-driven learned clauses arising from conflict analysis are typically much larger than those arising from two-input gates. Adding a large learned clause as a gate tree can lead to a significant increase in the size of the circuit. This in turn, can increase the number of circuit implications, thereby negating any potential gains obtained from circuit-based BCP. As an example, illustrated in Figure 4.1(a), the learned clause in CNF would be (-H + D + -A + -B). An equivalent circuit-learning [39] would add a circuit structure as shown in Figure 4.1(b), increasing the overall circuit structure. In the hybrid SAT approach [41], such clauses are maintained as CNF clauses and are processed during BCP by using CNF-based 2-literal watching scheme with lazy update.

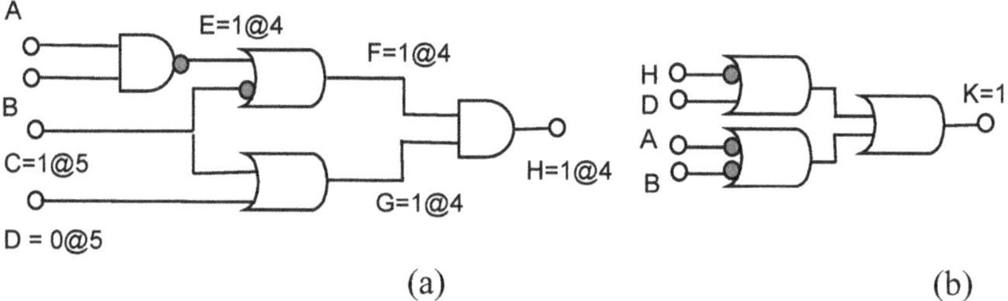

Fig. 4.1: a) Circuit-based SAT, b) Circuit-based learning

Improvements in CNF- and Circuit-based Solvers

There have been many ongoing research activities [106] to improve SAT solvers further. Often, the heuristics are very specific to either CNF-based or circuit-based SAT solvers, and not easy to incorporate in the other domain without incurring additional overhead. A Hybrid SAT solver overcomes this problem very easily, and benefits from any continuing and future improvements in SAT solvers of either domain.

In our later discussion, we use Chaff solver (version: *zChaff-2001*) [38] as a representative CNF-based solver for comparative evaluation. We performed experiments on a Linux workstation with 750 MHz Intel PIII and 256 MB. The examples used are large logic formulas, with 25K to 0.5 million gates, derived from the application of BMC on three large industrial circuits (bus, arbiter, and controller) and some public domain benchmarks [148] for which circuit descriptions were available. We consider 45 non-trivial examples for which Chaff took more than 40 seconds to solve. The formulas are distributed between 34 unsatisfiable and 11 satisfiable instances. In the following sections, we use this set to discuss our experimental results.

4.2 BCP on Circuit

BCP forms the computational core of a SAT solver, typically constituting 80% of solver time. Improvement in BCP directly benefits the overall performance of SAT solver. Existing circuit-based SAT solvers [39, 143-145] use a circuit representation based on AND and OR gate vertices, with INVERTERs either as separate vertices, or as attributes on the gate inputs. Constant propagation across AND/OR gates is, of course, well known, but the speed tends to be very implementation dependent. A circuit-based BCP implementation [39], algorithm *Imply* based on a table lookup, is shown in Figure 4.2.

The algorithm *Imply*, shown for a generic vertex type, iterates over the circuit graph, determining new implied values and the forward and/or backward directions for further processing. It takes specific advantage of the single vertex function for efficient propagation of logic implications. Based on the current values of the inputs and output of the vertex, the lookup table determines the next "state" of the gate where the state encapsulates any implied values and the next action to be taken for the vertex (line 14). As an example, Figure 4.3 shows some cases from the implication lookup table for a two-input AND gate. For Boolean logic, only one case, a logical *0* at the output of an AND vertex, requires a new case split to be scheduled for justification (line 19). All other cases cause a conflict and backtracking (line 16), or further implications (line 21), or a return to process the next element to be justified. Due to its low overhead, this implication algorithm is highly efficient. As an indication, on a 750MHz Intel PIII with 256 MB, it executed over one million implications per second.

```
1.  Synopsis:      BCP forward and backward
2.  Input:         Node p, Value val
3.  Output:        1: NO_CONFLICT, 0: CONFLICT
4.  Procedure:     Imply
5.
6.  Set_value(p,val); //assign val to p
7.  if (Is_var(p)) return 1; //o is variable
8.  //get children of p
9.  l = Left_child(p);
10. r = Right_child(p);
11. lval = Get_value(l);
12. rval = Get_value(r);
13.
14. next_state = Lookup(val, lval, rval);
15. switch (next_state→action) {
16.   Case CONFLICT:
17.     return 0;
18.   Case CASE_SPLIT:
19.     Add_justify_queue(p); //need to justify p
20.     return 1;
21.   Case PROP_LEFT_AND_RIGHT:
22.     if (Imply(l,next_state→lval) &&
23.         Imply(r,next_state→rval))
24.       return 1;
25.     }
26.     return 0;
27.   ...
28. }
29. return 1;
30. }
```

Fig. 4.2: Circuit-based BCP

4.2.1 Comparing CNF- and Circuit-based BCP Algorithms

There is an inherent overhead built into the translation of circuit gates into clauses. A two-input gate translates to three clauses in the CNF approach, while in the circuit-based approach a gate is regarded as a monolithic entity. Therefore, in the circuit approach an implication across a gate requires a single table lookup, while in the CNF approach it requires processing multiple clauses. In addition, the CNF-based BCP, such as in Chaff, does not keep track of the clauses that have been satisfied in order to reduce overheads. However, there is an inherent cost associated with visiting the satisfied clauses. Specifically, even if a clause gets satisfied due to an assignment to some un-watched literal, the watched literal pointers could still get updated whenever one of the watched literal is assigned false. Overall, for the generally small clauses arising from circuit gates, these differences translate to significant differences in BCP time, usually in favor of the circuit-based approach.

Current	Next	Action
1, X → X	1, X → X	STOP
0, X → 1	0, X → 1	CONFLICT
X, X → 0	X, X → 0	CASE_SPLIT
0, X → X	0, X → 0	PROP_FORWARD
X, X → 1	1, 1 → 1	PROP_LEFT_RIGHT
...	...	...

Fig. 4.3: 2-input AND Lookup Table for Fast Implication Procedure

We compared the CNF-based and Circuit-based BCP times, measured in seconds *per million implications* in a controlled setup, shown in Figure 4.4. The CNF-based solver uses lazy clause update BCP scheme, with two watched literals. The circuit-based solver uses fast table lookup as discussed before. In this controlled experiment, the implications of only gate clauses were considered, ignoring the effect of the learned clauses. One can clearly notice that circuit-based BCP is 50% faster than CNF-based BCP. In the hybrid approach, the circuit-based logic expressions are maintained using a uniform-gate data structure, and the learned clauses are maintained as CNF. Appropriate BCP is used to process them separately.

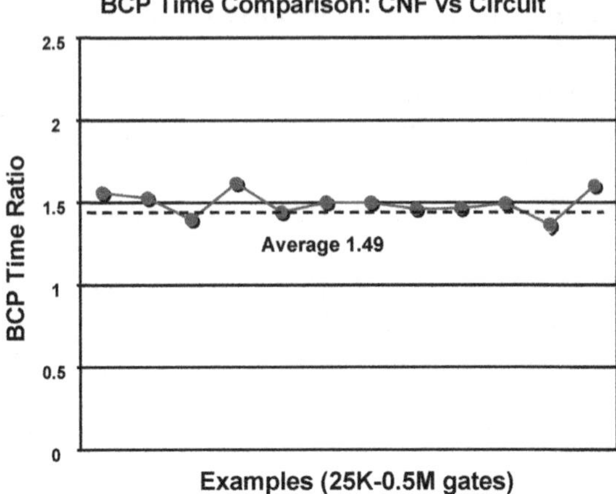

Fig. 4.4: Comparison of BCP time per million implications of gate clauses between CNF-based solver using *lazy update* and Circuit-based Solver using *fast table lookup*

4.3 Hybrid SAT Solver

A Hybrid SAT solver uses a hybrid representation for the logic expressions, i.e., short gate clauses are represented as 2-input gate structure (i.e., *Reduced AIG*) and long clauses such as those generated from conflict-driven learning are represented as CNF. We now discuss the three engines of a hybrid SAT solver, i.e., deduction engine, diagnosis engine and decision engine, that process the hybrid representation.

BCP engine involves separate processing of each representation, combining the best of solvers of each domain. For short gate clauses, circuit-based BCP scheme based on table lookup is used. For large clauses, CNF-based BCP scheme based on lazy update using two watched literals is used.

The diagnosis engine in the hybrid solver is similar to that in advanced CNF-solvers [34]. In the hybrid approach, conflict driven learning records both *clauses* and *circuit nodes* as the reasons for a conflict.

In addition to the VSIDS decision strategy, the hybrid approach can also benefit from circuit-based decision heuristics (described later in Section 4.4), which are otherwise not available to pure CNF-based approaches. Information regarding gate fanout/fanin, paths, and signal direction becomes readily available from the circuit structure, without incurring any overhead.

4.3.1 Proof of Unsatisfiability

We describe the procedure *SAT_get_refutation* to obtain *resolution proof tree* from a given unsatifiable instance in circuit form. Conflict analysis in a hybrid SAT solver traverses back from the conflicting nodes in an implication graph. However, edges in such a graph may correspond to hybrid representations of constraints. For example, while performing BCP directly on a circuit netlist, edges in the implication graph may correspond to nodes in the circuit. We record the reasons for the conflict, in their hybrid representations, and associate them with the learned constraint (corresponding to the conflict clause). When the final conflict is found, indicating the unsatisfiability, a marking procedure is started from its antecedents. Again, reasons for any learned constraints are marked recursively. At the end of this procedure, the marked constraints from the original problem constitute an unsatisfiable core. For example, given an unsatisfiable circuit problem due to external constraints, this procedure identifies a set of nodes in the circuit that are sufficient for implying the unsatisfiability.

4.3.2 Comparison with Chaff

We present comparison results for the hybrid SAT approach and Chaff. Apart from the different BCP algorithms on the circuit formula, the two solvers use identical algorithms for order of processing of implications, conflict-based learning, backtracking, and decision variable selection. We call this hybrid approach as the basic Hybrid SAT algorithm (*H1*).

We first compare the BCP time of Chaff and hybrid SAT (*H1*) on learned clauses and original circuit formula. The result is shown as the ratio of Chaff to *H1* in Figure 4.5. In spite of the same heuristics, a minor difference can creep in due to the uncontrolled choice of the conflict node when several nodes are in conflict. This difference may have a pronounced effect in satisfiable instances, for which one of the two solvers may get lucky in hitting upon a solution early. However, it has relatively less effect in unsatisfiable instances since the entire search space must be explored. With this in mind, we consider only the unsatisfiable instances for this controlled experiment. It is clear from the results that a hybrid BCP approach is consistently better than CNF-based BCP on these large problems, with an average improvement of 1.33.

In Figure 4.6, we compare the total SAT solver time of Chaff and hybrid solvers, shown as a ratio of the two approaches, respectively. Clearly, the overall performance of the hybrid solver is much better than Chaff, with an average 1.3 and the maximum 3.75. The ratio of the total time spent in Chaff to the total time spent in the hybrid solver for all the unsatisfiable instances is 1.48. Due to possibly a bad choice for a conflict node, we do see a larger time with the hybrid solver in a few cases. For the satisfiable instances shown on the right side, the Chaff to Hybrid ratio is distributed evenly on either side of 1.0 with a large standard deviation. This is most likely due to the randomness in choice of a conflict node, as explained earlier.

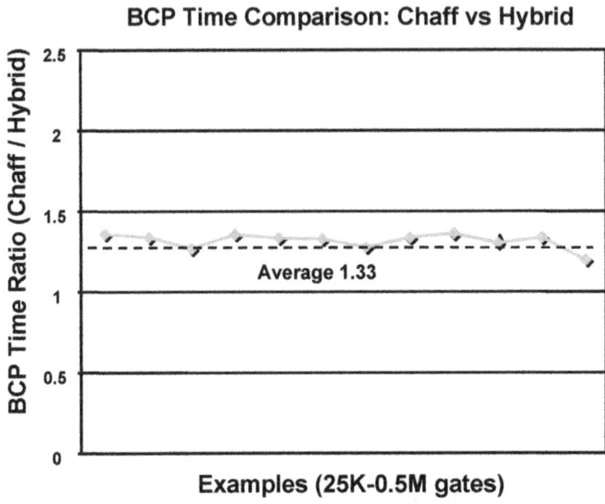

Fig. 4.5: Comparison of BCP time per million implications of gate and learned clauses between Chaff and hybrid solver

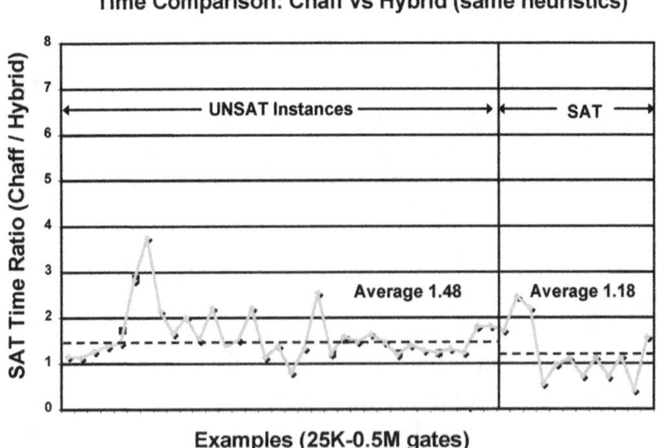

Fig. 4.6: Comparison of total solver time of Chaff and hybrid solver (both using same decision heuristics)

4.4 Applying Circuit-based Heuristics

Once it is established that the hybrid approach is, in fact, faster, it is possible to apply low overhead circuit-based heuristics that would be unavailable in the CNF domain to obtain an even higher speedup. Some of the circuit-based heuristics such as justification frontiers, gate fanout counts, *FIFO/LIFO* implication queue, and learning *XOR/MUX* gates, can easily be applied in this hybrid framework. We find that such an approach works much better than the circuit-oblivious decision heuristics used in the pure CNF-based methods. In the following, we present details of the circuit-based heuristics. We provide a summary of experimental results for each level of the heuristic to give the insight of this strategy in improving the performance of basic hybrid SAT with same heuristics as Chaff (*H1*). We used the same set of benchmarks and experimental setup as discussed previously.

4.4.1 Justification Frontier Heuristics

In propagation-justification type of Boolean reasoning in a circuit-based solver [39, 143-145], decisions are restricted to the nodes that justify the values on their fanout nodes. For example, in Figure 4.1, for justifying the OR node G=1, decision is restricted to C or D nodes. Variables that need to be justified are called *frontier* variables. Thus, we call this decision strategy as the *justification frontier* heuristic, where the frontier changes dynamically. Note, if there are no frontier variables when the decision engine is invoked, i.e., no more justification required, the problem can be declared satisfiable. Note, while some of the variables may still remain unassigned, one can show that any assignment to such variables will not affect the satisfying solution. Not surprisingly, this heuristic has the benefit of leading to a satisfying assignment faster when it exists. Such a satisfying solution with a partial assignment on variables has several benefits in applications such as SAT-based quantification, localizing circuit faults, test-vector compression in ATPG.

The stopping criterion of a CNF-based approach, such as [38], for a given satisfiable instance is when there are no free variables. There have been some attempts [147] to improve the stopping criteria by incorporating the circuit information dynamically. Clauses that do not determine the satisfaction of the formula are termed inactive clauses. These clauses arise from gates that become unobservable. Processing of these clauses and decisions on variables in these clauses is wasted effort. Dynamic detection and removal of these clauses requires marking and unmarking of these clauses. Even though these run time operations lead to a pruned search space, the repeated marking and unmarking leads to a loss in the overall performance. In contrast, justification frontier heuristic can be effectively implemented in circuit-based BCP using the directionality of the circuit structure.

Experimental results for using the justification frontier heuristic are shown in Figure 4.7. The use of this heuristic further improves performance in 24 of the 34 unsatisfiable instances. The ratio of the total time spent in Chaff to the total time spent in the hybrid solver for the unsatisfiable instances is now 1.89 versus the earlier 1.48, with the maximum ratio being 5.94 versus the earlier 3.75. Now consider the satisfiable instances (shown in the right part of Figure 4.7). Where there was wide variation in the speedups (or lack thereof) earlier, as shown in Figure 4.6, we now have consistent and significant speedup over the CNF approach. The ratio of the total time spent in Chaff to the total time spent in the hybrid solver is now 3.24 for the satisfiable instances, with the maximum ratio being 7.07. Clearly, the circuit heuristics lead to a satisfying solution much faster than the decision heuristics in Chaff. Comparing the hybrid solver with and without this circuit-based heuristic, we observed that this heuristic is useful in 80% of the examples.

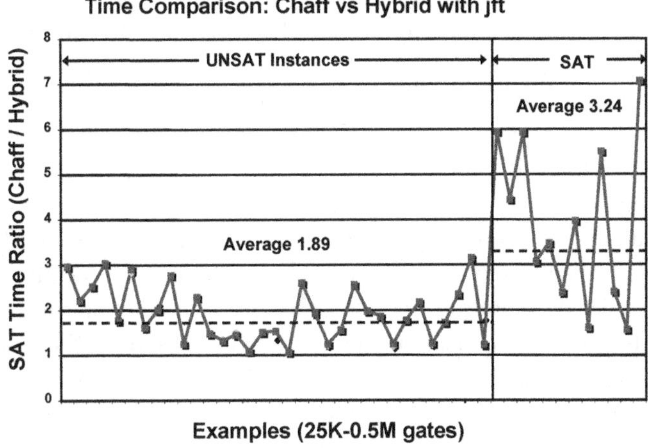

Fig. 4.7: Comparison of SAT time of Chaff and hybrid solver using justification frontier heuristics (jft)

4.4.2 Implication Order

As discussed earlier, most CNF solvers uses a FIFO queue to store the new assignments, i.e., implications are processed in the order in which they are generated. With the knowledge of fanouts and directionality in the hybrid approach, it is possible to follow implications based on circuit paths, i.e., implicitly in a LIFO order. We found that an approach in which the implications generated from gates are followed in FIFO order, while the implications generated from the learned clauses are followed in LIFO order, works very well. Intuitively, when multiple antecedents exist for an implication, such a strategy gives preference to conflict-driven learned clauses over gate clauses as antecedents. We refer to this heuristic as *hybrid implication* (*H2*). In our experiments with this heuristic, we observed a speed up over basic hybrid SAT (*H1*) on over 60% of the examples.

4.4.3 Gate Fanout Count

The VSIDS decision strategy, as discussed earlier, uses an initial literal count to establish scores of the literals, updates literals selectively when conflict-diven learned clauses are added, and decays the score of all literals to give higher weight to recently updated literals. We now discuss how circuit information can be used to enhance this basic mechanism.

A CNF-based solver such as Chaff [38] uses positive and negative literals count in clauses to prioritize the decision variables. The objective is to satisfy maximum number of clauses by the chosen decision variable. Unfortunately, for CNF clauses derived from circuit with 2-input gates, literal counts do not address the objective as shown in Figure 4.8. The OR gate *C* has fanins *A* and *B*, and fanouts *X*, *Y*, and *Z*. The clauses generated are shown on the right of the figure. Note that there is equal number of positive and negative literals of any fanin in the gate clauses for the corresponding gate. Also, the positive literal count is always one more than the negative literal count of the fanout in the corresponding gate clauses. The positive and negative literal count for C in these gate clauses are 5 and 4, respectively. In short, the literal count actually does not provide any useful information for clauses generated from gates. With the gate fanout information, on the other hand, one can accurately determine the number of positive and negative fanouts of a gate for decision making. We call this strategy the *fanout* (*fs*) heuristic. Combining *hybrid implication* with *fanout* heuristic (*H2+fs*), we found that *H2+fs* approach gives better results on 60% of the examples compared to *H2*. When we add *justification frontier* heuristic as well, i.e., *H2+ft+fs*, we obtain a speed up over *H2* on 73% of the examples.

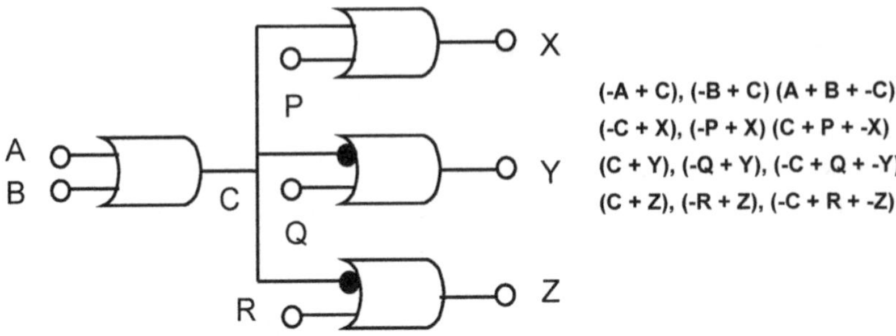

Fig. 4.8: Example of counting literals in Clauses

4.4.4 Learning XOR/MUX Gates

Circuits with large numbers of *XOR/MUX* gates are known to degrade the performance of SAT solvers [149, 150]. Given *Reduced AIG* representation, one can easily identify *XOR/MUX* gates due to their identical structures. From the input-output relation of these gates, one can learn clauses and add them as CNF in the hybrid representation as shown in Figure 4.9(a-b). For *XOR* circuit node, $C = \neg(A \vee B) \vee \neg(\neg A \vee \neg B)$, we learn four clauses as shown in Figure 4.9(a). Similarly for *MUX* node $C = \neg(S \vee B) \vee \neg(\neg S \vee A)$, we learn six clauses [151] as shown in Figure 4.9(b); out of which 2 are independent of mux select signal *S*. This learning combined with *H2+ft+fs* heuristics gives a speed up over *H2* in more than 62% of the examples.

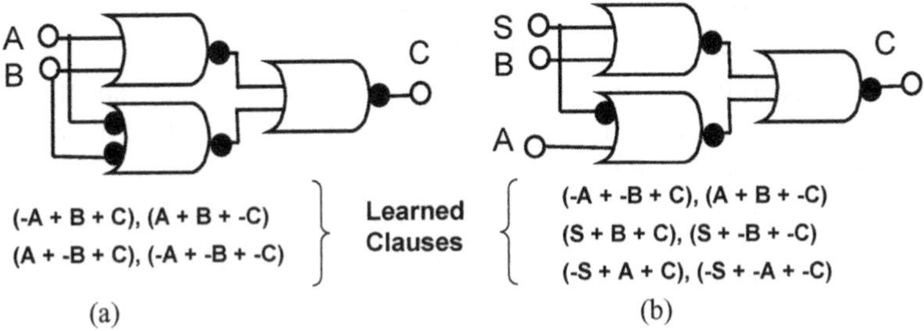

Fig. 4.9: Learning of CNF-clauses in Hybrid SAT (a) XOR gate, (b) MUX gate

Based on these data, there appears to be a clear benefit in using the circuit-based heuristics for both the satisfiable and unsatisfiable instances. In a sense, the hybrid approach allows us to exploit the benefits of the circuit-based as well as the CNF-based heuristics. Comparing the best results of the hybrid solver (among all the heuristics mentioned above) with that of Chaff, we found a speed up of at least 50% in all of the examples and greater than a factor of 2 in 69% of the examples.

4.5 Verification Applications of Hybrid SAT Solver

A Hybrid SAT solver is the primary workhorse and ideally suited for SAT-based model checking algorithms due to its robust and adaptable framework. A Hybrid SAT solver has several nice features that find use in various verification applications. We list some of them:

- *SAT-based Bounded Model Checking:* SAT is the core engine in SAT-based BMC. In fact, it is typical for a long BMC run to last for multiple days, requiring thousands of applications of the SAT solver. A speed up by a factor of two in the core engine, therefore, can prove to be very significant since it would lead to a large absolute saving in the run time. Further, it is unnecessary to incur the overhead of copying the entire circuit into a CNF data structure. This has the benefit of almost halving the memory requirement of these applications, allowing them to scale to large circuits, or large number of time frames.
- *SAT-based BMC with Efficient Memory Modeling:* A hybrid problem representation allows modeling of memory forwarding semantics efficiently in both CNF and circuit form, using the knowledge of underlying BCP schemes [50, 51].
- *SAT enumerations:* For SAT-based quantification, enumerating all SAT solutions is an essential step. A Hybrid SAT solver with justification frontier heuristics is ideally suited for the enumeration, as it generates a satisfying solution with a partial assignment [47]. In contrast, this effect is captured as cube enlargement in the post-processing step of a CNF-based solver.
- *Unsatisfiable SAT core:* Proof-based abstraction methods rely on extraction of unsatisfiable SAT cores. Using various circuit heuristics, one can direct the branching so as to obtain a smaller unsatisfiable core that leads to a smaller abstract model [48].
- *Distributed SAT:* As a hybrid SAT solver uses directionality of signals for BCP, one can very efficiently use this feature to distribute SAT in BMC application on a network of workstations [152].

4.6 Summary

We discussed a hybrid SAT solver that combines the strengths of circuit-based and CNF-based SAT solvers. It is demonstrably the highest performing SAT solver for circuit-based application domains. The approach is based on an analysis of the source of efficiencies in state-of-the-art techniques for BCP, decision variable selection, and backtracking. With this understanding, hybrid SAT techniques are developed to leverage off the circuit nature of these application domains while maintaining advantages arising from the CNF representation. Moreover, any future improvements in the performance of circuit-based and CNF-based SAT solvers can directly translate into improvements of the hybrid SAT solver as well.

4.7 Notes

The subject material described in this chapter are based on the authors' previous work [41] © ACM 2001.

PART II: FALSIFICATION

In Part II (Chapters 5—8), we discuss various accelerated techniques to find bugs and witnesses.

In Chapter 5, SAT-BASED BOUNDED MODEL CHECKING, we present a customized SAT-based Bounded Model Checking approach for verifying simple safety, reachability, and nested liveness properties. Customization involves property-specific problem partitioning, incremental formulation, and incremental learning. We also discuss various bug-finding accelerated techniques such as dynamic circuit simplification, SAT-based incremental learning, lightweight and effective BDD-based learning integrated with hybrid SAT solver.

In Chapter 6, DISTRIBUTED SAT-BASED BMC, we describe techniques to distribute hybrid SAT and SAT-based BMC over a network of workstations in order to overcome the memory limitation of a single server, for performing deeper search for very large industry designs. For the sake of scalability, at no point in the BMC computation does a single workstation have all the information. Such an approach scales well in practice, with acceptable communication cost.

In Chapter 7, EFFICIENT MEMORY MODELING IN BMC, we describe an abstract interpretation of embedded memory called Efficient Memory Modeling used in BMC. It provides a sound and complete abstraction that preserves the memory semantics, thereby, augmenting the capability of SAT-based BMC to handle designs with large embedded memory without explicitly modeling each memory bit.

In Chapter 8, BMC FOR MULTI-CLOCK SYSTEMS, we describe an integrated solution within the SAT-based BMC framework for verification of synchronous multi-clock systems with nested clocked specifications.

5 SAT-BASED BOUNDED MODEL CHECKING

5.1 Introduction

Bounded Model Checking (BMC) has been gaining ground as a falsification engine, mainly due to its improved scalability compared to other formal techniques. In BMC, the focus is on finding counterexamples (bugs) of a bounded length k. For a given design and correctness property, the problem is translated effectively to a propositional formula such that the formula is true if and only if a counterexample of length k exists [66, 67]. Such a translation basically involves unrolling the circuit of the transition relation for the required number of time frames as illustrated in Figure 5.1 (with k=d). Essentially, d copies of circuit are made and then clauses are built at each time frame for the unrolled circuit and the property to be checked, which is then fed to a SAT-solver. In practice, the bound k can be increased incrementally to find the shortest counterexample. A separate reasoning is needed to ensure completeness when no counterexample can be found up to a certain bound [66, 67]. However, with increasing depth the problem size, comprising the unrolled circuit and translated property, grows linearly [111] with the size of the model, thereby the making the Boolean reasoning increasingly difficult for large bounds, in general. The standard translation for these properties is *monolithic*, i.e., the entire propositional formula is generated for a given k and then the formula is checked for satisfiability using a standard SAT solver. Such a monolithic translation provides little scope for an incremental formulation within a k-instance BMC problem. Therefore, the past efforts [153, 154] on incremental learning for BMC have been limited to sharing of constraints across k-instances only.

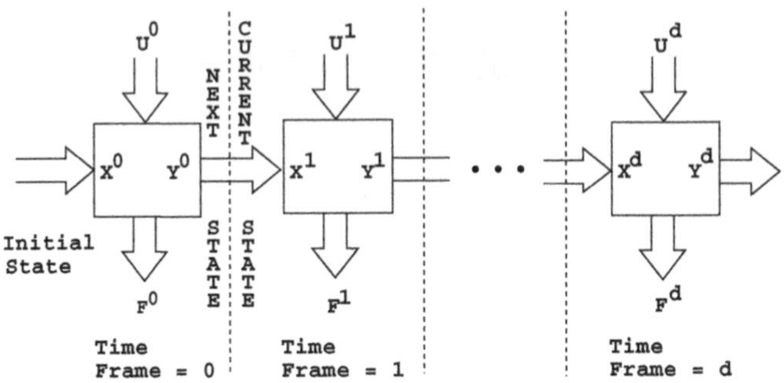

Fig. 5.1: Iterative array model of synchronous circuit

Outline

In this chapter, we discuss several enhancements proposed in the last few years to make the standard BMC procedure [66, 67] scale with large industry designs. The first key improvement that we discuss is dynamic circuit simplification [45] in Section 5.2. This is performed on the iterative array model of the unrolled transition relation, where an on-the-fly circuit reduction algorithm, discussed in Chapter 3, is applied not only within a single time frame but also across time frames to reduce the associated Boolean formula.

The second key enhancement we discuss is SAT-based incremental learning in Section 5.3. This is used to improve the overall verification time by re-using the SAT results from the previous runs.

Next, we discuss a lightweight and goal-directed effective BDD-based learning scheme in Section 5.4, where learned clauses generated by BDD-based analysis are added to the SAT solver on-the-fly, to supplement its other learning mechanisms. We also discuss several heuristics for guiding the SAT search process to improve the performance of the BMC engine.

Based on the above simplification and learning techniques, we discuss the use of customized property translations, in Section 5.5, for commonly occurring LTL formulas such as safety and liveness, with novel features that utilize *partitioning*, *learning*, and *incremental* formulation [46]. Such customized translations allow not only incremental learning across *k*-instances, but also within *k*-instances of BMC problems and thereby, greatly improve overall performance. Moreover, in comparison to the standard translation, customized translations provide additional opportunities for deriving completeness bounds for nested properties (this is discussed in more detail in Chapter 10).

In Section 5.6, we present controlled and comparative experimental results, and summarize our discussion in Section 5.7.

5.2 Dynamic Circuit Simplification

We discuss the main motivation and basic procedure for applying dynamic circuit simplification in the context of BMC. BMC, in some sense, can be considered as performing a symbolic simulation where symbolic expressions are used to evaluate functions that encode circuit behavior. As pointed out by Bryant [155], expressing simulation operation using Boolean algebra also leads directly to a symbolic formulation. A symbolic domain becomes in turn a set of Boolean functions over these variables. Therefore, a symbolic evaluation at every step can be expressed as logic operations — OR, AND, and INVERTER over such functions rather than over the binary values 0 and 1.

For illustration, consider the iterative model of a synchronous circuit as shown in Figure 5.1. The simulator starts from an initial state at time frame 0. At each successive time frame, a new symbolic variable is introduced for each primary input and then the primary outputs and the next state nodes are computed in terms of these symbolic variables. The outputs may correspond to checkers for safety properties, coverage goals, Miter [134] outputs for sequential equivalence checking. If the design has an unexpected behavior (i.e., a bug) at some output at d^{th} time frame, then one can determine the entire input sequence from the function at that primary output at the dth time frame. The downside of this technique is that the symbolic expressions grow exponentially for successive tests. The use of BDDs further limits the capability of symbolic simulators [156-158] due to memory intensive nature of BDDs.

More lately, use of non-canonical circuit graphs and reduction algorithms have been proposed [45, 109, 159] to represent the iterative array model succinctly during BMC. In one such approach [109], a Reduced Boolean Circuit (RBC) is used to represent the transition relation and formulas in a reduced form, however circuit simplification is still restricted to individual time frames. We call such BMC unrolling as *explicit* unrolling. This results in significant increase in formula size with each unrolling and therefore, becomes increasingly difficult for the SAT-solver to solve as shown in [45]. In another approach [159], authors have used Boolean Expression Diagrams (BEDs) [43], which are similar to RBC, for representing transition relation and formulas. They use quantification-by-substitution for (pre)-image computation, instead of explicit unrolling. Translation of a BED to a BDD for a tautology check is found to be expensive, in general, as BDDs are sensitive to variable ordering. Moreover, it was observed [159] that SAT-solvers gave poor results on the clauses built from the Boolean formulas represented using BEDs.

We focus here on the dynamic simplification approach [45] where the simplification is carried across time frames in addition to simplification within each time frame, by using on-the-fly logic reduction algorithms discussed in Chapter 3. Furthermore, we use local learning (i.e., within time frames and partitioned sub-problems) and global learning (i.e., across time frames and partitioned sub-problems) to further simplify the unrolled structure. We refer to such an approach as *implicit* unrolling.

5.2.1 Notation

Consider a synchronous single-clock design with m primary inputs $U_1,\dots,U_m$; n latches (state holding elements) $X_1,\dots,X_n$ and corresponding next state nodes $Y_1,\dots,Y_n$. Let U, X and Y represent the set of primary input, state and next state variables. We use $U^d=\{U_1^d,\dots,U_m^d\}$ to denote the set of primary input variables introduced in the dth time frame. Let $V^d=\cup_{0\le i\le d}U^i$ be set of all primary input variables introduced up to and including the time frame d. Note that each node in the expanded (or unrolled) design has an associated function $f: V^d \to \{0,1\}$ which arises by recursively substituting the fanins of the gates with their respective functions, stopping at the primary inputs. In our later discussion, we refer to a gate g in the transition relation as *ckt_node* and a logic node g^d, corresponding to the dth unrolling of g, as *uckt_node*. Let X_i^d represent the variable for the latch X_i at the time frame d and define X^d to be the set $\{X_1^d,\dots,X_n^d\}$. Note that as we have a single initial state, we can set the function associated with X_i^0 to the constant initial state value I_i of latch X_i in the design.

We illustrate these definitions with the following example as shown in Figure 5.2. The design has two latches X_1 and X_2, one input U_1, and one output F. The transfer function of the design can be expressed as follows:

$NEXT(X_1) = G_1 = \neg(U_1 \vee X_1)$ *//NEXT is next state node*
$NEXT(X_2) = X_1$
$F = \neg X_1 \wedge X_2$

In the following, we use a 4-tuple $UC^i = <U_1^i;X_1^i;X_2^i;F^i>$ to denote the ith $(i \le 2)$ unrolled circuit nodes (combinational logic functions) for U_1, X_1, X_2, and F, respectively, where $X_1^0=0$, and $X_2^0=0$ denote the respective initial states.

$UC^0 = <U_1^0;\ X_1^0=0;\ X_2^0=0;\ F^0=0>$
$UC^1 = <U_1^1;\ X_1^1=\neg U_1^0;\ X_2^1=0;\ F^1=0>$
$UC^2 = <U_1^2;\ X_1^2=\neg(U_1^1 \vee X_1^1);\ X_2^2=X_1^1;\ F^2=\neg X_1^2 \wedge X_2^2>$

Note that these symbolic expressions grow very large with successive unrolling of the time frames. In the next section, we discuss how we can reduce the growth of these expressions by allowing dynamic simplification across the time frames.

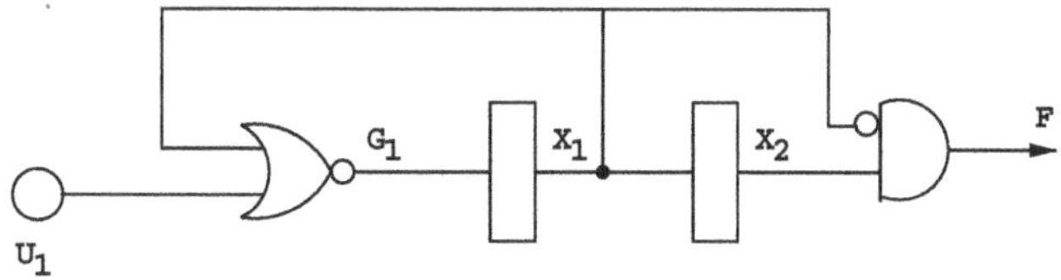

Fig. 5.2: A small design example. The initial values for X_1 and X_2 are both 0

5.2.2 Procedure *Unroll*

We present the procedure *Unroll* and sub-procedure *Compose* in Figure 5.3. These are applied on efficient non-canonical circuit graph representations, i.e., *Reduced AIG*. The procedure *Unroll* is invoked at every depth d and uses previously computed results at depth d-1. Note, that symbolic expression for latch X_i at depth d, i.e., *uckt_node* X_i^d, is equivalent to the symbolic expression Y_i^{d-1} for the corresponding next state node Y_i at depth d-1. At each successive time frame d, a new symbolic variable U_i^d is created for each input U_i^d.

We now define a mapping (*ckt_node*, *uckt_node*) in a hash table H^d with (X_i, X_i^d) for each latch X_i (lines 6-10), and (U_k, U_k^d) for each input U_k (lines 11-13). Note that the *uckt_node* X_i^0 for latch X_i is a *const_0*, *const_1*, or *const_X*, when the latch initial state I_i is the Boolean constant zero, the Boolean constant one, or a symbolic constant, respectively (lines 7-8). Using the mapping in H^d, we compute the *uckt_node* F^d for the output node F by invoking the sub-procedure *Compose* (line 14). We also compute the symbolic expression, in advance, for the each latch input Y^i which in turn becomes the symbolic expression X_d^{i+1} for the next unroll at depth d+1 (lines 15-17).

Now, we describe the sub-procedure *Compose* as shown in Figure 5.3. Given a *ckt_node* p and hash table H with the (*ckt_node*, *uckt_node*) mapping for primary inputs and latches, the *Compose* procedure recursively computes the symbolic expressions for the left and right children of the node and then invokes the procedure *And*, as described in Figure 3.4, on the results of the children nodes (line 29). It also inserts the computed symbolic expression of the internal nodes in the hash table (line 30). The procedure then returns the resulting *uckt_node P* that represents the symbolic expression for *p* for the mapping in *H* (line 31).

```
1.  Synopsis:    Compute symbolic expression at d depth
2.  Input:       ckt_node F and depth d
3.  Output:      symbolic expression, uckt_node F^d
4.  Procedure:   Unroll
5.
6.  forall latch i=1 to n {
7.    //For d=0,
8.    //X_i^0 = I^i=1? const_1: I^i=0? const_0 : const_X
9.    Hash_insert(H^d, X_i, X_i^d);
10. }
11. forall input k=1 to m {
12.   Hash_insert(H^d,U_k,U_k^d);
13. }
14. F^d = Compose(H^d,F);
15. forall latch i=1 to n {
16.   X_i^{d+1} = Compose(H^d,Y_i);
17. }
18. return F^d;
19.
20. Synopsis:    Compose using mapping in hash table
21. Input:       ckt_node p and mapping in Hash table H
22. Output:      symbolic expression, uckt_node P
23. Procedure:   Compose
24.
25. P = Hash_lookup(H,p); //mapping exists?
26. if (P != NULL) return P;
27. l = Left_child(p);
28. r = Right_child(p);
29. P = And(Compose(H,l),Compose(H,r)); //Fig 3.4
30. Hash_insert(H,p,P);
31. return P;
```

Fig. 5.3: Pseudo-code for *Unroll* Procedure

5.2.3 Comparing *Implicit* with *Explicit* Unrolling

We compare the *implicit* unrolling with *explicit* unrolling, where both use the same reduction algorithm for circuit simplification. As discussed above, in *implicit* unrolling, the reduction algorithm is applied not only on each time-frame but also across the time-frames; while in *explicit* unrolling, the reduction algorithm is applied only on the transition relation, and iterative copies of the transition relation are used for unrolling.

We conducted our experiments on a 600MHz P-III with 256M RAM running Red Hat Linux 6.2. We present results on three large designs obtained from IBM; these are part of the Gigahertz Processor Project. The component *m_ciu* is an intricate store queue, the logic *larx* is part of load-store unit, and the component *ifpf* is a portion of control logic for IFU (instruction fetch unit) and L2 cache communication. We also present results on a large design, *b17*, from the ITC99 benchmark series [160].

For each of these designs, we chose an output node corresponding to an invariant that was impossible to verify using BDD-based symbolic model checking [20, 161]. For each of the designs, we compare the graph sizes achieved corresponding to *explicit* and *implicit* unrolling as shown in Figure 5.4.

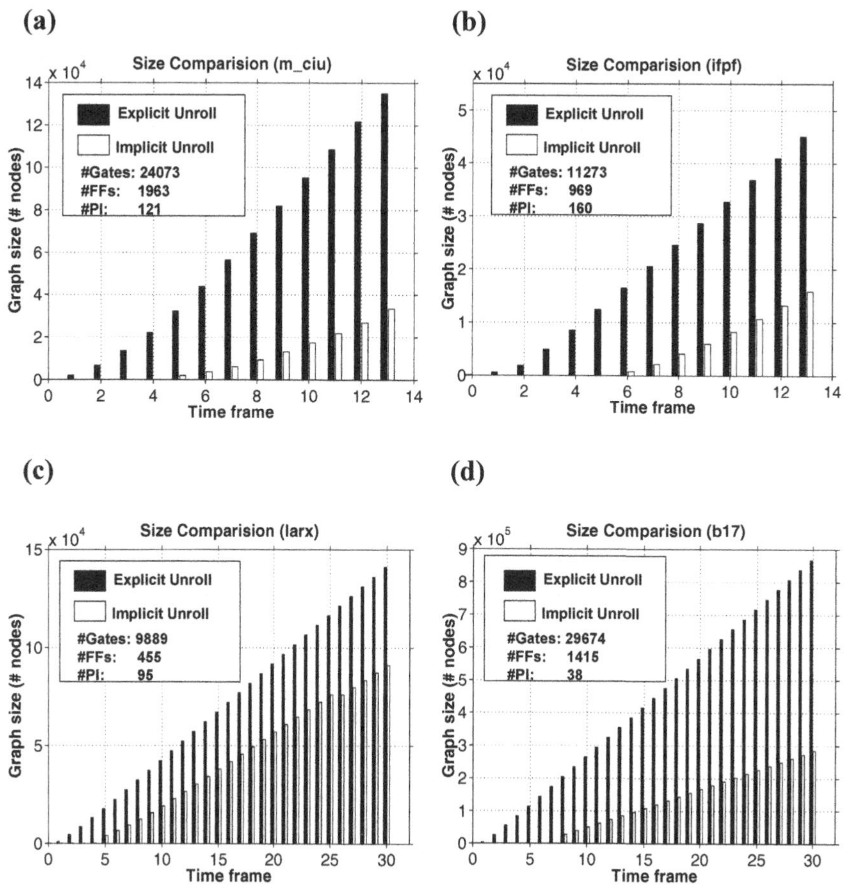

Fig. 5.4: Comparison of unrolled graph sizes between *explicit* and *implicit* approach on following examples: (a) *m_ciu*, (b) *ifpf*, (c) *larx*, (d) *b17*

These graphs represent the symbolic expressions of the output nodes at each time frame. Each vertical bar represents the number of AND nodes in the *Reduced AIG*; a solid bar represents the *explicit* unrolling approach, and a hollow bar represents the *implicit* unrolling approach. Note that in the implicit approach, graph representations at the first few initial time frames correspond to *const_0* node. This is attributed to the on-the-fly reduction algorithm, which maps the symbolic expression to the constant 0 function. This implies that we do not have to do further Boolean reasoning for the output node at this time frame. Note that we do not achieve a similar reduction in the *explicit* unrolling approach, since the reduced circuit is simply replicated at each time frame. The simplification achieved using the *implicit* unrolling approach results in a significantly smaller circuit graph size, in comparison to that achieved by the *explicit* unrolling approach. This simplification is attributed to the propagation of symbolic expressions across time frames. As the number of clauses and variables in the associated SAT problem are directly proportional to the circuit graph size, the search space for SAT-solvers is, therefore, also reduced in the *implicit* unrolling approach.

5.3 SAT-based Incremental Learning and Simplification

In a standard BMC procedure, there is a considerable overlap of circuit constraints due to unrolled transition relations between a k-instance and a $(k+1)$-instance of the BMC problem. Researchers have exploited the constraint sharing technique for speeding up the $(k+1)$- instance [153, 154], thereby leading to a reduction in the overall verification time. As shown in [45], one can also learn from previous unsatisfiable results to simplify the next overlapping problem. In the customized BMC translations [46] (discussed in Section 5.5), sharing occurs not just between the circuit constraints due to the unrolled transition relation, but also between the constraints arising from the property translations and constraints learned from unsatisfiable SAT sub-problems. Such translation schemes heavily rely on incremental SAT techniques. We now describe three kinds of SAT-based incremental learning that can be exploited in BMC by the customized translations. Later, in this section, we also discuss an efficient implementation of such learning techniques.

Learning from shared constraints (L1)

Given two SAT instances S_1 and S_2, as shown in Figure 5.5, conflict clauses that are deduced solely from the set of common constraints Y that is shared between S_1 and S_2 can be used as learned clauses while solving S_1 *or* S_2 [153, 154]. The constraint sharing technique in a SAT solver works as follows: first, the Y clauses are marked when solving S_1 *or* S_2. Then, every conflict clause, generated by a conflict analysis, is also marked if all clauses leading to the conflict are marked.

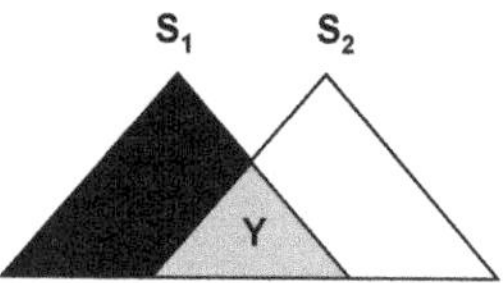

Fig. 5.5: Constraint sharing of SAT instances S_1 and S_2

Learning from satisfiable results (L2)

Suppose $\varphi_{1,\ldots,}\varphi_n$ represents a series of n SAT problems where problem φ_i is built incrementally by adding and removing constraints to and from φ_{i-1} respectively. The satisfying solution for φ_{i-1} (if it exists) can be used to guide the SAT decision engine when solving φ_I [154]. Basically, the SAT decision engine uses the satisfying assignment as initial branching literals, with high possibility to satisfy the overlapped problem quickly.

Learning from unsatisfiable results (L3)

Let a SAT problem $\Phi = \varphi_1 \vee \ldots \vee \varphi_n$ be a disjunction of n sub-problems. Instead of solving Φ as a monolithic problem, one can solve φ_i starting from $i=1$ with the additional constraint $C_i = \neg\varphi_1 \wedge \ldots \wedge \neg\varphi_{i-1}$ where each φ_j is unsatisfiable for $1 \leq j < i$. More benefit is potentially obtained when the problem φ_i shares more constraints with the additional constraint C_i, thereby, allowing *L1* learning.

For safety properties, one can apply *L3* learning [45] to simplify the *(k+1)*-instance problem as shown in Figure 5.6, in addition to dynamic circuit simplification. When the symbolic expression *uckt_node* F^k is unsatisfiable, i.e., a constant zero function, one can merge, i.e., structurally connect the F^k to *const_0* and propagate the effect structurally using the procedure *Merge*(F^k, *const_0*) (see Figure 3.15).

One can derive *circuit truths* using a separate learning process, and use them to simplify the BMC instances further [137]. Note that *circuit truths* can be regarded as invariant relations of circuit nodes in the given circuit structure, without external constraints. A clause representing *circuit truth* can be *replicated* at different time frames of BMC unrolling. Clause replication enables reuse of learning in SAT applications involving time frame expansion, such as BMC. Essentially, a learned clause, consisting of literals from specific time frames, is replicated by substituting corresponding literals from other allowable time frames. This was originally applied to clauses learned by conflict analysis [162]. For example, a clause representing *circuit truth* (V_1 + V_2 + -V_3) can be replicated, i.e., added as learnt clause (V_1^d + V_2^d + -V_3^d) at time frame *d*, where V_i represents *ckt_node* and V_i^d represents corresponding *uckt_node* at depth *d.* In our experience, overhead of replication outweighs its benefits in comparison with L1—L3 learning.

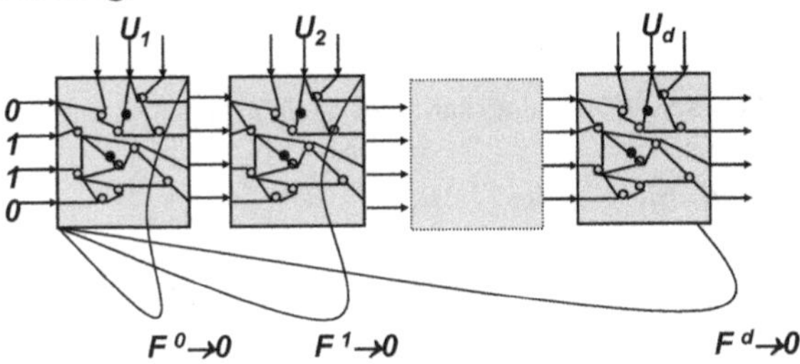

Fig. 5.6: *L3* Learning: Merging unsatisfiable F^k to *const_0* for $0 \leq k \leq d$

Before we discuss the implementation of an incremental SAT-based technique, we first classify the clauses into three types: *constraint, circuit,* and *conflict clauses. Constraint* clauses arise from properties and environmental constraints. *Circuit* clauses comprise gate clauses only in the transitive fanin cone of the property and externally constraint signals. Note, this is an optimization step, as gates clauses in unrolled transition relation, not in the transitive fanin cone of the constraint nodes, are not included. *Conflict* clauses are generated by the SAT solver during conflict-driven learning. In our implementation, each clause has a bit vector field, called a *gflag*. Each bit of *gflag* denotes whether the clause belongs to the group corresponding to that bit position. A *constraint* clause belongs to at most one group. A *circuit clause* does not belong to any group, since it always represents a *circuit truth.* A *conflict* clause that is added during conflict analysis becomes a member of a group if any clause leading to the conflict belongs to that group. Note that *conflict* clauses derived only from circuit clauses can be always replicated [162]. On the other hand, the *conflict* clauses derived from *constraint* and *circuit* clauses can be reused, though not necessarily replicated.

We describe a procedure *Inc_solve* that uses incremental SAT techniques for solving a formula $\Phi = \varphi_1 \vee (\neg\varphi_2 \wedge \varphi_3)$, as shown in Figure 5.7. Here, φ_1, φ_2, and φ_3 represent the gate outputs with constraint 1, 0, and 1, respectively. Thus, circuit clauses are the gate clauses of the fanin cone of the outputs, and the constraint clauses are $(\varphi_1=1)$, $(\varphi_2=0)$, and $(\varphi_3=1)$.

```
1.  Synopsis:      SAT-based Incremental Learning
2.  Input:         Formula Φ = φ1∨(¬φ2∧φ3)
3.  Output:        SAT/UNSAT
4.  Procedure:     Inc_solve
5.
6.  root_gid = Alloc_group_id(); //root clause group
7.  gid = Alloc_group_id(); //new clause group
8.  Add_constraint(φ1=1,gid); //add constraint
9.  status = SAT_solve();
10. if (status == SAT) return SAT;
11.
12. // delete conflict and constraint clauses
13. // that belong to the group gid. Note that
14. // conflict clauses derived only from the
15. // circuit clauses are never deleted as part
16. // of L1 learning
17. Delete_clauses(gid);
18. // L3 Learning: Add the learned clause
19. //(φ1=0) to the group root_gid
20. Add_constraint((φ1=0),root_gid);
21. //circuit simplify using circuit truth
22. Merge(φ1,const_0);
23.
24. Add_constraint((φ2=0),gid);
25. if (SAT_solve()==UNSAT) return UNSAT;
26. dvar_ord = SAT_get_decision_var_order();
27. Add_constraint((φ3=1),gid);
28. //L2 Learning
29. SAT_apply_orders(dvar_ord);//Guide decision engine
30. return SAT_solve();
```

Fig. 5.7: Use of Incremental SAT techniques on an example formula $\Phi = \varphi_1 \vee (\neg\varphi_2 \wedge \varphi_3)$

To solve the problem incrementally, we first partition the problem into φ_1 and $(\neg\varphi_2 \wedge \varphi_3)$ and pick φ_1 to solve first based on its smaller size. We start with an empty clause group *root_gid.* We add the constraint clause ($\varphi_1=1$) to a new group *gid* (line 8). If the result is SAT, we are done (line 10); otherwise, we delete the clauses in group *gid* (line 17), as it will not be reused later. With this deletion, all conflict clauses and constraint clauses that are members of the group *gid* are removed. However, conflict clauses that are deduced only from the circuit clauses still remain, and can be shared across the time frames, as part of *L1* learning. Furthermore, due to an UNSAT result, we use *L3* learning to add the clause $(\varphi_1=0)$ to the clause group *root_gid*. Any conflict clause deduced from such globally learned clauses can also be reused in later time frames. Since, $\varphi_1=0$ is a *circuit truth*, we can also simplify the circuit structure using the procedure *Merge*(φ_1, *const_0*) (line 22) Note, the clause $\varphi_1=0$ is still useful when the backward simplification of *Merge* is not deep enough. Next, we add the constraint *(φ_2=0)* to the group *gid* (line 24) If the problem is UNSAT, we are done; otherwise, we obtain the decision variable order *dvar_ord* that led to the satisfying solution as part of L2 learning. We then add constraint *(φ_3=1)* to the group *gid* and invoke the SAT procedure. Using the decision variable order, we apply *L2* learning using the procedure *SAT_apply_orders* in the SAT procedure in order to guide the search.

5.4 BDD-based Learning

Learned clauses play a crucial role in determining the performance of SAT solvers, both by pruning the search space, and by dynamically affecting the choice of decision variables. At the same time, there is an overhead associated with the addition of each learned clause. Therefore, learning techniques must ensure a good tradeoff between the usefulness and the overheads of adding learned clauses. Here we explore the use of learning from BDDs, where learned clauses generated by BDD-based analysis are added to the SAT solver, to supplement its other learning mechanisms [146]. The idea of generating learned clauses from a BDD is relatively straight forward, and has also been mentioned by other researchers [163]. Here, we explore the associated tradeoff issues in integrating such learning within a hybrid SAT solver. We describe heuristics and parameters in the BDD Learning engine that are targeted at increasing the usefulness of the learned clauses, while reducing the overheads.

5.4.1 Basic Idea

Essentially, a BDD is used to capture the *circuit truth* i.e., relationship between Boolean variables of (a part of) the SAT problem, in the form of a characteristic function. In such a BDD, each path to a "0" (false) node denotes a conflict. A learned clause corresponding to this conflict is easily obtained by negating the literals that define the path. Since a BDD captures all paths to 0, i.e., all possible conflicts among its variables, the potential advantage is that *multiple* learned clauses can be generated and added to the SAT solver at the same time. In contrast, conflict-driven learning typically analyzes a single conflict at a time. An example with multiple learned clauses generated from a BDD is shown in Figure 5.8. (The figure shows multiple terminal nodes, and no inverted edges for exposition only; standard ROBDDs can be used otherwise.)

Using the input and output relation in the circuit structure of the given SAT problem, one can create useful and manageable sized BDDs for the bounded cone of logic. The main goal for our BDD Learning technique is to be *effective* but *lightweight*, i.e., it should improve the performance of the SAT solver, but without overwhelming the SAT solver heuristics. We perform learning selectively, i.e., by selecting *seed nodes* in the circuit graph around which to perform learning. We distinguish between the following kinds of learning:

- *Static learning*: seed nodes are selected using static information, and learned clauses are added statically, before the SAT search starts.
- *Dynamic learning*: seed nodes are selected using dynamic information, and learned clauses are added on-the-fly, during the SAT search.

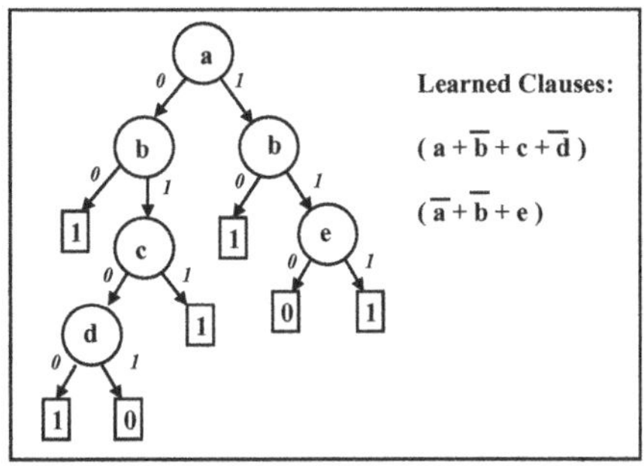

Fig. 5.8: Example for BDD Learning

We use a BDD Learning engine to encapsulate the essential tasks of seed selection, on-the-fly creation of BDDs, and generation of learned clauses. This engine is integrated with the hybrid SAT solver (see Chapter 4). A Hybrid SAT solver facilitates the integration with BDD learning by allowing BDD creation directly on the circuit structure, and by allowing long learned clauses to be represented efficiently in CNF. In particular, BDD Learning is performed in conjunction with other learning mechanisms in the SAT solver, e.g. conflict-driven learning [37]. Furthermore, a learned clause generated by the BDD Learning engine is treated similar to other learned clauses by the SAT solver. For example, scores of variables related to these learned clause are incremented [38, 40] when added, which can significantly impact the future decisions made by the SAT solver.

5.4.2 Procedure: *BDD_learning_engine*

The overall flow of the BDD-based learning technique is shown in Figure 5.9. The procedure *BDD_learning_engine* (lines 6-10) performs the essential tasks of seed selection using procedure *Select_seed*, on-the-fly creation of a BDD using procedure *Circuit_to_BDD*, and generation of learned clauses from the BDD using the procedure *Generate_learned_clauses*. It is integrated seamlessly with the hybrid SAT solver *HSAT_bddl*, as shown in lines 19-23.

The original problem is represented in circuit form, allowing the sub-procedures *Select_seed* and *Circuit_to_BDD* to directly operate on the *Reduced AIG*. Also, the learned clauses from BDD, like conflict-driven learned clauses, are represented in CNF efficiently. The remainder of this section describes details of the sub-procedures in BDD Learning engine, and the next section focuses on its integration with the SAT solver using the procedure *Add_clauses_dynamic*.

```
1.  Synopsis:   Learn circuit truths using BDDs
2.  Input:      Circuit Structure
3.  Output:     Learn CNF clauses
4.  Procedure:  Bdd_learning_engine
5.
6.  Update_engine_info();
7.  node = Select_seed();
8.  Bdd = Circuit_to_BDD(node,threshold_size);
9.  Cl_list = Generate_learned_clauses(Bdd);
10. return Cl_list;
11.
12. Synopsis:   Check Boolean Satisfiability
13. Input:      Boolean Problem in Circuit
14. Output:     SAT/UNSAT
15. Procedure:  HSAT_bddl
16.
17. if (Deduce()=CONFLICT) return UNSAT;
18. while(1) {
19.   if (Do_bdd_learn()=TRUE){//decide runtime
20.     CNF_cl_list = Bdd_learning_engine();
21.     if (Add_clauses_dynamic(CNF_cl_list)=UNSAT)
22.        return UNSAT;
23.   }
24.   if (Decide()=SUCCESS) {
25.     while(Deduce()=CONFLICT) {
26.       if (Diagnose()=FAILURE) return UNSAT;
27.     }
28.   }
29.   return SAT;
30. }
```

Fig. 5.9: BDD-based Learning Technique

5.4.3 Seed Selection

We explored many different heuristics in the procedure *Select_seed* for selecting seeds around which BDD Learning is performed. Since the goal is to improve the SAT solver performance, the seed selection heuristics are based on the decision ordering heuristics of the SAT solver itself. We assign a rank to each candidate seed node (not a primary input) based on criteria described below. We also keep track of which variables have already been chosen as seeds, to avoid adding duplicate learned clauses. The seed selection heuristics, listed *SSH1-SSH5*, are:

- *SSH1—Next decision rank*: The idea is to preempt the learning that would be performed by the SAT solver for a future decision. Rather than learn a single conflict clause, we learn all related conflict clauses simultaneously.
- *SSH2—Past decisions ranked back from the current one*: In this case, the idea is to learn some more clauses about variables that have been important in the past.
- *SSH3—Most frequent decisions*: The idea is that a variable chosen frequently as a decision variable, along different paths in the SAT search, is a good candidate for preempting future learning.
- *SSH4—Decisions at back-leap levels*: The back-leap technique identifies a good decision level to backtrack to, in the presence of many conflicts localized within a range of decision levels [107]. Like restarts, it allows jumping out of locally bad regions, but without having to backtrack all the way up to the starting decision level. The intuition here is that additional learning about decision variables at the back-leap levels is likely to be useful. The seeds are ranked from the backleap level down to the current decision level.
- *SSH5—Decisions at backtrack levels most often visited*: Since more number of backtracks signifies difficult SAT problems, we keep track of decision variables at those levels to which maximum number of backtracks have taken place. The intuition is that these variables are likely to be causing more conflicts, and additional learning might help.

We experimented with choosing a single seed at a time, versus choosing multiple seeds. In most cases, the latter incurred additional overhead, without helping improve the performance. Therefore, for all experiments described later in this chapter, we chose a single seed whenever BDD learning was invoked inside the SAT solver.

5.4.4 Creation of BDDs

Once seed selection is done, we create BDDs using sub-procedure *Circuit_to_BDD* that capture relationships among variables in the circuit region around the seed. We explored two different region heuristics – the *fanin cone of the seed*, and the *circuit region around the seed* including its fanins and fanouts as shown in Figure 5.10. Since the latter results in BDDs that may not include the seed variable at all, our experimental results were uniformly better with the fanin cone heuristic.

In both cases, we created BDDs across very few logic levels, typically 5-10, in order to avoid BDD size blowup. Keeping the number of logic levels any lower would likely result in local learning, which is relatively easy to infer from the circuit constraints. The potential benefit of BDD Learning is in its ability to perform non-local learning around the seed. Apart from avoiding memory blowup, keeping the BDD sizes small has the added benefit of creating shorter paths in the BDD, thereby resulting in shorter clause lengths. In general, shorter learned clauses are likely to be more beneficial than longer learned clauses, since they require less number of assignments before resulting in an implication in the SAT solver.

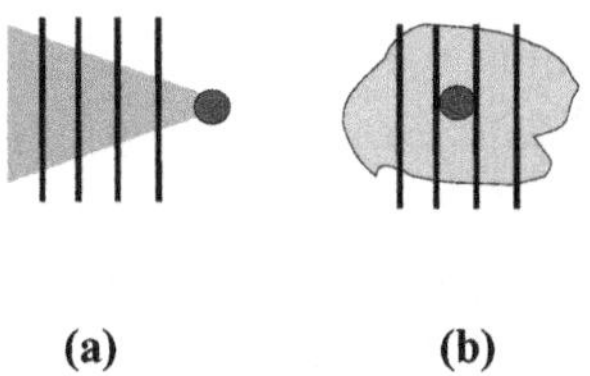

(a) **(b)**

Fig. 5.10: BDD from seed a) Fanin Cone, b) Region around the seed

5.4.5 Generation of Learned Clauses

After a BDD has been created, we generate learned clauses using the procedure *Generate_learned_clauses*. In order to directly use standard BDD packages [164], we can complement the given BDD, and simply enumerate its cubes, i.e., paths to the "1" (true) node. In order to favor shorter clauses, only those cubes that are shorter than a given maximum clause length, typically 5-10, are used for generating learned clauses. To avoid exploring all paths (potentially exponential) in the BDD, we enumerate only a fixed number of cubes.

An alternate method, which actually performed better in our experience, is to implement our own fixed-depth traversal for the complemented BDD. All paths leading to "1" that are shorter than the maximum clause length are enumerated. At the same time, all paths that are greater than the maximum clause length are changed to lead to "0", thereby resulting in an under-approximated BDD. Since the maximum clause length varies from 5 to 10, this traversal is very fast. An additional strategy is to perform a universal quantification on heuristically chosen variables in the under-approximated

BDD. This corresponds to performing a resolution on the learned clauses, and results in less number of learned clauses. However, in our experimental results, this typically performed worse than the fixed-depth traversal without quantification.

5.4.6 Integrating BDD Learning with a Hybrid SAT Solver

This section describes details of integrating a BDD Learning engine with a hybrid SAT solver [41], and addresses various issues. We invoke the BDD Learning engine at selective decision intervals using the procedure *Do_bdd_learn* (line 19, Figure 5.9). For static learning, we perform learning at decision level 0 only, and disable learning at all other levels. For dynamic learning, we invoke the BDD Learning engine after certain decision intervals during which backtracks have exceeded a certain threshold, indicating a difficult SAT problem. The procedure *Update_engine_info* (line 6, Figure 5.9) updates information regarding seed selection heuristics. One can also increase the number of backtrack threshold (*threshold_size*) adaptively, so as to not over-burden the SAT solver. This worked better than a fixed parameter for more difficult problems.

5.4.7 Adding Clauses Dynamically to a SAT Solver

We describe the procedure *Add_clauses_dynamic* (line 21, Figure 5.9) that adds clauses to a SAT solver *while the SAT search is in progress*. For static learning, the learned clauses generated by the BDD Learning engine are added to the SAT solver before any decisions are made (but after preprocessing). This is relatively straight forward, since all implications due to the learned clauses occur at the starting level. However, the situation is somewhat more complicated for dynamic learning, i.e., when learned clauses are added dynamically to the SAT solver.

Note that a conflict clause, i.e., a clause learned from conflict analysis by the SAT solver, is also added dynamically. However, a conflict clause is guaranteed to be either conflicting (or unit, depending upon the implementation) when it is added, which causes an immediate backtrack (or an implication at the current decision level). When multiple clauses are added after BDD Learning, or any other kind of learning performed externally with respect to the SAT solver, extra work may be required to maintain the decision level invariants in the SAT solver.

Consider the effect of a newly added clause on the existing state of the SAT solver. The effect depends on the status of the clause, as computed with respect to the current variable assignment stack in the SAT solver, as follows:

- If a learned clause is *conflicting*, i.e., all its literals are false, then the clause can be added immediately. As soon as it is added, conflict analysis will take place, resulting in appropriate action in the SAT solver.
- If a learned clause is *unsatisfied*, but has at least two unassigned literals, then it can be added immediately without changing the decision level of the SAT solver.
- If a learned clause is *unit*, i.e., all but one of its literals is false, and the remaining one is unassigned, then adding it would cause an implication. This might require backtracking up to the level where the implication should be made. Therefore, a choice exists between adding the clause immediately, followed by potential backtracking, or delaying its addition until the SAT solver goes back to the level where the implication should be made. Note that this case also applies to non-conflicting 1-literal clauses, for which backtracking up to the starting decision level (a restart) would be required.
- Finally, if a learned clause is *satisfied*, i.e., at least one of its literals is *true*, we can add it immediately to the SAT solver without changing the decision level in most cases. The exception is the case when all literals but one are assigned false, and the true literal is the only literal assigned at the highest decision level, i.e., the learned clause would have caused an implication on the true literal at a lower decision level. We call this a *pseudo-satisfied* learned clause. It is similar to the case of a unit clause, and is handled in the same way.

To summarize, learned clauses that are conflicting, or are unsatisfied with at least two unassigned literals, can be added immediately to the SAT solver. Satisfied clauses, but not pseudo-satisfied clauses, can also be added immediately. Finally, unit clauses and pseudo-satisfied clauses require implications to be made. For these, we have a choice between adding them immediately followed potentially by backtracking, or waiting to add them later at the correct implication level.

5.4.8 Heuristics for Adding Learned Clauses

We use some additional filters to determine whether to add a clause generated by the BDD Learning engine to the SAT solver. First, we prefer those learned clauses that capture non-local learning, in order to avoid duplication of circuit constraints. A heuristic that we use is to check if assignments to its literals took place at different decision levels. Though this does not capture non-locality precisely, it is a good indicator. We use the term *non-local* to describe such clauses.

We also use a relevance number to determine the potential usefulness of a learned clause, defined as the sum of its true and unassigned literals. If this number is large, the learned clause is unlikely to be useful, since it will not cause an implication. Therefore, we prefer clauses with a *relevance number* less than a certain threshold – we call these the *relevant* clauses. Typically, SAT solvers [38, 40] use a similar figure of merit to delete their own learned clauses when needed. We found a relevance threshold of 5 to give good performance.

Finally, for unit and pseudo-satisfied learned clauses, we heuristically choose when to add them to the SAT solver. We add them immediately if the difference between the current decision level and the implication level is less than a threshold parameter, but delay adding them otherwise. We typically used a level difference of 5 as the threshold in our experiments.

Based on these additional filters and the status of a learned clause, we have organized the following levels of learning:

- *Level 1* : adds only the conflict clauses and 1-literal unit clauses
- *Level 2*: adds all level 1 learned clauses, and all unit and pseudo-satisfied clauses
- *Level 3*: adds all level 2 learned clauses, and all non-local, relevant clauses (satisfied, as well as unsatisfied).

Note that these levels are organized intuitively, according to the projected usefulness of a learned clause. Furthermore, since each level includes clauses added by previous levels, we can easily investigate the effect of adding more clauses. We study the effect of various learning in our experimental Sections 5.6.4-5.6.6.

5.4.9 Application of BDD-based Learning

The distinguishing feature of the BDD Learning method as described here is that it directly adds learned clauses to the SAT solver. Its use is orthogonal to other BDD-based simplifications for SAT problems, such as for simplifying the goals [163], or for simplifying the problem using BDD sweeping [135]. It is also possible to combine BDD Learning techniques with clause replication techniques [162]. Furthermore, dynamic seed selection can also be used for *selective* replication, to increase its effectiveness in practice.

For BMC application, BDDs can be created for seed nodes in the unrolled transition relation (starting from the initial state), or a single copy of the transition relation (without the initial state) called the template. In the former case, learned clauses are generated directly in terms of the different

SAT variables in the expanded time frames. In the latter case, learned clauses are generated in terms of template variables. With each successive unrolling of the transition relation, a learned clause can be generated for the new time frame by substituting corresponding SAT variables for the template variables, similar to clause replication [162].

Though our focus here is on BMC application, the BDD Learning technique can be used to potentially improve performance in other circuit-based SAT applications as well, such as equivalence checking [163], and automatic test pattern generation (ATPG) [165].

5.5 Customized Property Translation

Most BMC methods use the standard monolithic translation originally proposed by Biere *et al.* [66]. Rather than generating a monolithic SAT formula, we describe a customized property translation [46] that can be viewed as building the formula incrementally, by lazily indexing over the bounded conjunctions/disjunctions, terminating early when possible. The main motivation is to use an incremental formulation, employing learning and partitioning to generate multiple simpler SAT sub-problems.

While the standard BMC procedure partitions the overall problem into separate k-instances, the customized translation goes further in partitioning the k-instance problem into multiple, smaller, SAT sub-problems. This can potentially make each SAT sub-problem easier to solve than the monolithic SAT problem. Furthermore, the partitioning is based on formulating sub-problems in a way that enables learning, when these sub-problems are found to be unsatisfiable. Such opportunities for learning are missed by the general translation, because when a monolithic SAT problem is found to be unsatisfiable, it may not be clear which part of the problem led to the unsatisfiability. Though implicit learning is performed by conflict analysis within the SAT solver, it is not as beneficial as the explicit learning performed by these translation schemas. In order to mitigate the overheads of partitioning into multiple SAT sub-problems, the translation schemas are geared toward an incremental formulation, i.e., reuse of variables and constraints wherever possible. This allows use of incremental SAT learning techniques in the SAT solver very effectively. In contrast, the standard translation does not directly provide an incremental formulation of the property constraints. Before we describe the details, it is useful to classify the constraints that define the SAT sub-problems into the following types:

- Circuit constraints
- Constraints due to the property sub-formulas
- Constraints due to the environment
- Loop-check constraints for considering only loop-free path skeletons

- Constraints learned from unsatisfiable SAT instances (*L3* learning)
- Constraints due to re-use of learned conflict clauses (*L1* learning)
- Constraints corresponding to "circuit-truths" learned using BDD learning engine (L1 learning)

Circuit constraints refer to gate clauses in the transition relation of the transitive fanin cone of the property and environmental variables. Note that the general translation uses only the first two types of constraints, i.e., circuit and property constraints. Constraints learned from unsatisfiable instances [45], and loop-check constraints [67], have been shown useful in the context of verifying simple safety properties. Other types of constraints such as those arising from conflict-driven learning and BDD-based learning have been shown useful especially in the customized translations [46]. Here we discuss how the incremental formulation of property translation facilitates the sharing of such constraints also.

In the presented BMC engine, we combine the incremental SAT-based techniques (Section 5.3) and BDD-based learning (Section 5.4) with circuit simplification (Section 5.2) based on detecting structural isomorphism as discussed in Chapter 3. At the backend, the SAT problems are solved by a hybrid SAT solver as described in Chapter 4, which combines the benefits of CNF-based and circuit-based logic representations, and includes use of circuit-based SAT heuristics.

To highlight the partitioning, learning, and incremental aspects of our customized property translations, we describe the pseudo code for handling commonly used LTL properties. In particular, we present the procedures for handling $\boldsymbol{F}(p)$, $\boldsymbol{G}(q)$, and $\boldsymbol{F}(p \wedge \boldsymbol{G}(q))$, in the presence of fairness constraints $B=\{f_1,\ldots,f_b\}$, in Figures 5.11-5.14. Here, property nodes f and g denote any Boolean combination of propositional atoms nested with the next-time operator $\boldsymbol{X}$. We construct a tree expression for the property nodes where each sub-tree node *prop_tree_node* represents a sub-expression of types *AND* ($\wedge$), *NOT*($\neg$), *LEAF*, or $\boldsymbol{X}$, where *LEAF* corresponds to the propositional atom. Recall, we use *ckt_node* to denote a propositional atom or gate in the transition relation, and *uckt_node* such as p^i to denote a propositional logic node corresponding to the ith unrolling of p. Further, *Is_sat(C)* returns *true* if the Boolean formula C is satisfiable; *L{1-3}* denote the kind of incremental SAT-based learning; ${}_jL_i$ denotes the loop transition constraint, i.e., ${}_jL_i = T(s_i, s_j)$; FC^{ij} denote the fairness constraints, i.e. $FC^{ij} = \wedge_{1\leq r\leq b} (\vee_{i\leq k\leq j} f_r^k)$; M, N denote the loop and non-loop bound, both under user's control. Due to incremental formulation, learning *L1* is always active. For ease of description, the circuit constraints are not shown in these translations – they are always added to the SAT sub-problems, using the simplified circuit representation.

The procedure *Prop_node*, shown in Figure 5.11, returns p^i using the procedures *Unroll* (line 9), *And* (line 11), and *Invert* (line 14). If $p \rightarrow type$ is X, the current unroll depth is advanced by one (line 16). The correctness of the translation for unclocked LTL is based on the re-write rules:

$\neg Xf \equiv X\neg f$, and

$X(f \wedge X(g)) \equiv Xf \wedge XXg$

Note, we choose to build a *uckt_node* p^i, instead of partitioning the problem into separate conjunctions/disjunctions of X operators and propositional atoms. In our experience, too many SAT sub-problems add a performance overhead and thus, we restrict our sub-problem partitioning at conjunctions/disjunctions with one operand being F or G. Also, the procedure *Prop_node* allows sharing of common sub-expressions in p by mapping identical structures of *uckt_node* logic nodes, using on-the-fly simplification procedures during unroll (as discussed in Section 5.2).

```
1. Synopsis:   Create property circuit node at depth i
2. Input:      prop_tree_node p, depth i
3. Output:     uckt_node p^i
4. Procedure: Prop_node
5.
6. j = i;
7. switch(p→type) {
8.    case LEAF://ckt_node
9.      return Unroll(p,j);//uckt_node at j
10.   case AND://p = p1 ∧ p2
11.     return And(Prop_node(p1,j),
12.                Prop_node(p2,j));
13.   case NOT: //p = !p1
14.     return Invert(Prop_node(p1,j));
15.   case X: //p = X(p1)
16.     j = j+1;
17.     return Prop_node(p1,j+1);
18.}
```

Fig. 5.11: Unrolling of property tree node *p*

Example 5.1

For the given LTL formula $F(a \wedge \neg X(b \wedge X(c)))$ with proposition atoms *a*, *b*, and *c*, the procedure *Prop_node(p,i)* with $p = a \wedge \neg X(b \wedge X(c))$, at depth *i* returns the Boolean expression $p^i = a^i \wedge \neg(b^{i+1} \wedge c^{i+2})$.

5.5.1 Customized Translation for *F(p)*

We describe the customized translation and various learning aspects for a negated (unclocked) safety property, i.e., $F(p)$ in the procedure *BMC_solve_F* as shown in Figure 5.12. (In our discussion of multi-clock system in Chapter 8, we discuss how we handle clocked properties.) Note that the outer while-loop on index i correspond to incrementing the bound k in BMC, up to the user-specified maximum limit N. It incrementally builds up the property database C, which initially consists of the initial state constraints I (line 6). At a given depth i, we obtain *uckt_node* p^i using the procedure *Prop_node*. We then check the satisfiability of $(C \wedge p^i)$ (in line 9), i.e., if the current clause database is satisfiable while p is true in the current time frame. If satisfiable, we return *TRUE*, indicating witness for ***F****(p)* is found. If not, then the fact that p is always false in time frame i is *learned* (*L3*) and added to the clause database C (line 10). We also merge p^i to *const_0* using the procedure *Merge* (see Figure 3.15), to simplify subsequent BMC instances (line 11). The inner for-loop on index j (lines 12—15) is optional, and is used selectively to derive proofs. It adds pair-wise constraints to C in order to ensure that there is no loop from current state s_i to any previous state s_j on the path. If the current path cannot be extended to remain loop-free, the procedure returns *FALSE*, indicating that no witness is possible and provides the *completeness* argument for ***F****(p)*. Such a path is also referred to as *longest loop-free path* [67]. Note, that the pair-wise loop constraints are added incrementally in order to obtain early termination. When $(i \geq N)$, the procedure aborts (line 18) without finding either a witness or a proof.

```
1.  Synopsis:   Customized BMC solve for F(p)
2.  Input:      prop_tree_node p, bound N, initial state I
3.  Output:     TRUE/FALSE/ABORT
4.  Procedure:  BMC_solve_F
5.
6.  C = I; i=0;
7.  while (i<N) {  //L1 is active always
8.    p^i = Prop_node(p,i);
9.    if (Is_sat(C ∧ p^i)) return TRUE; //wit found
10.   C = C ∧ ¬p^i; //L3 Learning
11.   Merge(p^i, const_0); //Simplify
12.   for (j=i;j>=0;j--) {//loop-free check
13.     C = C ∧ ¬_jL_i;
14.     if (!Is_sat(C)) return FALSE; //no wit exists
15.   }
16.   i = i+1; //increment depth
17. }
18. ABORT("Bound Reached"); //wit not found
```

Fig. 5.12: BMC Customization for ***F****(p)*

5.5.2 Customized Translation of *G(q)*

We describe the translation of *nested* ***G(q)*** with fairness constraints, $B=\{f_1,..,f_b\}$ using the procedure call *BMC_solve_G(IC,q,start),* shown in Figure 5.13.

```
Synopsis:   Customized BMC solve for nested G
Input:      prop_tree_node q, bound M,
            fairness constraints B={f_1,...,f_b},
            start depth s_depth,
            initial path constraint IPC
Output:     TRUE/FALSE/ABORT
Procedure:  BMC_solve_G

// L1 is active always
C = IPC; t = 0; C''= 1;
for (i=s_depth;i<M;i++) {
  q^i=Prop_node(q,i); C = C ∧ q^i;
  if (!Is_sat(C)) return FALSE;//L2
  for (j=i;j>=s_depth;j--) {
    if (!Is_sat(C ∧ FC^ij)) continue;//L2
    if (Is_sat(C ∧ _jL_i ∧ FC^ij)) return TRUE;
    C = C ∧ ¬_jL_i; //L3
  }
  for(j=s_depth-1;j>=t;j--) {
    C' = C & C''; q^j=Prop_node(q,j)
    if (!Is_sat(C' & q^j)) {
      t = j+1; C'' = C'' ∧ ¬q^j ; //L3
      break;
    }//else L2
    C' = C'∧q^j;
    if (!Is_sat(C' ∧ FC^ij)) break;//L2
    if (Is_sat(C'∧ _jL_i ∧ FC^ij)) return TRUE;
    C = C ∧ ¬_jL_i;     //L3
  }
  for(l=t-1;l>=0;l--){//loopFree check
    C = C ∧ ¬_lL_i; //L3
    if (!Is_sat(C)) return FALSE;
  }
}
ABORT("Bound Reached");
```

Fig. 5.13: BMC Customization for *G(q)* with fairness constraints *B*

Here *IPC* denotes the *Initial Path Constraint,* and *s_depth* denotes the start time frame. A non-nested ***G(q)*** (not shown) is a special case with *IPC=1* and *s_depth=0.* The procedure *BMC_solve_G*, essentially, searches for a path satisfying the constraint *IC* starting at s_{s_depth}, such that *q=TRUE* at each state in the path, and the path loops back to some previous state. The outer *for-loop* on index i (line 11) corresponds to incrementing the time-frames for BMC up to the user-provided bound M. It incrementally builds up the property database C, which is initially set to *IPC* (line 10). For each i, we first check the satisfiability of $(C \wedge q^i)$ (line 12). If it is unsatisfiable, it returns false, as clearly there is no way to find a witness loop. If it is satisfiable, we collect satisfying decision variables for *L2* learning (not shown). We then check the satisfiability of $(C \wedge FC^{ij})$ (line 15), to check for fairness constraints. If it is unsatisfiable, we skip the witness loop check; else we proceed to check for witness loops incrementally (line 16) so that we can terminate early at the first detection. However, if such a loop is not found, this fact is learned (*L3* at line 17).

For non-zero i, additional checks for a loop- back state at time frame $j < i$ are also performed (lines 19-29). In this case, it must additionally prove that q=*TRUE* at each such state. If it is found to be unsatisfiable at any j, $\neg q^j$ is learned (*L3* at line 22) and t is updated accordingly (line 22) so that we do not need to loop back to a state before t (line 19). If the current path cannot be extended to remain loop-free (lines 30-33), we conclude that no witness exists.

5.5.3 Customized Translation of $F(p \wedge G(q))$

Nested properties like $\boldsymbol{F}(p \wedge \boldsymbol{G}(q)$ with fairness constraints provide the opportunity to partition further, allowing greater learning and sharing across the partitions. The algorithm for $\boldsymbol{F}(p \wedge \boldsymbol{G}(q)$ is shown in Figure 5.14. There are two user-provided bounds N and M, for $\boldsymbol{F}$ and $\boldsymbol{G}$ checks, respectively. The outer *for-loop* on i corresponds to incrementing the bound k in BMC. The first SAT sub-problem (line 10) is to check $(C \wedge p_i)$. If it is unsatisfiable, $\neg p^i$ is learned (*L3*) and added to C (line 12); otherwise, *L2* learning is attempted. The procedure, then, checks for nested $\boldsymbol{G}(q)$ using *BMC_solve_G*$(C \wedge p^i, q, i)$ (line 14), which looks for a path starting at s_i, such that q=*TRUE* at each state in the path, and the path loops back to a previous state. If the current path cannot be extended to remain loop-free in lines 18-21, we stop further search since no witness exists. Moreover, if the G checks in line 20 have been *FALSE* at all depths strictly less than i, $lfc=1$, we conclude that no witness exists. Note the large number of SAT calls made in the process of analyzing this property, and how the databases *(C, C', C'')* are incremented between successive calls. Besides *L1* learning due to sharing of constraints and conflict clauses, SAT calls on sub-problems provide additional *L2* and *L3* learning opportunities. Also, due to reuse of variables and clauses in our incremental formulation, there is a large sharing of constraints that make *L1-L3* learning very effective. The analysis of properties translated with our approach is much faster than with the standard translation as observed in our experiments described in the next section.

```
Synopsis:  Customized BMC solve for nested F(p∧G(q))
Input:     prop_tree_node p,q, bound M,N, initial
           state I, fairness constraints B={f_1,..,f_b},
Output:    TRUE/FALSE/ABORT
Procedure: BMC_solve_FG

//L1 is active always
C = I; lfc=1;
for (i=0;i<N;i++) {
  p^i = Prop_node(p,i); C = C ∧ p^i;
  if (!Is_sat(C ∧ p^i)) {
    C = C ∧ ¬p^i; //L3
  }else {//L2
    R = BMC_solve_G(C ∧ p^i,q,i));
    if (R=TRUE) return TRUE;
    else if (R==ABORT) lfc=0;
  }
  for (j=i;j>=0;j--) {//loop-free check
    C = C ∧ ¬_jL_i; //L3
    if (lfc &&!Is_sat(C)) return FALSE;
  }
}
ABORT("Bound Reached");
```

Fig. 5.14: BMC Customization for ***F(p∧G(q))*** with fairness constraints

5.6 Experiments

We implemented dynamic circuit simplification, SAT-based incremental learning, and customized property translation in the SAT-based model checking framework called *VeriSol* (formerly *DiVer* [28]) that uses a hybrid SAT solver [41]. For comparision, we used the BMC engine in VIS [166] (version: *VIS-2.0*). In the first set of controlled experiments, we study the effect of a combination of various techniques on various in-house designs HD1-HD11 with safety and liveness properties. In the second set, we study particularly the effect of SAT-based incremental learning and customized property translation on some designs HD12-HD14 with liveness properties. In the third set, we study the effect of BDD-based learning on our BMC engine, on large industry designs D1-D5 obtained from various industry sources.

5.6.1 Comparative Study of Various Techniques

We applied the BMC engine in *VeriSol* for verification of industry designs HD1-HD11 with safety and liveness properties. In this controlled experiment, we study the effect of various techniques and their combination in improving the performance of the BMC engine. We use the following notation: *T* denotes dynamic circuit (Ckt) simplification, *C* denotes customized property translation, *Inc* denotes SAT-based Incremental Learning (*L1+L2+L3*), and *H* denotes use of a Hybrid SAT solver. We experimented with the following different combinations of these techniques. with their inclusion/exclusion denoted respectively by +/-:

- *Option 1* (op1): *-T-C* (General BMC Translation)
- *Option 2* (op2): *-T+C* (w/o Ckt Simplification; w/ Customization)
- *Option 3* (op3): *+T-C* (w/ Ckt Simplification; w/o Customization)
- *Option 4* (op4): *+T+C* (w/ Ckt Simplification; w/ Customization)
- *Option 5* (op5): *+T+C+H* (op4 + Hybrid SAT solver)

Using the options *op1* to *op5* and +/- *Inc*, we worked with 10 different combinations. Note, for *op1* to *op4*, we used the Chaff SAT solver [38] (version: zChaff-2001), enhanced with SAT-based incremental learning. The option *op1* combined with *-Inc* corresponds to standard BMC, i.e., a monolithic translation with no dynamic circuit simplification, and no incremental learning. The customized translations always use an incremental formulation, whether or not the SAT solver uses incremental SAT techniques.

The experimental results are summarized in Table 5.1. For HD1-5, and HD10-11, the experiments were performed on a 750 MHz Intel PIII with 256 MB. For the remaining designs HD6-9, the experiments were performed on a Sun UltraSparc 440 MHz, 1GB workstation. (Note that experiments were conducted on non-uniform machines for logistics reasons). We used a time limit of 10K seconds (~3 hours). In Table 5.1, Column 1 indicates the design and property type i.e, safety (S) or liveness (L); Columns 2-3 report the number of flip-flops (#FFs) and 2-input gates (#gates) in the static fanin cone of the property; Column 4 indicates the various options *op1-5*; Column 5 reports the number of completed BMC instances within the time limit; Column 6 reports whether a witness was found for that *k*. The next two columns report the time taken (in seconds) *with* and *without* the use of incremental SAT techniques, denoted *+Inc* and *–Inc*, respectively. The last column reports the memory required (in Mbytes) for performing the verification with incremental SAT techniques. (The memory used by the verification without incremental SAT was about the same, for most examples.)

Table 5.1: Customized BMC Verification Results

D(L/S)	# FFs	# Gates	Options	k	Wit?	Time (s) (+Inc)	Time (s) (-Inc)	Mem (MB)
HD1(L)	1735	9467	op1: -T -C	18	N	6620	6709	136
			op2: -T +C	20	Y	91	99	28
			op3: +T -C	20	Y	294	326	45.5
			op4: +T +C	20	Y	14	15	10.5
			op5: +T +C +H	20	Y	7	8	7.1
HD2(L)	1997	11043	op1: -T -C	18	N	4724	4754	168
			op2: -T +C	23	Y	270	379	67
			op3: +T -C	23	Y	602	733	60.7
			op4: +T +C	23	Y	70	108	21
			op5: +T +C +H	23	Y	36	51	15
HD3(L)	2259	12614	op1: -T -C	18	N	6369	6357	205
			op2: -T +C	29	Y	7036	7502	200
			op3: +T -C	29	Y	2134	9874	87
			op4: +T +C	29	Y	693	3083	34.1
			op5: +T +C +H	29	Y	300	1894	37.5
HD4(L)	2703	16670	op1: -T -C	18	N	9320	9402	255
			op2: -T +C	18	N	1644	1657	130
			op3: +T -C	18	N	1207	1540	91
			op4: +T +C	18	N	170	398	38
			op5: +T +C +H	18	N	87	159	34
HD5(S)	12	328	op1: -T -C	72	N	3880	3971	11
			op2: -T +C	72	N	1969	1961	9
			op3: +T -C	200	N	2	3	8
			op4: +T +C	200	N	1	1	8
			op5: +T +C +H	200	N	6	6	8
HD6(S)	234	2495	op1: -T -C	20	N	3436	5718	27
			op2: -T +C	20	N	811	1367	19
			op3: +T -C	50	N	97	578	29
			op4: +T +C	50	N	37	205	21
			op5: +T +C +H	50	N	28	242	19
HD7(S)	256	1582	op1: -T -C	16	Y	996	1010	22
			op2: -T +C	16	Y	49	55	6
			op3: +T -C	16	Y	3	3	7
			op4: +T +C	16	Y	1	1	3
			op5: +T +C +H	16	Y	1	1	2
HD8(S)	256	1582	op1: -T -C	24	Y	8808	9107	42
			op2: -T +C	24	Y	818	1076	18
			op3: +T -C	24	Y	7	8	10
			op4: +T +C	24	Y	3	5	6
			op5: +T +C +H	24	Y	3	4	5
HD9(S)	256	1582	op1: -T -C	17	Y	1741	1809	26
			op2: -T +C	17	Y	127	164	7
			op3: +T -C	17	Y	4	5	7
			op4: +T +C	17	Y	1	1	3
			op5: +T +C +H	17	Y	1	1	3
HD10(S)	256	1582	op1: -T -C	17	Y	694	725	24
			op2: -T +C	17	Y	32	37	7
			op3: +T -C	17	Y	2	2	7
			op4: +T +C	17	Y	0	0	2.8
			op5: +T +C +H	17	Y	0	0	1.9
HD11(S)	974	10940	op1: -T -C	21	Y	653	727	45
			op2: -T +C	21	Y	8	8	22.5
			op3: +T -C	21	Y	12	11	41
			op4: +T +C	21	Y	4	4	23.5
			op5: +T +C +H	21	Y	3	3	21

The effectiveness of the various enhancements can be seen clearly from the consistent performance improvements of option sets 2 through 5, in comparison to basic BMC (*op1*). The bigger examples show a performance improvement of upto two orders of magnitude for the same bound *k*. Note that in many cases, basic BMC could not complete the search for as many time frames as BMC with enhancements within the allotted time (10k seconds). Among the enhancements, the use of only dynamic circuit simplification (*op3*) gave better performance than use of only customized translations (*op2*) for the safety properties; however, for liveness properties, it was the other way around. However, the combination of using both (*op4*) is always better than using only one or the other. Finally, use of the Hybrid SAT solver (*op5*) enhances this performance in most examples.

In terms of use of incremental SAT techniques, note that the performance gains in using incremental SAT techniques are the lowest with basic BMC (*op1*), where only circuits constraints are shared across different *k*-instances of BMC. The real benefit of incremental SAT is demonstrated by our customized property translations, which offer additional opportunities for sharing constraints within each *k*-instance of BMC as well. Therefore, the improvement factors are generally greater with use of customized translations (*op2*, *op4*, *op5*) than without (*op1*, *op3*). This can be seen clearly in the bigger examples – HD1-4, HD11 – where the SAT solving time is found to be non-negligible. Note also that the comparative gain from using incremental SAT techniques in the SAT solver is less than that from using the customized property translations. This indicates the benefits of our incremental formulations as a partitioning strategy, even when the constraint-sharing is not exploited by the SAT solver.

For the design HD3, we show the individual gains and cumulative effect in improving BMC due to various techniques in Figure 5.15. Overall, we obtain 100x improvement over standard BMC.

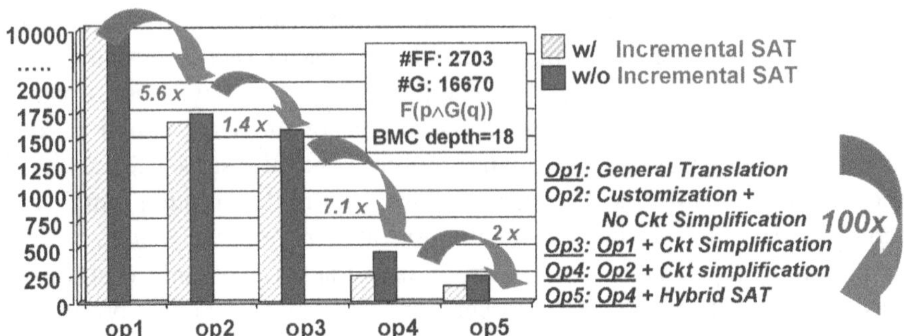

Fig. 5.15: Cumulative improvements of BMC using various techniques on HD3

5.6.2 Effect of Customized Translation and Incremental Learning

We show the benefits of our customized translation on liveness properties in the (negated) form $F(p \wedge G(q))$ on industry designs HD12-14. These are bus core designs with multiple masters and slaves. The properties to falsify are "*request should be eventually followed by acknowledge or error*". We compare various learning schemes, i.e., *NL* (no learning), *L1+L3*, *L2+L3*, and *L1+L2+L3* in the customized property translation approach, called *custom.* We also compare them against standard BMC in VIS [166], called *standard.* For fair comparison, we use the same SAT heuristics in our SAT solver as those used in the backend SAT solver (*zChaff* [38]) in VIS BMC. We experimented on a workstation with 2.8 GHz Xeon Processors with 4GB running Red Hat Linux 7.2. We present the comparison results in Table 5.2. Columns 2-4 show the characteristics of the designs in Column 1, i.e., number of flip-flops, primary inputs and gates, respectively; Column 5 shows the length of counter example (CEX); Columns 6-9 and Column 10 show time taken by our customized translation with different learning schemes, and the standard translation, respectively. Clearly, our customized property translation is able to find witnesses far quicker by 1-2 orders of magnitude than the standard monolithic translation [66, 67]. Moreover, the various learning schemes improve the performance of BMC significantly.

Table 5.2: Effect of Incremental Learning and BMC Customization

Design	#FF	#PI	#G	CEX (D)	Custom (VeriSol)				Standard (VIS)
					NL	L1,3	L2,3	L1,2,3	
HD11	2316	76	14655	19	2.3	2.2	2.3	2	77
HD12	2563	88	16686	22	11.2	8.9	11.7	8	201
HD13	2810	132	18740	28	730	290	862	240	2728

5.6.3 Effect of BDD-based Learning on BMC

We used six industrial designs D1-D6, ranging in gate count from 20K to half million gates, and flip-flop count from few hundreds to ~10K in the cone of influence of the correctness property. We used BMC to check safety properties, i.e., the search was for simple counterexamples without loops. Experiments for all designs except for D1, were performed on a 2.2 GHz Dual Xeon processor machine, with 4 GB memory, running Linux 7.2. Experiments for D1 were performed on a 900 MHz Dual Sun 220R machine, with 4 GB memory, running Solaris 5.8. For these examples, customized BMC (using *op5*) in *VeriSol* has orders of magnitude improved performance over VIS as shown in Table 5.3. Our goal is to show additive improvement due to BDD-based learning over a good baseline BMC.

Table 5.3: Effect of Incremental Learning and BMC Customization

Design	#FF/#G	VIS BMC		VeriSol BMC	
		k	Time(s)	k	Time(s)
D1	12.7k/416.1k	8	1906	96	10230
D2	4.2k/37.8k	30	802	64	7519
D3	5.2k/46.4k	29	8092	32	8667
D4	910/18k	57	10462	89	9760
D5	377/19.4k	12	5868	12	109
D6	952/18.1k	56	9134	56	29

5.6.4 Static BDD Learning

We first experimented with static BDD Learning. For our experiments, we used a maximum clause length of 6, and the seeds were the top 20 variables ranked by the SAT scoring mechanism for ordering decision variables, before any decisions are made. For D1, D2, and D4 we use time limit of 3 hours and present max BMC depths reached for different heuristics (Levels 1—3) for adding learned clause (see Section 5.4.8), as shown in Figure 5.16(a). For D3, D5, and D6, we used bounded depths of 32, 12, and 56, respectively, as stopping criteria since the SAT problem is non-trivial or satisfiable at those depths. The times are normalized with respect to the no learning case, as shown in Figure 5.16(b).

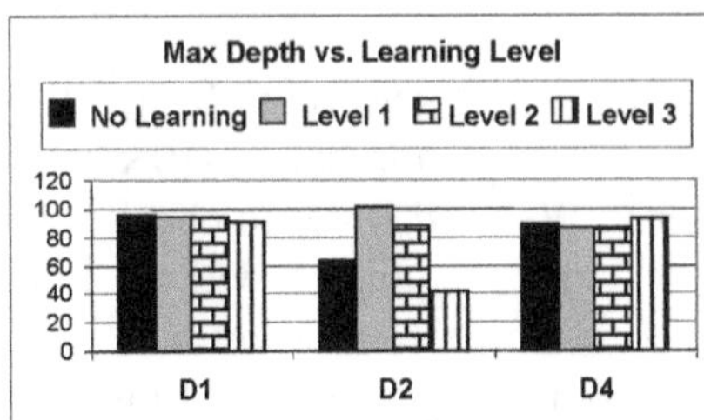

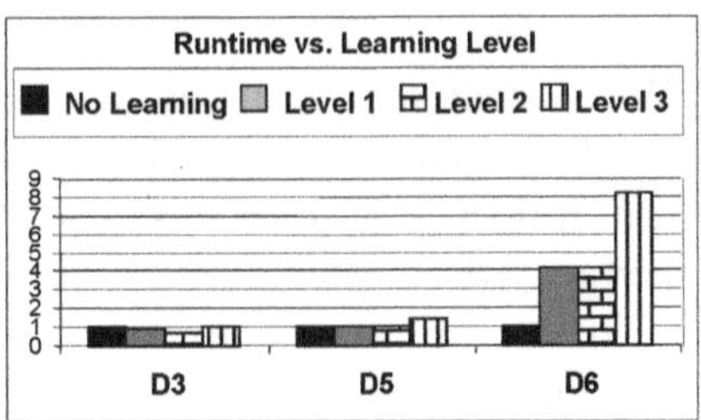

Fig. 5.16: Static Learning a) with time limit 3hrs for D1, D2, and D4, b) with BMC bound limit and normalized time for D3, D5, and D6

For four of six designs, static BDD Learning is better or the same as basic BMC, i.e., it improves the runtime, and in some cases allows a deeper search. The learning time is quite low in all cases. Comparing the different levels of learning, for five of six designs, the impact of adding more clauses is not encouraging at least for static learning. This can be seen clearly for design D2, where the maximum depth varied from 101 for Level 1, 87 for Level 2, down to 42 for Level 3. This shows the importance of learning "useful" clauses, while keeping their overheads low. On the other hand, for design D4, it was indeed more useful to add more clauses.

5.6.5 Dynamic BDD Learning

Next, we experimented with dynamic BDD Learning with different levels; learning after every 100 backtracks, and used a maximum clause length of 6. We observed that level 3 learning, in general, gave better performance, possibly due to dynamic seed selection.

We show the results using *Level 3* dynamic learning in Figure 5.17. We used BMC bounds of 96, 64, 32, 89, 12, and 56 for D1-D6 respectively, and compared the run times, normalized with respect to the no learning scheme, as shown in Figure 5.17a. Note, these bounds correspond to satisfiable instances, or time-outs using the no learning scheme for non-trivial SAT instances. For D1, D2, and D4, we used a time limit of 3 hours to compare the maximum depth reached.

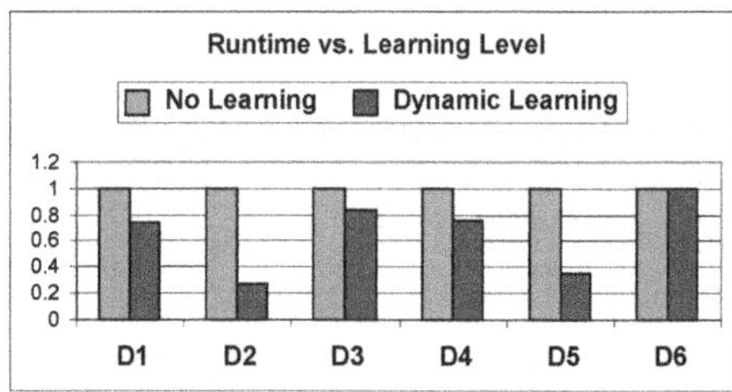

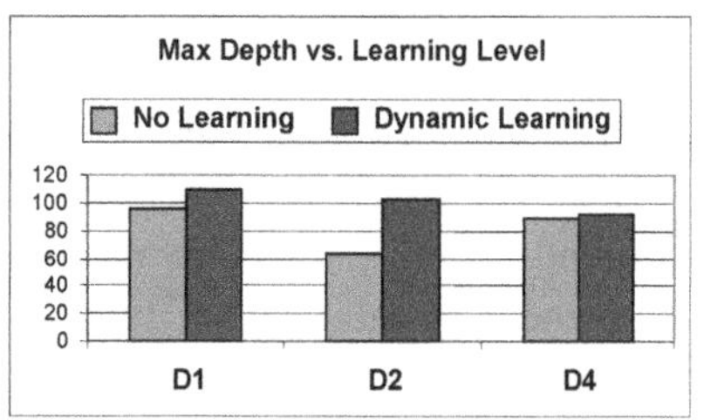

Fig. 5.17: Dynamic Learning: a) Depth limit with normalized time for D1-D6, b) Time limit on D1, D2 and D4

Compared with the no learning scheme, dynamic learning consistently reduces the runtime up to 73% as shown in Figure 5.17a, and allowed searches up to 60% deeper within the allotted time as shown in Figure 5.17(b). The learning time is again quite low. Interestingly, even a small number of added clauses can impact the overall performance significantly. For example, in design D1, the addition of just 15 learned clauses for 3 seeds, achieved a 25% reduction in run time for completing depth 96, and allowed an additional 13 time frames to be searched. This demonstrates the effectiveness of our heuristics for choosing seeds and learned clauses to be added. We also found that the past decision heuristic (SSH2) for seed selection provided good performance in general, although not consistently in each case.

In comparison to static BDD Learning, we obtained significantly better results (either less time, or increased depth, or both) with dynamic BDD Learning for four of six designs, and it was not much worse for the remaining two. While the time for learning itself is insignificant, the difference is due to the quality of the clauses added, and their impact on future decisions in the SAT solver. In particular, for the same level of learning (*Level 3*), less number of clauses is added by dynamic learning than by static learning. In a sense, this indicates that dynamic seed selection offers a better control than static seed selection over the tradeoff in adding learned clauses.

We also explored the effect of varying the maximum clause length from 5 to 9. Empirically, a maximum clause length of 6 or 7 gave the best results on all designs, where the gap between the worst and the best was about 10-20% on average.

5.7 Summary

SAT-based BMC is an effective technique for finding bugs in large designs. We discussed several techniques, individually as well as in combination, for improving the performance of the BMC engine for safety and liveness properties. Of these, customized property translation schemas and dynamic circuit simplification offer a significant performance improvement on the industry designs, as much as two orders of magnitude over standard BMC originally proposed [66]. This benefit is realized effectively due to incremental SAT problem formulation, exploiting partitioning, learning, and use of incremental SAT techniques in the SAT solver. When combined with the hybrid SAT solver techniques, our BMC engine shows another factor of two improvement in performance. Using lightweight and effective BDD Learning technique, the runtime of BMC engine is further reduced, allowing deeper searches to be performed within the allotted time. The various heuristics and parameters in the BDD Learning engine are targeted at increasing the usefulness of learned clauses, while reducing the inherent overheads.

5.8 Notes

The subject material described in this chapter are based on the authors' previous works [45] © 2002 IEEE, [146] © ACM 2003, [46] © ACM 2005.

6 DISTRIBUTED SAT-BASED BMC

6.1 Introduction

SAT-based Bounded Model Checking (BMC), though a robust and scalable verification approach, is still computationally intensive, requiring large memory and time. Even with the many enhancements discussed in Chapter 5, sometimes the memory limitation of a single server, rather than time, can become a bottleneck for doing deeper BMC search on large designs. The main limitation of a standard BMC application is that it can perform search up to a maximum depth allowed by the physical memory on a single server. This limitation stems from the fact that as the search bound k becomes large, the memory requirement due to unrolling of the design also increases. Especially for memory-bound designs, a single server can quickly become a bottleneck in doing deeper search for bugs.

Distributing the computing requirements of BMC (memory and time) over a network of workstations can help overcome the memory limitation of a single server, albeit at an increased communication cost. Here we describe such an approach [152], along with details that help to control the communication overhead to make it feasible.

The main intuition is that a BMC problem originating from an unrolling of the sequential circuit provides a *natural disjoint partitioning* of the underlying SAT problem and thereby, allows the computing resources to be configured in a linear topology with one Master and several Clients, as shown in Figure 6.1. Each Client C_i hosts a part of the unrolled circuit, i.e., from time frame n_i+1 to time frame n_{i+1} where n_i represents the partition depth. Each C_i (except for the terminal clients) is connected to C_{i+1} and C_{i-1}. The Master is connected to each of the Clients. Using a *linear topology*, we can *dynamically* distribute parts of the unrolled circuit over additional

Clients, as and when memory resources on the current set of Clients get close to exhaustion.

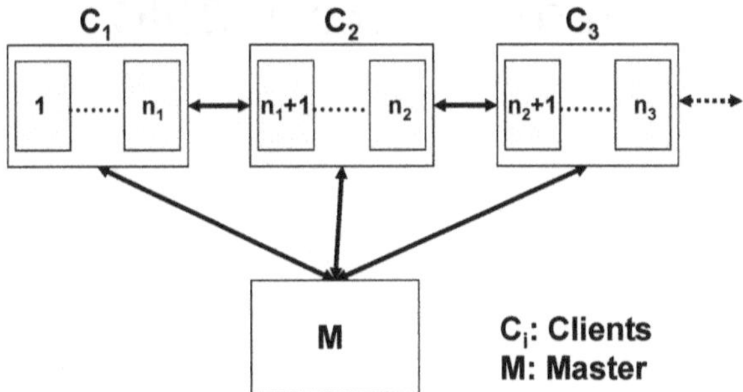

Fig. 6.1: Partitioning of Unrolled Circuit

To check the satisfiability of a Boolean problem originating from BMC, wherein the unrolled circuit is distributed over several servers, we must identify the parts of a SAT solver that may be delegated to each processor without requiring any processor to have the entire problem data. Since Boolean Constraint Propagation (BCP) on clauses can be done independently on an exclusive problem partition, it can be delegated to each processor. Moreover, since about 80% of SAT time involves BCP, one could achieve some level of parallelism by doing distributed-BCP. Note that any approach similar to SAT-based BMC can use a similar concept to exploit parallelism.

Outline

In Section 6.2, we provide an overview of distributed SAT-based BMC; in Section 6.3, we discuss topology cognizant distributed-BCP; in Section 6.4, we discuss distributed-SAT based on distributed-BCP; in Section 6.5, we present distributed SAT-based BMC algorithm; in Section 6.6, we several performance and memory optimizations; in Section 6.7, we provide experimental results; in Section 6.8, we discuss related work, and summarize in Section 6.9.

6.2 Distributed SAT-based BMC Procedure

We provide an overview of method *d-SAT_solve* that distributes SAT over a network of workstations (NOW) using a Master/Client model where each Client worsktation has an *exclusive partition* of the SAT problem. In other words, clauses and variables are exclusively owned by a client except for some shared variables that are on the interface of the partition. Such variables (clauses) are refered to a *local variables* (clauses). In contrast, any clause that spans variables from different partitions are referred to as *global clause*. We briefly describe the fine grain parallelization in *d-SAT_solve* implementing the three engines of a DPLL SAT algorithm [35], on a Master/Client distributed memory environment as shown in Figure 6.2. The Master controls the execution of d(istributed)-SAT. The *d-Decision* engine is distributed in such a way that each Client selects a good local variable and the Master then chooses the globally best variable to branch on. During the *d-Deduction* phase, each Client performs BCP on its exclusive partition, and the Master performs BCP on the (global) learned conflict clauses. *d-Diagnosis* is performed by the Master, and each Client performs a local backtrack when requested by the Master, in the procedure *d-backtrack*. The Master does not keep all problem clauses and variables; however, the Master maintains the global assignment stack and the global state for diagnosis. This requires much less memory than the entire problem data. To ensure proper execution of the parallel algorithm, each Client is required to be synchronized. We describe details of the parallelization and different communication messages in the following sections.

```
Synopsis:   Check distributed Boolean Satisfiability
Input:      Boolean Problem
Output:     SAT/UNSAT
Procedure:  d-SAT_solve

if (d-Deduce()=CONFLICT) return UNSAT; //Pre-process
while(d-Decide()=SUCCESS) { //branch
  while(d-Deduce()=CONFLICT) {//distributed BCP
    blevel = d-Diagnose(); //conflict-driven learning
    if (blevel = 0) return UNSAT;
    else d-backtrack(blevel);//backjump to blevel
  }
}
return SAT;
```

Fig. 6.2: Distribtued DPLL-style SAT Solver

The distinguishing features of the the distributed SAT algorithm are as follows:

- During *d-Deduction* phase, each Client is *topology cognizant* i.e., it has the knowledge of the partitioning topology and uses it to communicate with other Clients. This ensures that the receiving Client has never to read a message that is not meant for it. Alternatively, if the variables sets are not disjoint and Clients are not topology cognizant, as in [167], the Clients, after completing BCP, have to broadcast their new implications to all other Clients. In such case, after decoding the message, each receiving Client either reads the message or ignores it. In a communication network where BCP messages dominate, broadcasting implications can be an overkill when the number of variables runs into millions.

- The algorithm described here uses easily available existing networks of workstations, connected by Ethernet LAN. We discuss several efficient optimization schemes to reduce the effect of communication overhead on performance in general-purpose networks by identifying and executing tasks in parallel while messages are in transit.

We discuss how to use the *topology cognizant distributed-SAT* to obtain a SAT-based distributed BMC over a distributed-memory environment. For the sake of scalability, the method makes sure that at no point in the BMC computation does a single workstation have all the information. We have developed our distributed algorithms for a network of processors based on standard Ethernet and using the TCP/IP protocol. We can also potentially use dedicated communication infrastructures such as a clustered system for high performance computing that has low latency and high bandwidth communication [168]. Such a system may yield better performance, but for this work, we wanted to use an environment that is easily available, and whose performance can be considered a lower bound. We used a socket interface message passing library to provide standard bidirectional communications primitives.

We experimented on a network of heterogenous workstations interconnected with a standard Ethernet LAN. As an illustration, on an industrial design with ~13K FFs and ~0.5M gates, the non-disributed BMC on a single workstation (with 4 Gb memory) ran out of memroy after reaching a depth of 120; on the other hand, our SAT-based distributed BMC over 5 similar workstations was able to search up to 323 steps with a communication overhead of only 30%.

6.3 Topology-cognizant Distributed-BCP

BCP is an integral part of any DPLL-style SAT solver. We distribute BCP on multiple processes that are cognizant of the topology of the SAT-problem partition running on a network of workstations. In [167], during the distributed-SAT solve each Client broadcasts its implications to all other processors. After decoding the message, each receiving process either reads the message or ignores it. We improve this approach in the following way. Each process is made cognizant of the disjoint partitioning. The process then sends out implications to only those processes that share the partitioning interface variables with it. Each receiving process simply decodes and reads the message. This ensures that the receiving process has to never read a message that is not meant for it. This helps in two ways:

1. the receiving buffer of the process is not filled with useless information;
2. receiving process does not spend time in decoding useless information.

We use a *distributed model* with one Master and several Client processors. The Master's task is to distribute BCP on each Client that owns an exclusive partition of the problem. A bi-directional FIFO (First-in First-out) communication channel exists *only* between the process and its known neighbor, i.e., each process is cognizant of its neighbors. The process uses knowledge of the partitioning topology for communication so as to reduce traffic in the receiving buffer. A FIFO communication channel ensures that the channel is in-order, i.e., the messages sent from one process to another will be received in the order sent. Each message sent by a processor is *non-blocking*, i.e., process need not wait for the send to finish to process next task. Besides distributing BCP, the Master also records implications from the Clients as each Client completes its task.

6.3.1 Causal-effect Order

The main challenging task for the Master is to maintain *causal-effect* ("happens before") ordering of implications in a distributed-BCP algorithm since we cannot assume channel speeds and relative times of message arrivals during parallel BCP. *Maintaining such ordering is important because it is required for correct diagnosis during the conflict analysis phase of the SAT solver.* In the following we discuss the problem in detail and techniques to overcome it.

Consider the Master/Client model as shown in Figure 6.1. Client C_i can communicate with C_{i-1} and C_{i+1}, besides the Master M. The Master and Clients can generate *implication requests* to other Clients; however, Clients can send *replies* to the Master *only,* for any requests made to it. Along with the reply message, a Client also sends the message *ids* of the requests, if any, it made to the other Clients. This is an optimization step to reduce the number of redundant messages. To minimize the reply wait time, the Master is allowed to send requests to the Clients even when there are implications pending from the Client, provided that the global state (maintained by the Master) is not in conflict.

Let $p \rightarrow q$ denote an *implication request* from p to q and $p \leftarrow q$ denote an *implication reply* from q to p. Consider a sequence of three events E1-E3, shown below. Note that though the channel between C_i and the Master is in-order, what actually happens at Event E3 cannot be guaranteed.

E1: M→C1
E2: C1→C2
E3: M←C2 or M←C1

If M←C2 "happens before" M←C1, then we consider it an *out-of-order* reply since the implications due to M←C2 depend on C1→C2, which in turn depend on M→C1. Moreover, any out-of-order reply from a Client makes subsequent replies from that Client out-of-order until the out-of-order reply gets processed.

We propose a simple solution to handle out-of-order replies to the Master. For each Client, the Master maintains a *FIFO* queue where the out-of-order replies are queued. Since the channel between a Client and Master is in-order, this model ensures that messages in the *FIFO* will not be processed until the front of the *FIFO* is processed. We illustrate this with a short event sequence. For simplicity we show the contents of the *FIFO* for the Client C2.

E1: M→C1		FIFO(C2): -
E2: C1→C2		FIFO(C2): -
E3: M→C2		FIFO(C2): -
E4: M←C2	(in response to E2)	FIFO(C2): E4
E5: M←C2	(in response to E3)	FIFO(C2): E4, E5
E6: M←C1	(in response to E1)	FIFO(C2): -

Note: (E4 is processed before E5)

Note that in the reply event E6, the Client C1 also notifies the Master of the event E2. The Master enqueues the E4 reply as an out-of-order reply, since it is not aware of the responsible event E2 until E6 happens. The E5 reply is also enqueued as out-of-order, since the earlier out-of-order reply E4 has not been processed yet. When E6 occurs, the Master processes the messages from the events E6, E4, and E5 (in this order). This maintains the ordering of implications in the global assignment stack for correct diagnosis.

6.4 Distributed-SAT

We use fine grain parallelism in our distributed-SAT algorithm similar to the one proposed in [167]. However, we use the topology-cognizant distributed-BCP (as described in the previous section) to carry out distributed-SAT over a network of workstations. First, we describe the task partitioning between the Master and Clients as shown in Figure 6.3.

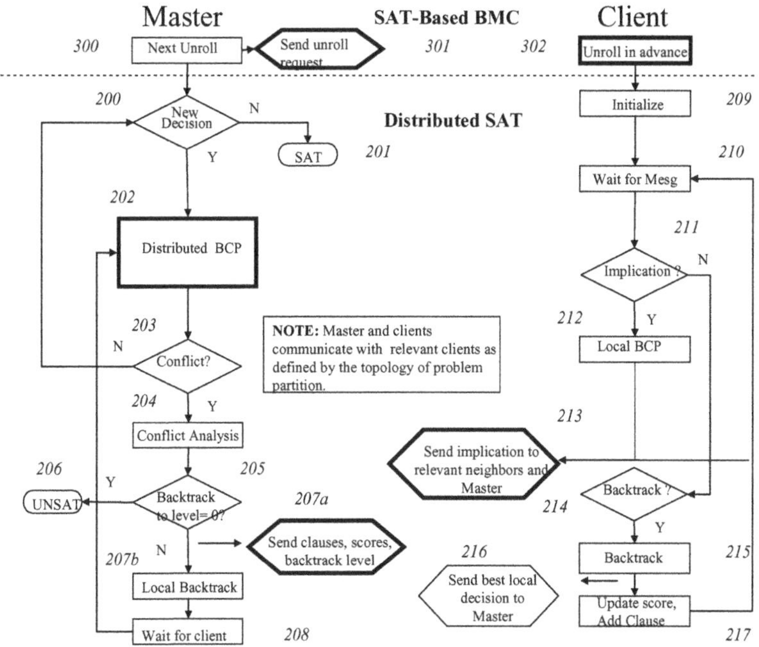

Fig. 6.3: Distributed-SAT and SAT-based Distributed-BMC

6.4.1 Tasks of the Master

Master controls the execution of *d-SAT_solve* procedure with the help of participating Clients. Specific tasks of Master are listed below, corresponding to the numbered boxes (as shown in Figure 6.3) in brackets.

- Maintains list of constraints, global assignment stack, learned clauses, antecedents
- Selects a new decision variable from the best local decision sent by each Client (box 216)
- Global conflict analysis using the assignments and antecedents (box 204)
- BCP, on globally learned clauses; manages distributed-BCP (box 202)
- *Receives from* C_i: New implications with antecedents and best local decision (box 213)
- *Sends to* C_i: Implication on variables local to C_i variables, backtrack request, learned local clauses, and update score request (box 207a)

6.4.2 Tasks of a Client C_i

Clients follow instructions from Master, but carries out BCP and conflict detection in their exclusive partition. Specific tasks of each Client are listed below, corresponding to the numbered boxes (as shown in Figure 6.3) in brackets.

- Maintains the ordered list of variables, scores, local assignment stack, local learned clauses
- Keeps the exclusive partition of the problem and topological information
- Executes on request: Backtrack, decay score, update variable score, local BCP (boxes 215, 217)
- *Receives from Master:* Implications, backtrack request, update score, conflict-driven learnt clause with all literals owned by C_i
- *Receives from neighbor C_j:* Implications on interface variables
- *Sends to Master:* New Implications with antecedents and best local decision, best local decision when requested, conflict node when local conflict occurs during BCP, request *id* when an implication request comes from another Client (boxes 213, 216)
- *Sends to neighbor C_j:* New implication requests on interface variables (box 213)

6.5 SAT-based Distributed-BMC

A SAT-based BMC problem originating from an unrolling of the sequential circuit over different time frames has a natural linear partition, and thereby allows configuring the computing resources in a linear topology. The topology using one Master and several Clients is shown in Figure 6.1. Each Client C_i is connected to C_{i+1} and C_{i-1}. The Master *controls the execution of the SAT-based distributed BMC algorithm.* The BMC algorithm in Figure 5.11-5.12 remains the same, except for the following changes. The *Unroll* procedure is now replaced by a *d-Unroll* procedure in which the procedure *Unroll* is actually invoked on the Client that hosts the partition for the depth *i.* Note that depending on the memory availability, the host Client is decided dynamically to be the same as that for depth *i-1*, or a new Client. After the unrolling, the *d-SAT_solve* algorithm (Figure 6.2) is invoked (in place of *SAT_solve)* to check the satisfiability of the problem on the unrolled circuit that has been partitioned over several workstations. The tasks for BMC are distributed as follows:

Tasks of the Master

- Allocates an exclusive problem partition to each host Client (box 300 in Figure 6.3)

- Requests an unrolling to the terminal Client (box 301 in Figure 6.3)
- *Controls* distributed-SAT.

Tasks of a Client

- Handles current unroll request and also advances by one (box 302 in Figure 6.3)
- Initiates a new Client as defined by the topology when the new unroll size is too large
- *Participates* in distributed-SAT.

6.6 Optimizations

We discuss several memory and performance optimization techniques in distributed-SAT and distributed SAT-based BMC. We also discuss a tight estimation of communication overheads in order to improve the parallelization of a single task.

6.6.1 Memory Optimizations in Distributed-SAT

The book-keeping information kept by the Master grows with the unroll depth. The scalability of our distributed-BMC is determined by how low is the *scalability ratio*, i.e., ratio of the memory utilized by the Master to the total memory used by the Clients. We take the following steps to lower the *scalability ratio*:

- By delegating the task of choosing the local decision and maintaining the ordered list of variables to the Client, we save the memory otherwise required by the Master.
- The Master does not keep the entire circuit information at any time. It relies on the Clients to send the reasons of implications that are used during diagnosis.

In our experiments, we observed that the scalability ratio for large designs is close to 0.1, which implies that we can hope to do a 10 times deeper search using our distributed-BMC in comparison to a non-distributed (monolithic) BMC. (In our observation, the global learned clauses maintained by the Master do not grow prohibitively large).

6.6.2 Tight Estimation of Communication Overhead

The inter-workstation communication time can be significant and adversely affects the overall performance. We can mitigate this overhead by hiding execution of certain tasks behind the communication latency. To have some idea of the communication overhead, we first need some strategy to measure the *communication overhead* and actual processing time. This is non-trivial due to asynchronous clock domains of the workstations. In the following, we first discuss a simple strategy to make a tight estimation of the wait time incurred by the Master due to inter-workstation communication in Distributed BMC.

Consider a *request-reply* communication scenario, where the time stamps are local to the Master and Client in the time line as shown in Figure 6.4. At time T_s, the Master sends its request to the Client. The Client receives the message at its time t_r. The Client processes the message and sends the reply to the Master at time t_s. The Master, in the meantime, does some other tasks and then starts waiting for the message at time T_w. The Master receives the message at time T_r.

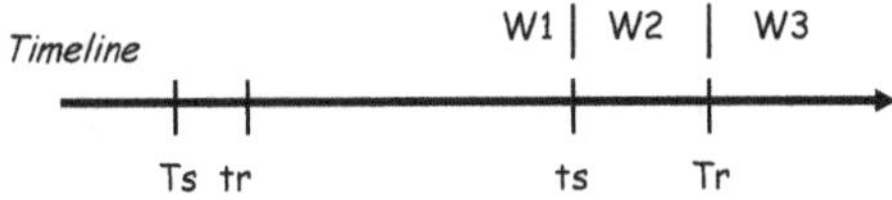

Fig. 6.4: Timeline for a tight estimation of communication overhead

Without accounting for the Client processing time, the wait time for Master is:

$$Wait\_Time = T_r - T_w \text{ for } T_r > T_w \text{ (= 0 otherwise)} \tag{6.1}$$

The above wait time, so calculated, does not discount the Client processing time, while the Master is waiting for the reply message from the Client. Therefore, the calculated wait time is an over-estimation of the actual *communication latency.* To obtain a wait time that truly represents the communication latency, discounting the Client processing time, we propose the following steps:

- Master sends the request with T_s embedded in the message.
- Client replies back to the Master with the time stamp $(T_s+(t_s- t_r))$.
- The Master, depending on the time T_w, calculates the actual wait time as shown in the follows three cases W1, W2, and W3, shown in Figure 6.4

1. Case W1: $T_w < (T_s+(t_s- t_r))$
   $Wait\_Time = T_r-(T_s+(t_s- t_r))$

2. Case W2: $(T_s+(t_s- t_r)) < T_w < T_r$
   $Wait\_Time = T_r - T_w$

3. Case W3: $T_r < T_w$
   $Wait\_Time = 0$

6.6.3 Performance Optimizations in Distributed-SAT

Now we discuss several performance optimizations in the distributed-SAT algorithm.

1. A large number of communication messages tend to degrade the overall performance. We propose several means to reduce the overhead:

   a) The Master waits for all Clients to stabilize before sending a new implication request. This reduces the number of implication messages sent.
   b) Clients send their best local decision along with every implication and backtrack reply. At the time of decision, the Master, then, only selects from the best local decisions. It is not required to make explicit requests for a decision variable to each Client separately (box 213).
   c) For all implication requests, Clients send replies to only the Master. This reduces the number of redundant messages in the network.

2. Each Client sends active variables to the Master before doing the initialization. While the Master waits and/or processes the message, the Client does its initialization in parallel.

3. When Master requests each Client to backtrack, it has to wait for the Clients to respond with a new decision variable. The following overlapping tasks can mitigate the wait time:

   a) Local backtrack (box 207b in Figure 6.3) by the Master is done after the remote request is sent (box 207a in Figure 6.3). While the Master waits for the decision variable from the Client, the Master also sends the local learned conflict clauses to the respective Client.
   b) The function for adjusting variable score (box 217 in Figure 6.3) is invoked in the Client after it sends the next decision variable (during backtrack request from the Master) (box 216 in Figure 6.3). Since *message-send* is *non-blocking*, potentially the function is executed in parallel with *send.* On the downside, the decision variable that is chosen may be a stale decision variable. However, note that the local decision variable that is sent is very unlikely be chosen as decision variable. The reason is that in the next step after backtrack most likely there will be an implication. Since the Client sends the decision variable after every implication request, the staleness of the decision variable will be eventually eliminated.

6.6.4 Performance Optimization in SAT-based Distributed-BMC

We list two main steps in performance optimization in *d-SAT_solve.*

1. The design is read and initialization is done in all the Clients to begin with. This reduces the processing time when the unrolling is initiated onto a new Client.
2. Advance unrolling is done in the Client while the Client is initially waiting for implication request from the Master. This includes invoking a new partition in a new Client.

6.7 Experiments

We conducted our evaluation of distributed-SAT and SAT-based distributed BMC on a network of workstations; each composed of dual Intel 2.8GHz Xeon Processor with 4Gb physical memory running Red Hat Linux 7.2, interconnected with a standard 10Mbps/100Mbps/1Gbps Ethernet LAN. We compare the performance and scalability of our distributed algorithm with a *non-distributed approach* referred to as *Mono*. We also measure the communication overhead using the accurate strategy described in Section 6.6.2.

We performed our first set of experiments to measure the performance penalty and communication overhead for the distributed algorithms. We employed our SAT-based distributed algorithm on 15 large industrial examples, each with a safety property. For these designs, the number of flip-flops ranges from ~1K to ~13K and number of 2-input gates ranges from ~20K to ~0.5M in the fanin cone of the property. Out of 15 examples, 6 have counterexamples, and the rest do not have any counterexample within the bound chosen. We used a model with one Master (referred to as M) and two Clients (referred as C1 and C2) where the Clients can communicate with each other. We used a controlled environment for the experiments in which, at each SAT check in the distributed-BMC, the SAT algorithm executes the tasks in a distributed manner as described earlier except at the time of decision variable selection and backtracking, when it is forced to follow the sequence that is consistent with the sequential SAT solver. We also used 3 different settings of the Ethernet switch to show how the network bandwidth affects the communication overheads. We present the results of the controlled experiments in Table 6.1 and 6.2.

In Table 6.1, the 1st Column shows the set of designs (D1-D6 have a counterexample), the 2nd Column shows the number of Flip Flops and 2-input Gates in the fanin cone of the safety property, the 3rd Column shows the bound depth limit for analysis, the 4th Column shows the total memory used by the non-distributed BMC, the 5th Column shows the partition depth when Client C2 took an exclusive charge of further unrolling, Columns 6-8 show the memory distribution among the Master and the Clients. In Column 9, we report the scalability ratio, i.e., the ratio of memory used by the Master to that of the total memory used by Clients. *We observe that for larger designs, the scalability factor is close to 0.1* although for comparatively smaller designs, this ratio was as high as 0.8. This can be attributed to the minimum bookkeeping overhead in the Master. Note that even though some of the designs have the same number of flip-flops and gates, they have different safety properties. The partition depth chosen was used to balance the memory utilization; however, the distributed-BMC algorithm chooses the partition depth dynamically to reduce the peak requirement on any one Client processor.

Table 6.1: Memory Utilization in distributed SAT-based BMC

Ex	FF (K)/ Gate (K)	D	M Mem (Mb)	Part D	P Mem (Mb)			S ratio
					M	C1	C2	
D1	4.2/30	16	20	5	8	5	16	0.4
D2	4.2/30	14	18	5	8	6	13	0.4
D3	4.2/30	17	21	5	9	5	17	0.4
D4	4.2/30	9	10	5	3	4	6	0.3
D5	4.2/30	15	18	5	8	5	15	0.4
D6	4.2/30	7	8	5	2	4	4	0.3
D7	4.2/30	21	24	5	7	4	20	0.3
D8	1.0/18	55	68	30	20	35	31	0.3
D9	0.9/18	67	124	30	65	33	49	0.8
D10	5.2/37	21	29	5	10	4	24	0.4
D11	13/448	61	1538	45	172	1071	480	0.1
D12	4/158	81	507	40	47	246	267	0.1
D13	4/158	41	254	20	24	119	141	0.1
D14	4/158	81	901	40	149	457	447	0.2
D15	4/158	81	901	40	135	457	443	0.2

In Table 6.2, the 1st Column shows the same set of designs (D1-D15) as in Table 6.1, the 2nd Column shows the cumulative time taken (over all steps) by non-distributed BMC, the 3rd Column shows the cumulative time taken (start to finish of Master over all steps) by the distributed-BMC method excluding the message wait time, Columns 4-6 show the total message wait time for the Master in a 10/100/1000Mbps Ethernet Switch setting. In Column 7, we calculate the performance penalty by taking the ratio of the time taken by distributed to that of non-distributed BMC (=Para Time/ Mono Time). In Column 8, we calculate the communication overhead for the 1Gbps switch setting by taking the ratio of the message waiting time to distributed BMC time (=wait time for 1 Gbps/ Para Time). We find that the performance penalty is 50% on average, and the communication overhead is 70% on average, with overall degradation by a factor of 2.55 (=1.5 * 1.7).

Table 6.2: Performance evauation in distributed SAT-based BMC

Ex	MT (sec)	PT (sec)	MWT (sec)			Perf Pntly	Com Ovr
			1 gbs	100 mbs	10 mbs		
D1	8.9	12.8	11.4	34.5	991	1.4	0.9
D2	4.2	6.7	10.5	24.2	699	1.6	1.6
D3	9.7	15.6	11.2	33.2	768	1.6	0.7
D4	0.8	1.9	1.8	3.8	108	2.4	0.9
D5	5.2	8.2	10	31.4	680	1.6	1.2
D6	0.3	1.1	0.6	1.6	45	3.7	0.5
D7	9.5	15	9	40	855	1.5	0.6
D8	37.9	52.1	22.1	109	1895	1.4	0.4
D9	314.6	454	130	702	12923	1.4	0.3
D10	23.4	38.4	17.8	71.8	764	1.6	0.5
D11	919	1261	1136	2403	5893	1.4	0.9
D12	130.5	89.1	0.1	65.1	63	0.7	0.0
D13	33.7	23.2	0.4	6.3	16	0.7	0.0
D14	452.8	361	87.4	654	1289	0.8	0.2
D15	442.2	345	97.2	680	1138	0.8	0.3

In some cases, D12-D15, however, we find an improvement in performance over non-distributed BMC. This is due to the exploitation of parallelism during the Client initialization step as described in Sections 6.6.3 and 6.6.4. Note that the message wait time is adversely affected by lowering the switch setting from 1Gbps to 10Mbps. This is attributed to the fact that Ethernet LAN is inherently a broadcast non-preemptive communication channel.

In our second set of experiments, we show that distributed-BMC is able to go deeper for examples where there are no witnesses, and is able to find a witness in one example, while non-distributed BMC runs out of memory on each example. We used the 5 largest (of 15) designs D11-D15 that did not have a witness. To show the performance of distributed BMC on a large design with a witness, we experimented on a hardware implementation (using VHDL [169]) of a quick sort algorithm (QS). The algorithm is recursively called on the left and right partitions of the array. We implemented the array as a memory module with AW (address width) = 13 and DW (data width) = 32. We implemented the stack (for bounded recursive function calls) also as a memory module with AW=13 and DW=30. The property we chose is that the sorting algorithm eventually terminates. The design including the memory modules has 270K latches and 2.7M 2-input OR gates in the fanin cone of the property. We used a partition of 25 time frames on each client.

For distributed-BMC, we configured 5 workstations into one Master and 4 Clients C1-C4; each connected with the 1Gbps Ethernet LAN. In this setting, the Clients are connected in a linear topology and the Master is connected in a star with the others. For design D11, we used a partition of 81 unroll depths on each Client and for designs D12-15, we used a partition of 401 unroll depths on each Client. For the design QS, we used a partition of 25 unroll depths on each Client. The results are shown in the Table 6.3. The designs D11-D15 with no witnesses are shown in the top part, and the design QS with a witness is shown in the bottom part of the table.

In Table 6.3, the 1st Column shows the set of large designs that were hard to verify, the 2nd Column shows the farthest depth to which non-distributed BMC could search before it runs out of memory, the 3rd Column shows the time taken to reach the depth in the 2nd Column, the 4th Column shows the unroll depth reached by distributed-BMC using the allocated partitions, the 5th Column shows the time taken to reach the depth in the 4th Column excluding the message wait time, Columns 6-10 show the memory distribution for the Master and Clients, the 11th Column shows the total message wait time. In Column 12, we report the communication overhead by taking the ratio of message wait time to the distributed-BMC time (=MWT time/Para Time). In Column 13, we report the scalability ratio by taking the ratio of memory used by the Master to that of the total memory used by the Clients.

We use design D11 with ~13K flip-flops and ~0.5Million gates to illustrate the performance comparison on designs with no witnesses. For the this design, we could analyze up to a depth of 323 with only 30% communication overhead, while using a non-distributed version we could analyze only up to a depth of 120 with the per-workstation memory limit. The low scalability factor, i.e., 0.1 for large designs indicates that for these designs our distributed-BMC algorithm could have gone 10 times deeper compared to the non-distributed version for similar machines. We also observe that the communication overhead for these designs D11-D15 was about 45% on average, *a small penalty to pay for deeper search.*

For the QS design, we were able to find the witness at depth 33 (using Clients C1 and C2) with communication overhead of 15%, while using the non-distributed BMC we ran out of memory at depth 29. For the depth 29, distributed BMC required 2507s (including the communication overhead time of 495s) as compared to 2098s by non-distributed BMC.

Table 6.3: Comparison of non-distributed (mono) and distributed BMC on Industry designs

Ex	Mono Depth	Mono Time (sec)	Para Depth	Para Time (sec)	Para Memory (in Mb)					MWT (sec)	Comm Ovrhd	S Ratio
					M	C1	C2	C3	C4			
D11: FF/Gate=13K/448K, Witness not found for the bound D12-15: FF/Gate=4K/148K, Witness not found for the bound												
D11	120*	1642	323	6778	634	1505	1740	1740	1730	1865	0.3	0.1
D12	553*	4928	1603	13063	654	1846	1863	1863	1863	5947	0.5	0.1
D13	553*	4899	1603	12964	654	1846	1864	1864	1864	5876	0.5	0.1
D14	567*	642	1603	2506	654	1833	1851	1851	1851	1585	0.6	0.1
D15	567*	641	1603	1971	654	1833	1851	1851	1851	879	0.4	0.1
Quick Sort (QS): FF/Gate: 270K/2.7M, Witness found at depth 33												
QS	29*	2098	33	17796	817	1868	1548	-	-	2696	0.15	0.24

(*: memory out after solving for the bounded depth)

6.8 Related Work

Parallelizing SAT solvers have been proposed by many researchers [167, 170-174]. Most of them target performance improvement of the SAT solver. These algorithms are based on partitioning the search space on different processors using partial assignments on the variables. Each processor works on the assigned space and communicates with other processors only after it is done searching its portion of the search space. Such algorithms are not scalable memory-wise due to high data redundancy as each processor keeps the entire problem data (all clauses and variables).

In a closely related work on parallelizing SAT [167], the authors partition the problem by distributing the clauses evenly on many application specific processors. They use fine grain parallelism in the SAT algorithm to get better load balancing and reduce communication costs. Though they have targeted the scalability issue by partitioning the clauses disjointedly, the variables appearing in the clauses are not disjoint. Therefore, whenever a Client finishes BCP on its set of clauses, it must broadcast the newly implied variables to all the other processors. The authors observed that over 90% of messages are broadcast messages. Broadcasting implications can become a serious communication bottleneck when the problem contains millions of variables.

Reducing the space requirement in model checking has been suggested in several works [93, 175, 176]. These studies suggest partitioning the problem in several ways. The work in [175] shows how to parallelize the model checker based on explicit state enumeration. They achieve it by partitioning the state table for reached states into several processing nodes. The work in [176] discusses techniques to parallelize the BDD-based reachability analysis. The state space on which reachability is performed is partitioned into disjoint slices, where each slice is owned by one process. The process performs a reachability algorithm on its own slice. In [93], a single computer is used to handle one task at a time, while the other tasks are kept in external memory.

6.9 Summary

For verifying designs with high complexity, we need a scalable and robust solution. SAT-based BMC is quite popular because of its robustness and better debugging capability. Although, SAT-based BMC is able to handle increasingly larger designs as a result of advancement in SAT solvers, the memory of a single server has become a serious limitation to carrying out deeper search. Existing parallel algorithms either focus on improving the SAT performance, or have only been used in explicit state-based model checkers or in unbounded implicit state-based model checkers. Here, we discussed a feasible solution for SAT-based distributed-BMC using an improved distributed SAT algorithm.

The distributed algorithm uses the normally available large pool of workstations that are inter-connected by a standard Ethernet LAN. For the sake of scalability, the distributed algorithm makes sure that no single processor has the entire data. Also, each process is cognizant of the partition topology and uses this knowledge to communicate with the other process; thereby, reducing the process's receiving buffer traffic due to unwanted information. We have also described several memory and performance optimization schemes to achieve scalability and decrease the communication overhead.

6.10 Notes

The subject material described in this chapter are based on the authors' previous works [152, 177], used with kind permission of Springer Science and Business Media.

7 EFFICIENT MEMORY MODELING IN BMC

7.1 Introduction

Designs with large embedded memories are quite common and have wide application. However, these embedded memories add further complexity to formal verification tasks due to an exponential increase in state space with each additional memory bit. In typical BMC approaches [45, 66, 109, 178], the search space increases with each time-frame unrolling of a design. With explicit modeling of large embedded memories, the search space frequently becomes prohibitively large to analyze beyond a small depth. In order to make BMC more useful, it is important to have some abstraction of these memories. However, for finding real bugs, it is sufficient that the abstraction techniques capture the memory semantics [179] without explicitly modeling each memory bit.

We discuss an *Efficient Memory Modeling* (EMM) approach for memory abstraction that preserves memory semantics, thereby augmenting the capability of SAT-based BMC to handle designs with large embedded memory without explicitly modeling each memory bit [50, 51]. EMM method does not require examining the design or changing the SAT-solver, and is guaranteed not to generate false negatives. The described method is similar, but with key enhancements, to the abstract interpretation of memory [179, 180] that captures its forwarding semantics, i.e., a data read from a memory location is the same as the most recent data written at the same location. In EMM method, we construct an abstract model for BMC by eliminating memory arrays, but retaining the memory interface signals and adding constraints on those signals at every analysis depth to preserve the semantics of the memory. The size of these *memory-modeling constraints* depends quadratically on the number of memory accesses and linearly on the bus widths of memory interface signals. Since the analysis depth of BMC bounds the number of memory accesses, these constraints are significantly smaller than the explicit memory model. The novelty of our memory-modeling constraints is that they capture the exclusivity of a read and write pair explicitly, i.e., when a SAT-solver decides on a valid read and write pair, other pairs are *implied invalid immediately*, thereby reducing the SAT solve time. We have implemented these techniques in our SAT-based BMC framework (*VeriSol*) where we demonstrate the effectiveness of such an abstraction on a number of hardware and software designs with large embedded memories. We show at least an order of magnitude improvement (both in time and space) using our method over explicit modeling of each memory bit. We also show that adding exclusivity constraints boosts the performance of the SAT solver (in BMC) significantly, as opposed to the conventional way of modeling these constraints as nested *if-then-else* expressions [59, 179].

Outline

In Section 7.2 we describe the basic idea used in our approach, in Section 7.3 we give relevant background on memory semantics, in Section 7.4 we discuss our *EMM* approach in detail for single and multiple ports, in Sections 7.5 and 7.6 we discuss our experiments on single and multiple port memory, respectively, in Section 7.7 we discuss related work, and in Section 7.8 we summarize our discussion

7.2 Basic Idea

For typical designs with embedded memory, a *Main* module interacts with the memory module *MEM* using the interface signals as shown in Figure 7.1. For a single-port memory at any given clock cycle, the following observations can be made: a) *at most one address is valid*, b) *at most one write occurs*, and c) *at most one read occurs*.

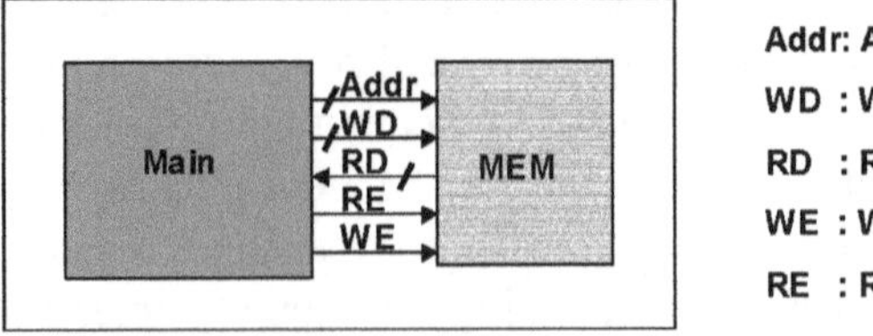

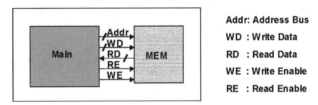

Fig. 7.1: Design with embedded memory

As BMC-search-bound k becomes larger, the unrolled design size increases linearly in the size of the memory module as shown in Figure 7.2. For designs with large embedded memories, this increases the search space prohibitively for any search engine.

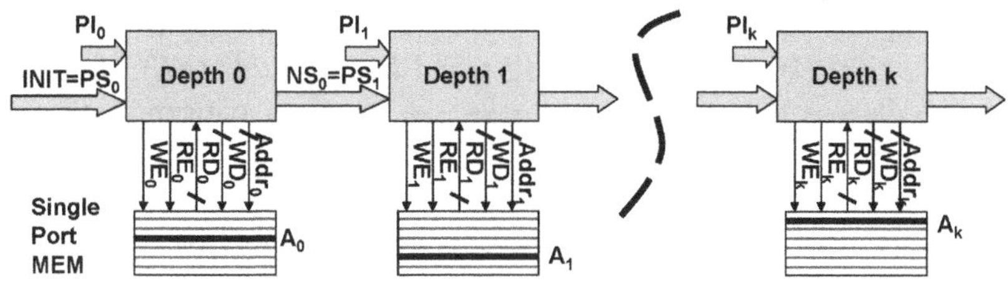

Fig. 7.2: Unrolled design with embedded memory

It can be observed in this memory model that memory read at any depth depends only on the most recent data written previously at the same address. Therefore, to enable the SAT-based BMC to analyze deeper on such designs, we replace an explicit memory model with an Efficient Memory Model (EMM) as follows:

1. We remove the *MEM* module but retain the memory interface signals and the input-output directionality with respect to the *Main* module.
2. We add constraints at every analysis depth k on the memory interface signals that preserve the forwarding semantics of the memory.

To improve the SAT solver (in BMC), we also do the following:

3. We *add constraints* such that when the SAT-solver decides on a valid read and write pair, other pairs are *implied invalid immediately*, i.e., additional branching by SAT solver.

Note that although steps (1) and (2) are sufficient to generate an efficient model that preserves the validity of a correctness property, step (3) improves the performance of SAT-based BMC as observed in our experiments. Moreover, we do not have to examine the *Main* module while adding these memory-modeling constraints. In our initial discussion, we consider a single-port memory for simplicity reasons. Later, we extend the method to discuss more general multi-port memories. In the next section, we discuss the formalization of the memory semantics.

7.3 Memory Semantics

Embedded memories are used in several forms such as RAM, stack, FIFO, etc., with at least one port for data access. For simplicity, we assume a single read/write port memory as shown in Figure 7.1. Such a memory has the following interface signals: *Address Bus* (*Addr*), *Write Data Bus* (*WD*), *Read Data Bus* (*RD*), *Write Enable* (*WE*), and *Read Enable* (*RE*). (In a later section, we discuss the more general case, i.e., memory with multiple read and write ports.)

We show a typical timing diagram for memory access operations in an example shown in Figure 7.3. The *write* operation requires two clock cycles. In the first clock cycle, the data value is assigned to the *WD* bus, the write address is assigned to the *Addr* bus, and the *WE* signal is made active. In the second clock cycle, the address location of the memory has the new data. Thus, written data is available after one clock cycle, i.e., after write enable and write address are valid. In the example shown, data 0 and 1 are written at address locations 5 and 4 in clock cycles 2 and 3, respectively. The *read* operation requires one clock cycle. When the read address is assigned to the *Addr* bus and the *RE* signal is made active, the read data is assigned to *RD* bus. Thus, data is available in the same cycle when the read enable and read address are valid. In the example shown, the data are read at clock cycles 4, 5, and 6, from address location 4.

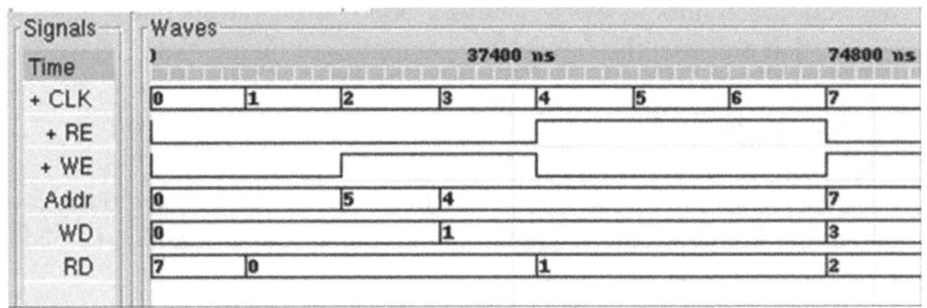

Fig. 7.3: Typical memory access timing diagram

The *data forwarding semantics* of memory states that read data equals the last written data (some time in the past) at the same address location. Formally, we define it as follows:

Assume that we unroll the design up to depth k (starting from 0). Let S^j denote a memory interface signal variable S at time frame j. Let the Boolean predicate variable $E^{i,j}$ denote the address comparison between time frames i and j, defined as $E^{i,j}=(Addr^i=Addr^j)$. Then, the data *forwarding semantics* is expressed formally as:

$$RD^k = \{WD^j \mid E^{j,k}{=}1 \wedge WE^j{=}1 \wedge RE^k{=}1 \wedge \forall_{j<i<k}(E^{i,k}{=}0 \vee WE^i{=}0)\}, \text{ for } j < k \quad (7.1)$$

In other words, data read at depth k equals the data written at depth j if

- addresses are equal at k and j, and
- write enable is active at j, read enable is active at k, and
- for all depths between j and k no data was written at $Addr^k$.

7.4 EMM Approach

We augment SAT-based BMC (see Chapter 5) with a mechanism to add memory-modeling constraints at every unroll depth of BMC analysis. We use our hybrid SAT solver (see Chapter 4) that uses hybrid representations of Boolean constraints, i.e., *Reduced AIG* to represent the original circuit problem, and CNF to represent the learned constraints. We use a hybrid representation to model memory constraints efficiently as desribed below

In order to add constraints for the data forwarding semantics of memory as in Eq 7.1, one can use a conventional approach based on the selection operator ITE, as shown in Figure 7.4. (Note, $\mathrm{ITE}(s,t,e) \equiv (s\wedge t)\vee(\neg s\wedge e)$, where s, t, e are Boolean variables.) Let the Boolean variable $s^{j,k}$ denote the *valid read signal,* defined as $s^{j,k} = E^{j,k} \wedge WE^j$. Then the data read at depth k (given RE^k=1) is expressed as:

$$RD^k = \mathrm{ITE}(s^{k-1,k}, WD^{k-1}, \mathrm{ITE}(s^{k-2,k}, WD^{k-2}, \ldots.\mathrm{ITE}(s^{0,k}, WD^0, WD^{-1}))\ldots) \quad (7.2)$$

where WD^{-1} denotes the initial data value.

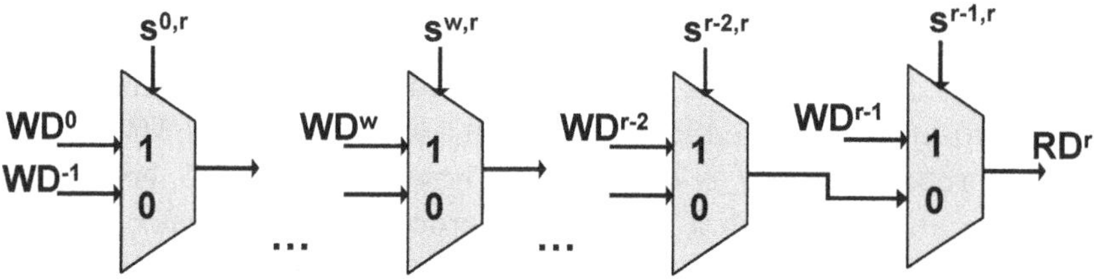

Fig. 7.4: Conventional way of adding constraints using ITE operator

To choose a read-write pair RD^k=WD^i; a SAT-solver has to make several decision to obtain the required constraints $s^{i,k}$=1, $s^{i+1,k}$=0, $s^{i+2,k}$=0,…,$s^{k-1,k}$=0. Instead we add explicit constraints to capture selection of the read-write pairs exclusively. Once a read-write pair is chosen by the SAT-solver, the other pairs are implied invalid immediately. Let the Boolean variable $S^{i,k}$

denote the *exclusive valid read signal* and the Boolean variable $PS^{i,k}$ denote the intermediate exclusive signal. They are defined recursively as follows:

$$\begin{aligned} \forall_{0\leq i<k}\ PS^{i,k} &= \neg s^{i,k} \wedge PS^{i+1,k} \quad (= RE^k \text{ for } i=k) \\ \forall_{0\leq i<k}\ S^{i,k} &= s^{i,k} \wedge PS^{i+1,k} \quad (= PS^{0,k} \text{ for } i=-1) \end{aligned} \tag{7.3}$$

Now, Eq. (7.1) can be expressed as

$$RD^k = (S^{k-1,k} \wedge WD^{k-1}) \vee (S^{k-2,k} \wedge WD^{k-2}) \vee \ldots \vee (S^{0,k} \wedge WD^0) \vee (S^{-1,k} \wedge WD^{-1}) \tag{7.4}$$

Note that $S^{i,k}=1$, *immediately implies* $S^{j,k}=0$ where $j \neq i$ and $i,j < k$.

7.4.1 Efficient Representation of Memory Modeling Constraints

As mentioned earlier, we use a hybrid representation for adding EMM constraints. We capture Eqs. (7.3)-(7.4) efficiently,

- by not representing the constraints as a large tree-based circuit structure, since such a structure adversely affects the BCP performance as observed in the context of adding large conflict clauses [41], and
- by not creating unnecessary 2-literal clauses since they too adversely affect a CNF-based SAT-solver that uses 2-literal watch scheme for BCP [107].

We refer to our hybrid representation as *hybrid Exclusive Select Signal* (hESS) representation – this is described in the following. We implemented the addition of memory modeling constraints in *hESS* as part of the procedure *EMM_constraints,* which is invoked after every unrolling. This is highlighted in the modified BMC algorithm, *BMC_solve_F_EMM* in Figure 7.5. The procedure *EMM_constraints,* as shown in Figure 7.6, generates the constraints at every depth *k* using 3 sub-procedures: *Gen_addr_equal_sig, Gen_valid_read_sig,* and *Gen_read_data_constraints.* It then returns the constraints *EMM_C* at each depth *k.* As we see in the following detailed discussion, these constraints capture the forwarding semantics of the memory very efficiently at depth *k.*

```
Synopsis:     Customized BMC solve for F(p)
Input:        prop_tree_node p, bound N, Initial state I
Output:       TRUE/FALSE/ABORT
Procedure:    BMC_solve_F_EMM

C = I; i=0;
while (i<N) {  //L1 is active always
 p^i = Prop_node(p,i);
 C = C ∧ EMM_constraints(i);//update constraints
 if (Is_sat(C ∧ p^i)) return true; //wit found
 C = C ∧ ¬p^i; //L3 Learning
 Merge(p^i, const_0); //Simplify
 for (j=i;j>=0;j--) {//loop-free check
   C = C ∧ ¬_jL_i;
   if (!Is_sat(C)) return false; //no wit exists
 }
 i = i+1; //increment depth
}
ABORT("Bound Reached"); //wit not found
```

Fig. 7.5: BMC augmented with EMM constraints

Gen_addr_equal_sig: Generation of Address Comparison Signals

Let m denote the bitwidth of the memory address bus. We implement the address comparison as follows: for every address pair comparison ($Addr^j{=}Addr^k$) we introduce new variables $E^{j,k}$ and $e^{j,k}_i$ $\forall_{0\le i<m}$ (for every bit i). Then we add the following CNF clauses for each i:

$$(-E^{j,k} + Addr^j_i + -Addr^k_i)\ ,\ (-E^{j,k} + -Addr^j_i + Addr^k_i),$$
$$(e^{j,k}_i + Addr^j_i + Addr^k_i)\ ,\ (e^{j,k}_i + -Addr^j_i + -Addr^k_i)$$

Finally, we use one clause to connect the relation between $E^{j,k}$ and $e^{j,k}_i$, *i.e.*,

$$(-e^{j,k}_0 + \ldots + -e^{j,k}_i + \ldots + -e^{j,k}_{m-1} + E^{j,k})$$

Note that these clauses capture the relation that $E^{j,k}{=}1$ if and only if ($Addr^j{=}Addr^k$). A naïve way to express the same equivalence relation structurally would be to use an AND-tree of X-NOR (⊗) gates as follows:

$$E^{j,k} = (Addr^j_0 \otimes Addr^k_0) \wedge \ldots \wedge (Addr^j_{m-1} \otimes Addr^k_{m-1})$$

Clearly, this representation would require *4m-1* 2-input OR gates, amounting to *12m-3* equivalent CNF clauses (3 clauses per gate). A hybrid representation, on the other hand, requires only *4m+1* clauses and does not require any 2-literal clause. Thus, at every depth k, we add only *(4m+1)k* clauses for address comparison, rather than *(12m-3)k* gates clauses required by the purely circuit structure approach.

```
Synopsis:     Generate EMM constraints
Input:        BMC depth k
Output:       EMM constraints in hybrid representation
Procedure:    EMM_constraints

 //Generate address equal signals
Gen_addr_equal_sig(k);
//Generate exclusive valid read signals
Gen_valid_read_sig(k);
//Generate constraints on read data
EMM_C = Gen_read_data_constraints(k);
return EMM_C;
```

Fig. 7.6: Efficient Modeling of Memory Constraint

Gen_valid_read_sig: Generation of exclusive valid read signals

To represent the exclusive valid read signals as in Eq (7.3), we use a *Reduced AIG* rather than CNF clauses. Since each intermediate variable has fan-outs to other signals, we cannot eliminate them. If we were to represent these using CNF clauses, it would introduce too many additional 2-literal clauses. This representation adds *3k* 2-input gates (or *9k* gate clauses) at every depth *k*.

Gen_read_data_constraints: Generation of constraints on read data signals

By virtue of Eq (7.3), we know that for a given *k*, at most one $S^{j,k}=1$, $\forall_{1\leq j<k}$. We use this fact to represent the constraint in Eq (7.4) as CNF clauses. Let *n* denote the bitwidth of the data bus. We add the following clauses

$$\forall_{0\leq i<n,}\ \forall_{-1\leq j<k}\ (-S^{j,k} + -RD^{k}_{i} + WD^{j}_{i}),\ (-S^{j,k} + RD^{k}_{i} + -WD^{j}_{i})$$

To capture validity of the read signal, we add the following clause:

$$(-RE^{k} + S^{-1,k}+\ldots+S^{j,k}+\ldots+S^{k-1,k})$$

Thus we add *2n(k+1)+1* clauses at every depth *k*. On the other hand, if we use a circuit gate representation, it would require *n(2k+1)* gates and therefore, *3n(2k+1)* gate clauses.

Overall, at every depth k, our *hybrid Exclusive Select Signal* representation adds $(4m+2n+1)k+2n+1$ clauses and $3k$ gates, as compared to $(4m+2n+2)k+n$ gates in a purely circuit representation. Note that although the size of these accumulated constraints grows quadratically with depth k, they are still significantly smaller than an explicit memory model.

7.4.2 Comparison with ITE Representation

If we were to use *Reduced AIG* to represent nested ITE representation, at every depth k, we would need $(4m-1)k$ gates for address comparison, k gates for the ITE control signals, and $3nk$ gates for implementing ITE operators, where total number of ITE operators is nk. Overall, we need $4mk+3nk$ gates.

If we use a hybrid representation for the nested ITE, at every depth k, we would need $(4m+1)k$ clauses for address comparison, k gates for the ITE control signals, $4nk+1$ clauses for implementing the ITE operators. We need to add the following clauses $\forall_{0\leq i<n}\ \forall_{0\leq j<k}$

$$
\begin{gathered}
(-s^{jk} + -rd^{j+1}{}_i + WD^j{}_i) \\
(-s^{jk} + rd^{j+1}{}_i + -WD^j{}_i) \\
(s^{jk} + -rd^{j+1}{}_i + rd^j{}_i) \\
(s^{jk} + rd^{j+1}{}_i + -rd^j{}_i)
\end{gathered}
$$

where $rd^k=RD^k$, and $rd^0=WD^{-1}$. Note that we have to introduce new intermediate variables $rd^j\ \forall_{0<j<k}$. In addition, we need to add a clause to capture the validity of the read signal. Overall we need $(4m+1+4n)k+1$ clauses and k gates.

We summarize the number of gates and clauses required to represent the constraints in each of the cases *A-D* as discussed in Table 7.1. Clearly, the growth in number of constraint clauses is quadratic in the analysis depth k.

For a meaningful comparison between different cases *A-D*, we use 3 equivalent CNF clauses for each 2-input gate. The growth curves of constraint clauses are shown in Figure 7.7 for n=32, m=12. As we can see, hESS representation D is 3 times more succinct than the gate representation of nested ITE (A), and 50% in size when compared with hybrid representation of nested ITE (B) at k=50. In other words, BMC would be able to do deeper searches using D representation on a limited memory machine, when compared with *A-C* representations. Moreover, as hESS representation captures exclusivity of valid read signals explicitly, the SAT-solver performance is improved, as observed in our experiments.

While the growth of memory modeling constraint clauses is quadratic with the analysis depth, one can observe that although the constraint clauses are sufficient, they may not be necessary in every time frame and can be added lazily, in a manner similar to lazy refinement approaches discussed later in Chapter 11.

Table 7.1: Comparison of # clauses and gates in EMM constraints

Modeling Styles	Nested ITE		Exclusive selection criterion	
	Gate (A)	**Hybrid (B)**	**Gate (C)**	**Hybrid (D)**
#Gates	$(4m+3n)k$	k	$(4m+2+2n)k+n$	$3k$
#Clause	0	$(4m+4n)k+1$	0	$(2n+4m+1)k+1+2n$

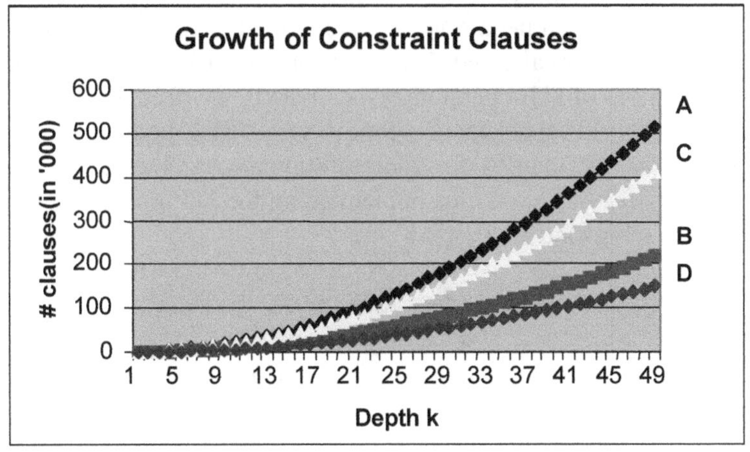

Fig. 7.7: Quadratic growth of EMM with BMC depth k

7.4.3 Non-uniform Initialization of Memory

Typically, memories are used with uniform initialization, i.e., the same initial value is considered stored at all memory locations (after some kind of power-on reset). However, in other applications of BMC such as semi-formal verification approaches [63, 64] where BMC can be applied at any state other than a reset state, all memory locations may not have uniform value. Handling such non-uniform initial values can be easily done within this framework.

Assume that the memory has p different initial values, i.e., $I^0,\ldots,I^{p-1}$ at address locations $Addr^0,\ldots,Addr^{p-1}$ for the given initial state of the design. Note that it is not important how we got that memory state as long as we can show that there exists a write sequence to get to that state. Such a write sequence can be constructed as follows: at t=-p all the memory locations are initialized with I^0; then $\forall_{1\leq i\leq p-1}$ a write cycle is issued at t=-p+i for writing data value I^i at address location $Addr^j$. We then need to add p-1 more terms to Equation (7.4) to capture these writes cycles.

We illustrate this for p=2. Assume, that the upper half of the memory array, i.e., $Addr_{m-1}$=1 (MSB) has initial value 2 and the lower half, i.e., $Addr_{m-1}$=0 has initial value 1. We construct a write sequence as follows:

- at t=-2, all address locations have value 1, i.e., WD^{-2}=1, WE^{-1}=1; and
- at t=-1, value 2 is written, i.e., $Addr^{\wedge -1}{}_{m-1}$=1, WD^{-1}=2, WE^{-1}=1. Now we need to add one more term in Eq (7.4), i.e., $S^{-2k} \wedge WD^{-2}$.

7.4.4 EMM for Multiple Memories, Read, and Write Ports

In modern designs, it is quite common to have a large number of diverse memories, each with multiple read and write ports. Here, we discuss how we extend EMM (that we have already discussed for a single read/write port single memory system) to more commonly occurring systems with multiple memories, having multiple read and write ports.

Before we delve into a discussion of efficient modeling, we first define memory semantics in the presence of multiple read and write ports. We assume there are no data races. In other words, a memory location can be updated at any given cycle through only one write port. (The approach can also be extended to check for data races.) Since each memory module is accessed only through its ports, the memory modules can be considered independent of each other. In the following discussion, we first consider a single memory with multiple read and multiple write ports.

Let the design be unrolled up to depth k (starting from 0). Let $X^{j,p}$ denote a memory interface signal variable X at time frame j for a port p. Let R and W be the number of read and write ports, respectively, for the given memory. Let the Boolean predicate variable $E^{j,i,w,r}$ denote the address comparison of the read port r at depth i, and the write port w at depth j, defined as =($Addr^{i,r}$=$Addr^{j,w}$). Then the *forwarding semantics* of the memory can be expressed as:

$$(E^{j,k,w,r} \wedge WE^{j,w} \wedge RE^{k,r} \wedge \forall_{0 \leq p < W}\ \forall_{j<i<k}(\neg E^{i,k,p,r} \vee \neg WE^{i,p})) \rightarrow (RD^{k,r} = WD^{j,w}) \tag{7.5}$$

In other words, data read at depth k through read port r, equals the data written at depth j through write port w, if

- the read addresses of port r at depth k is same as write address of port w at depth j, and
- write enable is active at j for the write port w, and
- read enable is active at k for the read port r, and

- for all depths strictly between j and k, no data was written at the address location $Addr^{k,r}$ through any write port.

Let the Boolean variable $s^{j,k,w,r}$ be defined as $s^{j,k,w,r}=E^{j,k,w,r} \wedge WE^{j,w}$. The decision $s^{i,k,w,r}=1$ does not necessarily imply $RD^{k,r}=WD^{i,w}$; other read-write pairs $RD^{k,r} = WD^{j,p}$ where $j \neq i$ or $p \neq w$, need to be established invalid through the decision procedure as well. Similar to the single read/write port approach, we add explicit constraints to capture the exclusivity of the matching read-write pair, in order to improve the SAT solve time. Let the Boolean variables $S^{i,k,w,r}$ and $PS^{i,k,w,r}$ denote the exclusive valid read signal and intermediate signal respectively for a given read port r and write port w. They are defined recursively as follows:

$$
\begin{aligned}
PS^{k,k,0,r} &= RE^{k,r} \\
\forall_{0\le i<k} \forall_{0\le p<W} PS^{i,k,p,r} &= \neg s^{i,k,p,r} \wedge PS^{i,k,p+1,r} \; (PS^{i,k,W,r} = PS^{i+1,k,0,r}) \\
\forall_{0\le i<k} \forall_{0\le p<W} S^{i,k,p,r} &= s^{i,k,p,r} \wedge PS^{i,k,p+1,r}
\end{aligned}
\tag{7.6}
$$

Now the forwarding semantics for multiple read and write ports can be expressed as:

$$RD^{k,r} = (\vee_{0\le p<W, 0\le i<k} S^{i,k,p,r} \wedge WD^{i,p}) \vee (PS^{0,k,0,r} \wedge WD^{-1}) \tag{7.7}$$

Note that $S^{i,k,p,r}=1$, immediately implies $S^{j,k,q,r}=0$ where either $q \neq p$ or $j \neq i$, and $i,j < k$. As before, we use a hybrid representation to add the memory modeling constraints as part of the procedure *EMM_constraints*, which is invoked after every unrolling as shown in Figure 7.6. Given $DW = n$ and $AW = m$, we give the sizes of EMM constraints added in terms of clauses and gates for each read port at a given depth k.

- *Address comparison:* We require $(4m+1)kW$ CNF clauses to represent address comparison signals.
- *Exclusive constraints:* We require $3kW$ 2-input gates to represent the exclusivity constraints in Eq (7.7).
- *Read data constraints:* We require 2nkW+$2n$+1 CNF clauses to represent read data constraints in Eq (7.5).

In total, we need $(4m+2n+1)kW+2n+1$ clauses and $3kW$ gates for embedded memory with a single read port and W write ports. For R read ports, we need $((4m+2n+1)kW+2n+1)R$ clauses and $3kWR$ gates. Note, the growth of constraints remain quadratic with analysis depth k and is WR times the constraints required for a single memory having a single read/write port. In the presence of multiple memories, we add these EMM constraints for each of them.

7.4.5 Arbitrary Initial Memory State

To model a memory with an arbitrary initial state, we introduce new symbolic variables at every time frame. Observe that for a (k-1)-depth analysis of a design, there can be at most k different memory read accesses from a single read port; out of which at most k accesses can be to un-written memory locations. Therefore, in total we need to introduce k symbolic variables for the different data words for each read port at analysis depth k-1. However, these variables are not entirely independent. Simply introducing new variables maintains soundness but introduces additional behaviors in the verification model. Therefore, we need to identify a sufficient set of constraints that models the arbitrary initial state of the memory more precisely.

Let $V^{i,p}$ and $V^{j,q}$ represent new data words introduced at depths i and j, for read ports p and q, respectively. Let $RA^{i,p}$ and $RA^{j,q}$ be the corresponding read addresses for the ports p and q (p and q need not be distinct). Let $N^{i,p}$ (and $N^{j,q}$) denote the condition that no write has occurred until depth i (and j) at address location $RA^{i,p}$ (and $RA^{j,q}$). We can then express the data read from the ports p and q at depths i and j, respectively, as:

$$N^{i,p} \rightarrow (RD^{i,p}=V^{i,p}),$$
$$N^{j,q} \rightarrow (RD^{j,q}=V^{j,q})$$

Note that, if read addresses $RA^{i,p}$ and $RA^{j,q}$ are equal, then $V^{i,p}$ and $V^{j,q}$ should also be equal. We add the following constraint to capture the same,

$$(RA^{i,p}=RA^{j,q} \wedge N^{i,p} \wedge N^{j,q}) \rightarrow (V^{i,p}=V^{j,q}) \tag{7.8}$$

For R read ports at (k-1)-depth analysis, we need to add $kR(R$-1) such constraints. We add these constraints using a hybrid representation in a separate sub-procedure call *EMM_arb_init_constraints,* within the procedure *EMM_constraints* (not shown). In discussion of our proof methods later in Part III of this book, we will show that the correctness of safety properties cannot be shown without adding these constraints.

7.5 Experiments on a Single Read/Write Port Memory

We first explain our approach using a typical implementation (in Verilog HDL) of a stack using random access memory (RAM) as shown in Figure 7.8.

```
1.  //random access memory
2.  `define AW 12//address width
3.  `define DW 32//data width
4.  `define SZ 4096//Mem Size

5.  module mem(Addr,WD,RD,WE);
6.  input [`AW-1] Addr;
7.  input WE;
8.  output [`DW-1:0] RD;
9.  input [`DW-1:0] WD;
10. reg [`DW-1] mem [`SZ-1:0];
11. assign RD = mem[Addr];
12. always @ (WE or Addr or WD)
13.   if (WE) mem[Addr] <= WD;
14. endmodule

15. //Stack model using RAM
16. module Stack(op,Din,Dout);
17. //op=2(PUSH),1(POP),0(NOP)
18. input op;
19. //push value
20. input [`DW-1:0] Din;
21. //pop value
22. output [`DW-1:0] Dout;
23. reg WE;
24. reg [`AW-1:0] SP;
25. initial SP=0;
26. always @(posedge clk) begin
27.   if (op==`PUSH)
28.     SP <= SP+1;
29.   else if (op==`POP)
30.     SP <= SP-1;
31.   WE <= (op==`PUSH);
32. end
33. mem(SP,Din,Dout,WE);
34. endmodule
```

Fig. 7.8: Stack implementation using RAM

We use the same notation of memory interface signals as discussed earlier in this chapter. Lines 5-14 show the implementation of RAM and lines 16-34 show the implementation of a stack. Note that if we were to model the memory explicitly we would require 130K state bits. Clearly, such a model would make SAT-based BMC (or any other kind of model checking) impractical. In our approach, we first remove lines 5-14 and line 33 and declare all the memory interface signals as input and output with respect to the stack. Then, we run the BMC algorithm with EMM constraints — *BMC_solve_F_EMM,* as shown in Figure 7.5 — on the resulting abstracted stack.

Next, we discuss our experimental results on several other software and hardware designs based on the above modeling of the embedded memories. For our experiments, we used three well-known recursive software programs *Fibonacci, 3n+1,* and *Towers-of Hanoi* with an embedded stack (bounded) as shown in Figure 7.9 and one hardware design with embedded RAM.

```
//Fibonacci
//cache and recursion
fib(n) {
  if (n<2) return n;
  //cache lookup
  if(lookup(n,&f)) return f;
  f = fib(n-1)+fib(n-2);
  //insert cache
  store(n,f);
  return f;
}

//3n+1
//period tracks # of
//calls required. to converge;
//initialized to 0
3nPlus1(n) {
  if (n==1) return;
  if ((odd(n))
    3nPlus1(3*n+1);
else
    3nPlus1(n/2);
  period++;
}

//Towers of Hanoi
//count tracks # of moves
//req.; initialized to 0
toh(n,s,d,a) {
  if (n==0) return;
  toh(n-1,s,a,d);
  count++;
  toh(n-1,a,d,s);
}
```

Fig. 7.9: Software programs with embedded stack used in experiments

In each of these cases, we chose a safety property that makes the memory modeling imperative, i.e., we cannot abstract away the memory from the design without affecting the validity of the property used. We translated each of the software programs into equivalent hardware models using Verilog HDL using a bounded but large stack model. For each of the software designs, we use an inverse function to describe the negated safety property that requires a non-trivial state space search, e.g., *given a certain value of Fibonacci number does there exist a corresponding n*? (Similar queries are made for a given number of recursive calls to terminate *3n+1*, and for a given number of legal moves required in *Towers-of-Hanoi*.)

We conducted our experiments on a workstation composed of dual Intel 2.8 GHz Xeon Processors with 4GB physical memory running Red Hat Linux 7.2, using a 3 hours time limit for each BMC run. We compared the performance of *BMC_solve_F_EMM* for handling embedded memory with *BMC_solve_F* (Figure 5.12) using *explicit* memory modeling. We also compared the performance of *BMC_solve_F_EMM* using hybrid exclusive select signal representation (*hESS*), with that of a hybrid nested ITE representation (*hITE*). In addition, we show the effect on their performance with increasing memory sizes for a given property and design.

We performed our first set of experiments on the hardware models of the software programs with several properties selected as described above. Each of the properties has a non-trivial witness and is listed in Tables 7.2-7.5 in the order of increasing search complexity. We used a fixed memory size in each of the models. We also used one industrial hardware design with a safety property that is not known to have a counterexample. For these properties, we show the performance and memory utilization comparison of the memory modeling styles, i.e., explicit memory modeling, memory modeling using our *hESS* representation and that using *hITE* representation in the Tables 7.2-7.5. In Tables 7.2-7.4, we show comparison results for *Fibonacci*, 3*n*+1, and *Towers-of-Hanoi,* respectively. We used address width (*AW*)=12, data width (*DW*)=32 for *Fibonacci, AW*=12, *DW*=2 for 3*n*+1, and *AW*=12, *DW*=22 for *Towers-of-Hanoi* models. In Table 7.5, we show comparison results for the industrial hardware design with a given safety property *S* for various intermediate analysis depths, since no counterexample was found within the resource limit. Without the memory, the design has ~400 latches and ~5k gates. The memory module has *AW*=12 and *DW*=12.

In Tables 7.2-7.5, the 1st Column shows the properties with increasing complexity, the 2nd Column shows the witness depth (intermediate analysis depth in Table 7.5), Columns 3-7 show the performance figures and Columns 8-12 show the memory utilization figures. Specifically in the performance columns, Columns 3-5 show the BMC search time taken (in seconds) for explicit memory modeling (P1), using *hITE* (P2), and using *hESS* (P3) respectively; Columns 6 and 7 show the speed up (ratio) using *hESS* over the explicit memory modeling and *hITE* respectively. For the memory utilization columns, Columns 8-10 show the memory used (in MB) by explicit memory modeling (M1), using *hITE* (M2), and using *hESS* (M3) respectively; Column 11 and 12 show the memory usage reduction (ratio) using *hESS* over the explicit memory modeling and *hITE* respectively (Note, *MO* ≡ Memory Out, *NA* ≡ Not Applicable).

Table 7.2: Comparison of memory modeling on *Fibonacci* model (AW=12, DW=32).

Prp	Wit Depth	Performance					Memory Utilization				
		Explicit P1(s)	hITE P2(s)	hESS P3(s)	Speed P1/P3	Speed P1/P3	Explicit M1(mb)	hITE M2(mb)	hESS M3(mb)	Red. M3/M1	Red. M3/M2
1-1	14	179	1	1	146	1.1	517	7	6	0.01	0.86
1-2	25	1050	5	4	248	1.3	1411	12	10	0.01	0.83
1-3	38	2835	20	15	184	1.3	2239	22	17	0.01	0.77
1-4	51	NA	79	47	NA	1.7	MO	41	34	NA	0.83
1-5	64	NA	125	100	NA	1.3	MO	63	52	NA	0.83
1-6	77	NA	252	311	NA	0.8	MO	100	75	NA	0.75
1-7	90	NA	587	362	NA	1.6	MO	175	92	NA	0.53
1-8	103	NA	625	557	NA	1.1	MO	163	161	NA	0.99
1-9	116	NA	1060	674	NA	1.6	MO	189	187	NA	0.99
1-10	129	NA	1674	1359	NA	1.2	MO	343	204	NA	0.59
1-11	142	NA	3782	2165	NA	1.7	MO	353	372	NA	1.05
1-12	155	NA	2980	2043	NA	1.5	MO	421	303	NA	0.72
1-13	168	NA	4349	4517	NA	1.0	MO	319	623	NA	1.95
1-14	181	NA	5573	4010	NA	1.4	MO	485	335	NA	0.69
1-15	194	NA	6973	4889	NA	1.4	MO	558	531	NA	0.95
1-16	**207**	**NA**	**> 3hr**	**7330**	**NA**	**NA**	**MO**	**541**	**461**	**NA**	**0.85**

Observing the performance figures in Column 6 of the Tables 7.2-7.5, we see that EMM approach using *hESS* representation improves the performance of BMC by 1-2 orders of magnitude when compared to explicit memory modeling. Similarly, as shown in Column 11 of these tables, there is a reduction in memory utilization by 1-2 orders of magnitude by the use of *hESS* approach. Moreover, our modeling style of using the hybrid exclusive select signals representation is better than the hybrid nested ITE, as shown in Column 7 and 12. Noticeably, in the last row of Tables 7.2 and 7.5, *hITE* times out, while *hESS* approach completes the analysis within the 3 hours time limit. Note that due to tail recursive nature of the 3*n*+1 program, the search complexity is not severe and therefore, we don't see a consistent benefit in using the exclusive select signals for this example in Table 7.3. On average, we see a performance improvement of 30%, and a reduction in memory utilization of 20%, noticeably more at higher analysis depths.

Table 7.3: Comparison of memory modeling on *3n+1* model (AW=12, DW=2).

Prp	Wit Depth	Performance					Memory Utilization				
		Explicit P1(s)	hITE P2(s)	hESS P3(s)	Speed P1/P3	Speed P2/P3	Explicit M1(mb)	hITE M2(mb)	hESS M3(mb)	Red. M3/M1	Red. M3/M1
2-1	44	2736	85	51	54	1.7	293	25	20	0.07	0.8
2-2	47	3837	109	138	28	0.8	314	30	37	0.12	1.23
2-3	50	2811	167	160	18	1.0	412	44	37	0.09	0.84
2-4	53	3236	205	207	16	1.0	407	42	58	0.14	1.38
2-5	56	5643	258	264	21	1.0	569	48	44	0.08	0.92
2-6	59	4518	312	277	16	1.1	432	56	49	0.11	0.88
2-7	62	9078	324	368	25	0.9	479	58	59	0.12	1.02
2-8	65	9613	426	483	20	0.9	585	72	85	0.15	1.18
2-9	68	10446	487	522	20	0.9	648	73	64	0.10	0.88
2-10	71	9903	562	590	17	1.0	668	82	74	0.11	0.90
2-11	74	> 3hr	674	692	NA	1.0	981	83	92	NA	1.11
2-12	77	> 3hr	910	746	NA	1.2	719	110	83	NA	0.75
2-13	80	> 3hr	820	861	NA	1.0	875	106	89	NA	0.84
2-14	83	> 3hr	969	990	NA	1.0	586	113	80	NA	0.71
2-15	89	> 3hr	1292	1201	NA	1.1	659	127	113	NA	0.89

Table 7.4: Comparison of memory modeling on *Towers-of-Hanoi* (AW=12, DW=22).

Prp	Wit Depth	Performance					Memory Utilization				
		Explicit P1(sec)	hITE P2(s)	hESS P3(s)	Speed P1/P3	Speed P2/P3	Explicit M1(mb)	hITE M2(mb)	hESS M3(mb)	Red M3/M1	Red. M3/M2
3-1	10	4	0	0	149	1.3	71	3	3	0.04	1.00
3-2	24	182	1	1	264	1.2	664	6	5	0.01	0.83
3-3	52	2587	13	10	255	1.2	2059	16	12	0.01	0.75
3-4	108	NA	229	129	NA	1.8	MO	68	43	NA	0.63
3-5	220	NA	1266	838	NA	1.5	MO	214	143	NA	0.67
3-6	444	NA	8232	6925	NA	1.2	MO	845	569	NA	0.67

In the second set of experiments, we used different memory sizes for the model 3*n*+1 and the property *2-1*. We varied the address bus width *AW* from 4 to 14 bits and compare the performance and memory utilization of the three approaches as shown in Column 2 of Table 7.6. The description of the remaining Columns in Table 7.6 is same as that in Tables 7.2-7.5. As shown in Columns 6 and 10, the performance improvement and memory usage reduction gets more pronounced, about 2 orders of magnitude, with increasing memory size. Clearly, the benefits of using memory-modeling constraints outweigh its quadratic growth cost. Moreover, *hESS* approach shows on average 50% performance improvement and 20% memory usage reduction over nested ITE expressions.

Table 7.5: Comparison of memory modeling on industrial design (AW=12, DW=12).

Prp	Inter Depth	Performance					Memory Utilization				
		Explicit P1(s)	hITE P2(s)	hESS P3(s)	Speed P1/P3	Speed P2/P3	Explicit M1(mb)	hITE M2(mb)	hESS M3(mb)	Red. M3/M1	Red. M3/M2
S	68	10680	1264	925	11	1.3	2049	91	64	0.03	0.7
	150	NA	9218	7140	NA	1.3	MO	770	261	NA	0.3
	178	**NA**	**>3hr**	**10272**	**NA**	**NA**	**MO**	**NA**	**908**	**NA**	**NA**

Table 7.6: Comparison of memory modeling (for *3n+1*) with DW=12 and varying AW.

Prp	AW	Performance					Memory Utilization				
		Explicit P1(s)	hITE P2(s)	hESS P3(s)	Speed P1/P3	Speed P2/P3	Explicit M1(mb)	hITE M2(mb)	hESS M3(mb)	Red M3/M1	Red M3/M2
2-1	4	64	85	60	1.1	1.4	23	22	20	0.87	0.9
	5	81	72	47	1.7	1.5	21	22	17	0.81	0.8
	6	110	77	79	1.4	1.0	25	23	24	0.96	1.0
	7	117	99	53	2.2	1.9	28	25	19	0.68	0.8
	8	146	87	78	1.9	1.1	41	24	24	0.59	1.0
	9	265	79	73	3.6	1.1	49	25	22	0.45	0.9
	10	767	86	59	12.9	1.4	95	27	22	0.23	0.8
	11	1490	89	56	26.4	1.6	153	27	20	0.13	0.7
	12	2736	85	51	54.1	1.7	293	25	20	0.07	0.8
	13	3759	83	54	69.6	1.5	569	24	21	0.04	0.9
	14	11583	81	46	249.2	1.7	1452	25	18	0.01	0.7

7.6 Experiments on Multi-Port Memories

We report our experiences on several case studies consisting of large industry designs and software programs that have embedded memory modules with multiple read and write ports. Two case studies correspond to industry designs with many reachability properties. Another case study involves a sorting algorithm with properties validating the algorithm. For each of the properties, we require modeling of the embedded memory, and the case studies were chosen to highlight the use of different approaches described. We compare hESS approach (labeled EMM), with explicit memory modeling (labeled Explicit Modeling) to show the effectiveness of efficient memory modeling. We experimented on a workstation with 2.8 GHz Xeon processors with 4GB running Red Hat Linux 7.2.

7.6.1 Case Study on Quick Sort

This case study makes use of EMM for multiple memories and EMM that models arbitrary initial state. We implemented a quick sort algorithm using Verilog HDL (Hardware Description Language). The algorithm is recursively called, first on the left partition and next on the right partition of the array (Note: a pivot partitions the array into left and right). We implemented the array as a memory module with AW=10 and DW=32, with 1 read and 1 write port. We implemented the stack (for recursive function calls) also as a memory module with AW=10 and DW=24, with 1 read and 1 write port. The design has 200 latches (excluding memory registers), 56 inputs, and ~9K 2-input gates. We chose two properties:

a) P1: the first element of the sorted array (in ascending order) cannot be greater than the second element,
b) P2: after return from a recursive call, the program counter should go next to a recursive call on the right partition or return to the parent on the recursion stack.

The array is allowed to have arbitrary values to begin with. This requires precise handling of the arbitrary initial memory state (Eq (7.8)) to show the correctness of the property.

For different array sizes N, we compared the performance of EMM and Explicit Modeling approaches, using the loop-free path proof checks in *BMC_solve_F_EMM* (see Figure 7.5) and *BMC_solve_F* (see Figure 5.12) respectively. We used a time limit of 3 hours for each run. We present the results in Table 7.7. Column 1 shows different array sizes N; Column 2 shows the properties; Column 3 shows the forward *longest loop-free path* length; Columns 4-5 and 6-7 show performance time and space used by EMM and *Explicit* Modeling, respectively. Note that using EMM we were able to reach the depth D for all properties in the given time limit, while *Explicit* Modeling times out on all of them.

Table 7.7: Comparision of EMM vs Explicit memory modeling on *Quick Sort*

N	Prop	D	EMM		Explicit	
			Sec	MB	Sec	MB
3	P1	27	64	55	>3hr	NA
3	P2	27	30	44	>3hr	NA
4	P1	42	601	105	>3hr	NA
4	P2	42	453	124	>3hr	NA
5	P1	59	6376	423	>3hr	NA
5	P2	59	4916	411	>3ht	NA

7.6.2 Case Study on Industry Design (Low Pass Filter)

The industry design is a low-pass image filter with 756 latches (excluding the memory registers), 28 inputs and ~15K 2-input gates. It has two memory modules, both with address width, AW = 10 and data width, DW = 8. Each module has 1 write and 1 read port, with memory state initialized to 0. There are 216 reachability properties.

EMM

We were able to find witnesses for 206 of the 216 properties, in about 400s requiring 50Mb. The maximum depth over all witnesses was 51. For the remaining 10 properties, we were able to reach the forward *longest loop-free path* in less than 1s, and 6Mb memory (thereby, obtaining the proofs by induction as discussed later in Chapter 9). Note that the introduction of new variables, to model arbitrary initial memory state without the constraints in Eq (7.8), was sufficient for the proofs although they capture extra behavior in the verification model.

Explicit Modeling

We required 20540s (~6Hrs) and 912Mb to find witnesses for all 206 properties. For the remaining 10 properties, we were able to reach the longest loop-free path using BMC in 25s requiring 50Mb.

7.7 Related Work

Burch and Dill introduced symbolic representation of memory arrays in their logic of equality with un-interpreted functions (EUF) [179]. Their abstract interpretation of memory uses interpreted *read* and *write* operations. These interpreted functions were used to represent the memory symbolically by creating a nested *if-then-else* (ITE) structure to record the history of writes to the memory. While this interpretation is limited to the control part of the model, the datapath is abstracted out completely.

In an effort to model the entire circuit with control and data memory, Velev *et al.* have introduced a behavioral model which allows the number of symbolic variables used to be proportional to the number of memory accesses rather than to the size of the memory [180]. This model replaces the memory and it interacts with the rest of the circuit through a software interface. Similar to [179], these reads and writes are defined in such a way that the forwarding property of the memory semantics—data read from a memory location is same as the recent data written at the same location—is satisfied. BDDs used to represent the Boolean expressions tend to blow up for processors with branch or load/store instructions. A good BDD ordering was impossible due to dependencies of address on the data and vice versa.

Bryant *et al.* proposed that for processors in which writes are not reordered relative to each other or to reads, it is sufficient to represent data memory as a generic state machine, changing state in some arbitrary way for each write operation, and returning some arbitrary value dependent on the state and the address for each read operation [181]. Such an abstraction is sound, but it does not capture all the properties of a memory. Velev, in subsequent work, automated the process of abstraction of data memory by applying a system of conservative transformation rules [182] and generating a hybrid memory model where forwarding semantics of memory is satisfied only for some levels of forwarding.

Bryant *et al.* also proposed a logic of Counter arithmetic with Lambda expressions and Un-interpreted functions (CLU) to model infinite-state systems and unbounded memory in the UCLID system [59]. Memory is modeled as a lambda function expression whose body can change with each time step. Similar to [179], the memory is represented symbolically by creating a nested *if-then-else* (ITE) structure to record the history of writes to the memory. In this restricted CLU logic, one can use symbolic simulation to verify safety properties.

7.8 Summary

Verifying designs with large embedded memories is typically handled by abstracting out (over-approximating) the memories. Such an abstraction is generally not useful for finding real bugs. Conventional SAT-based BMC efforts are incapable of handling designs with explicit memory modeling due due to the enormously increased search space. We discussed a practical approach to use memory-modeling constraints to augment SAT-based BMC in order to handle embedded memory designs without explicitly modeling each memory bit. This method does not require transforming the design, and is also guaranteed not to generate false negatives. This method is similar to abstract interpretation of memory [179, 180], but with key enhancements. We showed that our hybrid representation of added constraints boosts the performance of a SAT solver significantly, as opposed to the conventional way of modeling these constraints as nested ITE expressions [179]. We extended the EMM approach for a single memory with a single read/write port, to the more commonly occurring systems with multiple memories, and multiple read and write ports. We also discussed precise modeling of the arbitrary initial state of memory, for use in SAT-based proofs using BMC and non-uniform memory initialization. We demonstrated the effectiveness of the approach on a number of software and hardware designs with large embedded memories. We showed about 1-2 orders of magnitude time and space improvement with the EMM approach, in comparison to explicit modeling of each memory bit. While the growth of memory modeling constraint clauses is quadratic with the analysis depth, we also observed that although the constraint clauses are sufficient, they may not be necessary in every time frame and can be added lazily.

7.9 Notes

The subject material described in this chapter are based on the authors' previous works [50], used with kind permission of Springer Science and Business Media, and [51] © 2005 IEEE.

8 BMC FOR MULTI-CLOCK SYSTEMS

8.1 Introduction

Current System-on-Chip (SoC) designs are essentially *multi-clock systems* with multiple clock domains, gated clocks, and latches. Further, the intellectual property (IP) blocks building SoC, with their own clock generators, mandate the requirement for synchronous primitives between the resulting asynchronous clock domains. To meet the high performance and low power requirements [183], multi-clock systems have become the norm of current designs, taking over the single global clock synchronous designs. Routing single clock over a large die incurs large skew delays, unacceptable for high-performance designs. With low power stringent requirements, it is difficult to reduce the clock skews for a distributed clock simply by increasing the power of the clock drivers. For power-conscious designs, designers often use gated clocks to reduce or disable the switching activity of certain portions of the design. Each of these design styles increases the verification complexity in terms of increased number of state bits and deeper bug traces. The following design features and specification of clocked systems pose additional *challenges* to the existing verification efforts.

8.1.1 Nested Clock Specifications

Property variables that involve gates with support from state elements in multiple clock domains require the use of clocks in the formula to avoid ambiguities. The Property Specification Language (PSL) standardized by Accellera [83] has formal semantics for specifying clocked properties using the clock operator @, based largely on the work of Eisner *et al.* [120]. The general translation scheme for clocked properties tends to generate large nested LTL formulas that can limit the effectiveness of a standard BMC solver [66]. For example, a clocked LTL formula

$$\boldsymbol{F}(p\wedge (\boldsymbol{X}q@clk1)@clk)$$

gets translated into an equivalent unclocked LTL formula

$$\boldsymbol{F}(p\wedge(\neg clk\boldsymbol{U}(clk\wedge\boldsymbol{X}(\neg clk\boldsymbol{U}(clk\wedge(\neg clk1\,\boldsymbol{U}(clk1\wedge q))))))).$$

We discussed in Chapter 5 how customized translations, using partitioning and incremental formulation, of commonly occurring un-clocked properties in BMC, improve the performance of BMC. However, it is not practical to devise property-specific algorithms for each clocked property translation due to the following:

- large nesting in the resulting translation, and
- many possible ways of using clock operator @ in sub-formula of commonly occurring properties to imply specifications with subtle differences

8.1.2 Verification Model for Multi-clock Systems

Model checking algorithms [17, 66, 67] are typically applied on a single-clock synchronous model. To generate such a verification model for synchronous multi-clock systems, one typically derives a *global clock* that ticks with a frequency equal to the least common multiple (LCM) of various input clock frequencies. Further, multiple phases effectively multiply the clock frequencies. Observing that the system state in a multi-clock system can change only at the tick of some clock, and not at every tick of global clock, such a model could be very inefficient for model checking applications, if used naively, in general. In particular, for BMC, problem instances with no corresponding clock events become computational overhead due to redundant unrolling and unnecessary loop-checks. For example, let the frequency of clock *C1* be 2Mhz, and that of the Clock *C2* be 3Mhz. If the design is unrolled with clock frequency of 6Mhz (=LCM(2,3)), assuming 0 initial phases of *C1* and *C2*, then every 2nd (i.e., k=1)and 6^{th}(i.e., k=5) BMC unrollings and checks are redundant, as none of the clocks tick during those unrollings.

8.1.3 Simplification of Verification Model

In the past [184], simplification of verification model for multi-clock system is achieved by identifying the periodic clock signals, and choosing a suitable number of phases *C* that captures the periodicity of the clocks. Using that number of phases, the model is *phase abstracted* as follows: the transition relation is unrolled *C* times and then simplified using the phase values of the clocks. Such an approach is not practical when the number of phases C gets large due to the presence of non-integral or large clock frequency ratios. In such scenarios, one can however, limit C and restrict phase abstraction to limited clocks signals with frequency ratio less than C. Further, gated clocks can further limit such simplification of a verification model due to its non-periodic behavior. We refer to such *phase abstraction* as *static simplification* to contrast with the dynamic simplification (during BMC unrolling) as described later.

8.1.4 Clock Specification on Latches

For verification purposes [185], latches (level-sensitive) are modeled as flip-flops (edge-triggered) clocked on a global clock in synchronous designs. Clocked specifications with latch enabling clocks that are gated clocks pose further verification challenges.

In practice, it is important to address the scalability issues in verifying multi-clock systems with clocked specifications. With our focus on the significance of BMC customization and the difficulties in handling translated clocked properties in a multi-clock system, we discuss an integrated BMC-based solution [52] to verify multi-clock systems with clocked specifications. This integrated method builds on the recent advancements in SAT-based BMC (see Chapter 5) and several methods targeted for verifying multi-clock systems such as clocked LTL property translation [120], phase abstraction techniques [184-187], and reducing BMC unrolling [188].

Outline

In Section 8.2, we present a *uniform clock modeling scheme* to handle multiple clocks with arbitrary frequencies and ratios, gated clocks, multiple phases, latches and flip-flops in a multi-clock synchronous system, to obtain a single-clock verification model.

In Section 8.3, we discuss a method to reduce BMC unrolling. Given external clock characteristics, we automatically generate schedules and clocks constraints based on an *event queue semantics,* to eliminate redundant unrollings.

In Section 8.4, we discuss techniques to reduce the number of explicit loop-checks using a SAT-solver in BMC by identifying the repetition period of global clock states.

In Section 8.5, we describe techniques to perform dynamic simplification as opposed to static simplification. Since not all clock domains are active at each unrolling, we perform *on-the-fly simplification of the unrolled transition relation* using the clock constraints at each unrolling, where we re-use the current unrolled sub-circuit corresponding to an inactive clock-domain for the next unrolling; thereby reducing size of the BMC problem instance. The presence of gated clocks and clocks with large frequency ratios limit the effectiveness of phase abstraction techniques in static simplification of the verification model [184]. Our dynamic simplification overcomes this limitation, as it does not depend explicitly on the periodicity of clocking signals.

In Section 8.6, we discuss BMC customization for translation of clocked properties *directly* rather than customizing each translated unclocked property, and simultaneously offer the benefits of partitioning and incremental BMC formulation [58]. We discuss a customized translation for the clocked LTL formula $(\boldsymbol{F}(f))^{@}$ in detail. We also discuss how we can extend customized translation to other clocked LTL formulas such as $(\boldsymbol{F}(f \wedge \boldsymbol{G}(g)))^{@}$, where f and g are clocked expressions with propositional combinations of atoms and nested $\boldsymbol{X}$ operators.

In Section 8.7, we discuss our experiments on publicly available multi-clock systems in *OpenCores* [189] benchmark. In Section 8.8, we discuss some of the related work, and summarize in Section 8.9.

8.2 Efficient Modeling of Multi-Clock Systems

We consider modeling of synchronous multi-clock systems that have clocks with arbitrary but known fixed frequencies and fixed initial phases. Multi-clock systems with the clocks derived from a single source generator lead to synchronized clocks with fixed frequencies and known initial phases. However, if the clocks are generated from independent sources, they are generally unsynchronized, typically with fixed frequencies but unknown initial phases. For modeling such systems, we consider one representative at a time from the various combination scenarios of initial phases, unlike the work of Clarke *et al.* [188] where all possible scenarios are considered simultaneously in modeling. Our goal is to trade generality for scalability and practicality of BMC methods.

Consider a synchronous multi-clock example shown in Figure 8.1(a). We use the following notations:

- $\underline{X}$ is a set of FFs (flip flops) triggered on the positive edge (↑) of an input clock *C1*,
- $\underline{Y}$ is a set of FFs triggered on the negative edge (↓) of an input clock *C2*,
- $\underline{Z}$ is a set of level-sensitive latches triggered on active high of a gated clock *GC*,
- f_1-f_5 are combinational blocks,

- <u>PI</u> is the set of primary inputs, and
- <u>PO</u> is the set of primary outputs.

Note, "/" on the connectors (→) indicates multiple connecting wires.

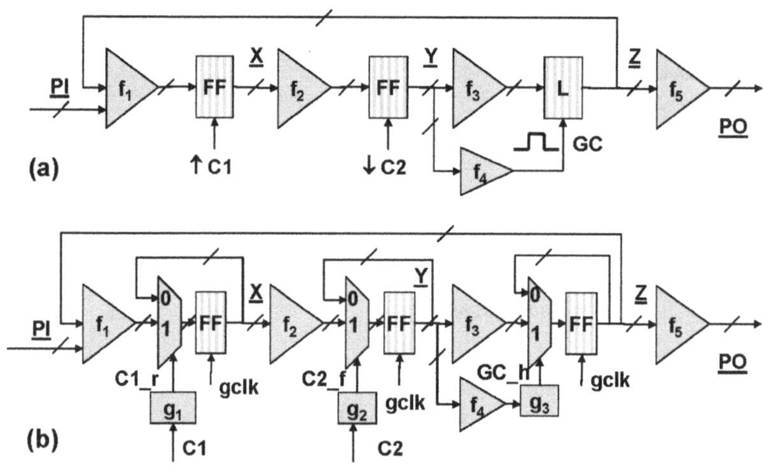

Fig. 8.1: (a) Multi-clock system, (b) Single-clock model

We derive a single-clock model, as shown in Figure 8.1(b), such that any value change on inputs, internal signals, and state elements occurs only at the tick of *gclk*. In order to do so, we add mux-circuits (multiplexers) in the next state transition logic of <u>X</u>, <u>Y</u>, and <u>Z</u>; and generator circuits g_1-g_3 (details shown in Figure 8.2) for *enable clocking signals C1_r*, *C2_f*, and *GC_h* that take value 1 at the *posedge* of *C1*, *negedge* of *C2,* and high value of *GC*, respectively. For example, the clocking signal *posedge C1* is modeled using the enable clocking signal $C1\_r = C1 \wedge \neg C1\_d$, where *C1_d* represents clock *C1* delayed by one *gclk*.

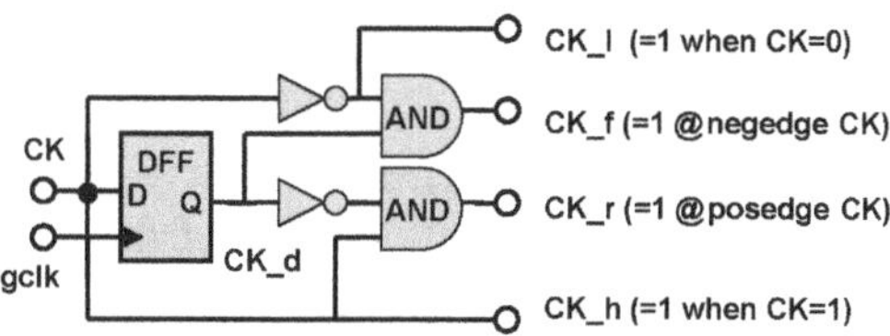

Fig. 8.2: Enable clock signal circuit

For all possible complex clock constructs, as shown in the Verilog HDL description in Figure 8.3, we generate the flip-flops outputs, clocked on global clock *gclk* with muxes in their next state logic.

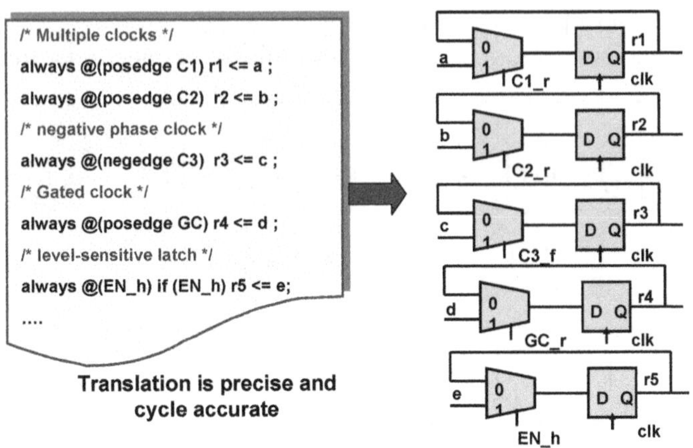

Fig. 8.3: Modeling complex clock constructs

8.3 Reducing Unrolling in BMC

For fixed input clock frequencies and initial phases, one can model a *clock-generator* ticking on *gclk* with frequency equal to LCM of the frequencies of all input clocks, and use the *clock-generator* to compute deterministically the values of input clocks at each tick of *gclk*. For SAT-based BMC, unrolling a single-clock model at every tick of *gclk* would add computational overhead, as there may be some ticks when no input clocks change values (and so the state elements). Instead, we use *event queue semantics* similar to the work of Clarke *et al.* [188], where only those ticks of *gclk* are considered when at least one input clock changes value. We now discuss the derivation of relevant ticks, i.e., *clock schedules*, and *clocking constraints* on the input clocks, using an example as shown in Figure 8.4. Let frequencies of input clocks *C1* and *C2* be *100Mhz* (time period, $T_{C1}=10ns$), and *62.5Mhz* ($T_{C2}=16ns$), respectively, and let the initial phases be *0ns* and *4ns,* respectively. An active edge of the input clock is indicated by the edge direction in the figure. Under the event queue semantics, events are recorded chronologically where each event corresponds to a value change on input clocks *C1* and *C2*. Let a *3*-tuple $S^i=<t^i,c_1^i,c_2^i>$ denote a configuration corresponding to the ith recorded event in the queue at time t^i, with c_1^i and c_2^i representing the values of clock signals *C1* and *C2,* respectively. In the event queue, we obtain the following configurations for $0 \le t \le 30$:

$S^{0}=<t^{0}=0,1,1>$,
$S^{1}=<t^{1}=4,1,0>$,
$S^{2}=<t^{2}=5,0,0>$,
$S^{3}=<t^{3}=10,1,0>$,
$S^{4}=<t^{4}=12,1,1>$,
$S^{5}=<t^{5}=15,0,1>$,
$S^{6}=<t^{6}=20,1,0>$,
$S^{7}=<t^{7}=25,0,0>$,
$S^{8}=<t^{8}=28,0,1>$,
$S^{9}=<t^{9}=30,1,1>$.

Observe that state elements cannot update between the consecutive configurations, i.e., S^{i} and S^{i+1}. We sometimes refer the configuration S^{i} as a *global clock state. Thus, we generate clock schedules for BMC unrolling by considering only those ticks of gclk that correspond to these configurations, with* S^{i} *occurring at* i^{th} *tick. We also generate clocking constraints at the* i^{th} *tick by constraining input clock signals such as C1 and C2 with the tuple values* $c_1^{\,i}$, *and* $c_2^{\,i}$. *During witness generation, we use* t^{i} *to time-stamp the* i^{th} *depth in the witness trace.*

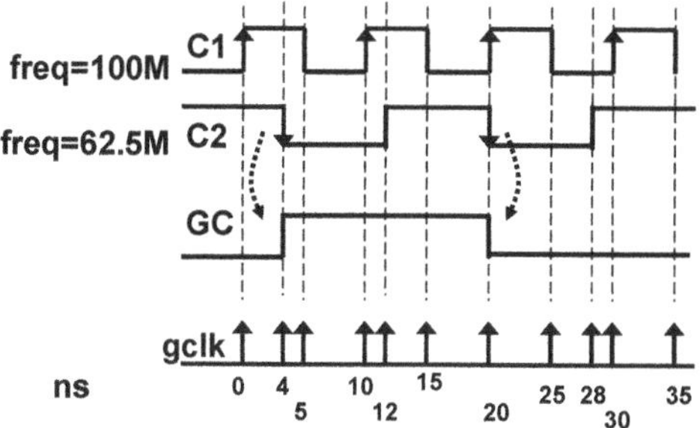

Fig. 8.4: Ticks of global clock *gclk*

8.4 Reducing Loop-Checks in BMC

Using the same example above, we now discuss the repetition of configurations and its significance in removing some loop-checks constraints in BMC between unrollings at the ith and j^{th} tick of *gclk.* Two configurations $S^{i}=<t^{i},c_1^{\,i},\ldots,c_n^{\,i}>$ and $S^{j}=<t^{j},c_1^{\,j},\ldots,c_n^{\,j}>$ are said to be *equivalent* (i.e., recur every R ticks of *gclk*), if and only if $\forall_{0\le k<R}\ (\forall_{0\le m<n}\ (c_m^{\,i+k}=c_m^{\,j+k}))$ and $\exists o(\forall_{0\le k<R}\ (t^{i+k}-t^{j+k}=o))$. In other words, the corresponding successive configurations have matching clock signal values and have a fixed time

difference. We call *R* the *recurrence length.* The *repetition period T* of the clock-generator (i.e., when clock states repeat) can be obtained by taking the LCM of the clock periods T_C. The equivalent configurations can be shown to correspond to equivalent clock states. For our running example, the repetition period *T=80ns* (LCM of *10ns* and *16ns*) and recurrence length *R=26. We use this information in BMC to consider loops-checks between unrolling depths i and j only if (i-j) mod R=0, i.e., when the clock states at the* i^{th} *and* j^{th} *ticks are the same. (Note, the clock states are not equivalent otherwise.)* In other words, a loop back ${}_jL_i$ from state s_i to s_j will be infeasible if the global clock states are not equivalent.

To summarize so far, we first generate a single-clock model from given multi-clock system. From the input clocking characteristics, we derive clocking constraints, automatic schedules, and the recurrence length, and then use them in BMC as described later.

8.5 Dynamic Simplification in BMC

We discuss dynamic simplification of the unrolled model to reduce size of the BMC problem instances with clock constraints generated as above. We use a multi-clock example and the corresponding single-clock model as shown in Figure 8.1(a) and 8.1(b), respectively, introduced in the previous section. We use the clocking characteristics of inputs clocks *C1* and *C2,* and assume the dynamic behavior of gated clock *GC* as shown in Figure 8.4, with initial states *C1_d*=0 in circuit g_1, and *C2_d*=1 in circuit g_2. Note, the dotted arrows indicate cause-effect relations, as *GC* is a (combinational) function of $\underline{Y}$ FFs, clocked by *C2.* The transition functions for the single-clock model are as follows:

$$
\begin{aligned}
NEXT(\underline{X}) &= (C1\_r)\ ?\ f_1(\underline{Z},\underline{PI}) : \underline{X}; \quad //\ s?b{:}c \equiv ITE(s,b,c) \\
NEXT(\underline{Y}) &= (C2\_f)\ ?\ f_2(\underline{X}) : \underline{Y}; \\
NEXT(\underline{Z}) &= (GC\_r)\ ?\ f_3(\underline{Y}) : \underline{Z}; \\
\underline{PO} &= f_5(\underline{Z});
\end{aligned}
$$

We use the scheduling of *gclk* as shown in Figure 8.4 to unroll the model in BMC. We constrain the input clocks *C1* and *C2* at the ith unrolling using the clocking constraints at the ith tick. In the following Table 8.1, we show the unrolled model, denoted as UC^i at different ticks. Columns 2-9, show $\underline{PI}^i$, $C1\_r^i$, $C2\_f^i$, GC_h^i, $\underline{X}^i$, $\underline{Y}^i$, $\underline{Z}^i$, and $\underline{PO}^i$ to denote the ith $(i \le 8)$ unrolled circuit nodes (combinational logic) for $\underline{PI}$, *C1_r, C2_f, GC_h,* $\underline{X}$, $\underline{Y}$, $\underline{Z}$, *and* $\underline{PO}$ respectively, with $\underline{X}^0$, $\underline{Y}^0$ and $\underline{Z}^0$ denoting respective initial states.

Note that by using dynamic simplification, unrolled circuit at i=8 UC^8 has far fewer copies of combinational blocks in its cone-of-influence *(COI)* than without its use, i.e., three f_1, two f_2, one f_3 and two f_5, compared to nine copies each of f_1, f_2, f_3, and f_5 without simplification. This is due to the mapping of circuit nodes of a clock-domain in one time frame to those of the previous time frame, if the clocking signal for that domain is inactive in the previous time frame. For example, circuit nodes $\underline{X}^3$ and $\underline{X}^2$ map to $\underline{X}^1$ as $C1\_r$=0 at $i=1,2$. We use *Reduced AIG* with on-the-fly circuit simplification procedures for further compacting the unrolled circuits.

Table 8.1: Unrolled circuit nodes for the multi-clock Example 8.1(a)

i	PI^i	$C1\_r^i$	$C2\_f^i$	GC_h^i	$\underline{X}^i$	$\underline{Y}^i$	$\underline{Z}^i$	$\underline{PO}^i$
0	PI^0	1	0	0	$\underline{X}^0$	$\underline{Y}^0$	$\underline{Z}^0$	$PO^0=f_5(\underline{Z}^0)$
1	PI^1	0	1	0	$\underline{X}^1=f_1(\underline{Z}^0,\underline{PI}^0)$	$\underline{Y}^0$	$\underline{Z}^0$	$\underline{PO}^0$
2	PI^2	0	0	1	$\underline{X}^1$	$\underline{Y}^2=f_2(\underline{X}^1)$	$\underline{Z}^0$	$\underline{PO}^0$
3	PI^3	1	0	1	$\underline{X}^1$	$\underline{Y}^2$	$\underline{Z}^3=f_3(\underline{Y}^2)$	$\underline{PO}^3=f_5(\underline{Z}^3)$
4	PI^4	0	0	1	$\underline{X}^4=f_1(\underline{Z}^3,\underline{PI}^3)$	$\underline{Y}^2$	$\underline{Z}^3$	$\underline{PO}^3$
5	PI^5	0	0	1	$\underline{X}^4$	$\underline{Y}^2$	$\underline{Z}^3$	$\underline{PO}^3$
6	PI^6	1	1	1	$\underline{X}^4$	$\underline{Y}^2$	$\underline{Z}^3$	$\underline{PO}^3$
7	PI^7	0	0	0	$\underline{X}^7=f_1(\underline{Z}^6,\underline{PI}^6)$	$\underline{Y}^7=f_2(\underline{X}^6)$	$\underline{Z}^3$	$\underline{PO}^3$
8	PI^7	0	0	0	$\underline{X}^7$	$\underline{Y}^7$	$\underline{Z}^3$	$\underline{PO}^3$

8.6 Customization of Clocked Specifications in BMC

Given a clocked specification of the form $(\boldsymbol{F}(p))@clk$, we present our BMC customization using the procedure *BMC_solve_F*$^{@}$ and sub-procedurs as shown in the Figures 8.5—8.7 Note, the procedure *BMC_solve_F*$^{@}$ is a modified version of the procedure *BMC_solve_F* (see Figure 5.12) for unclocked $\boldsymbol{F}(p)$. Recall, we allow property node p to be a Boolean combination of nested $\boldsymbol{X}$ operators with propositional atoms, where sub-expressions can have clocks specified with @. For our discussion, we only allow primary input clocks of the design to be in the support (i.e., *COI*) of clocks in the specification. For example, a clocked specification can be of the form

$$(\boldsymbol{F}(a \wedge \neg\boldsymbol{X}(b \wedge \boldsymbol{X}(c))@clk1))@clk$$

Here, we allow only those input clocks that are in the *COI* of *clk* and *clk1*. Note that the specification clock *clk* corresponds to the *enable clock signal* in our single-clock model.

We construct a tree expression for p where each node *prop_tree_node* represents a sub-expression. Each node is of type *AND* ($\wedge$), *NOT* ($\neg$), *LEAF*, or ***X***, where *LEAF* corresponds to a propositional atom. Note, we use *ckt_node* to denote a propositional atom or gate in the transition relation, and *uckt_node* such as p^i to denote a propositional logic node corresponding to the ith unrolling of p. Moreover, for a *prop_tree_node* p, we use $p \rightarrow clk$ to denote the associated specification clock.

We first discuss the procedures used in *BMC_solve_F*$^{@}$. The procedure *Get_clk_tick_depth*, shown in Figure 8.5, uses procedure *Is_clock_enable* to determine when *clk* ticks next, starting from depth *d*. The procedure *Is_clock_enable* returns *true* if *clk* evaluates to 1 at depth *d*, and returns *false* otherwise. Recall, as we allow input clocks only in the *COI* of *clk*, we obtain the value of *clk* by simulating the logic circuit in its *COI* until *d*, using the input clock constraints values. The procedure *Is_ckt_node_valid* returns *false* when *clk* is not valid in the case of unclocked property. (Note, for invalid *clk*, procedure *Get_clk_tick_depth* returns *d*.)

```
Synopsis:  Get next clock tick depth ≥ d
Input:     ckt_node clk, depth d, Bound N
Output:    next depth  nd
Procedure: Get_clk_tick_depth

nd = d;
if (!Is_ckt_node_valid(clk)) return nd;
while(nd < N) {//check upto the bound N
  if (Is_clock_enable(clk,nd)) return nd;
  nd = nd + 1;
}
ABORT("Bound Reached");
```

Fig. 8.5: Procedure to get next clock tick depth

The procedure *Prop_node*$^{@}$, shown in Figure 8.6, returns an unrolled node using the procedures *Unroll* (see Figure 5.3), and *And/Invert* (see Figure 3.4). Note, in the absence of a valid associated *clk*, it is identical to the procedure *Prop_node* (Figure 5.11). For a clocked property (i.e. with a valid *clk*) the nested rule (*R6:* $T^{clk}((f)@clk1) = T^{clk1}(f)$, refer Section 2.5.2, Chapter 2) is applied. Using the procedure *Get_clk_tick_depth,* the next *clk* tick depth $j \geq i$ is determined. If *clk* is associated with an ***X*** operator, another call to the procedure *Get_clk_tick_depth* returns the next *clk* tick depth after *j*. Again, correctness of the translation for a clocked LTL formula is based on the re-write rules:

- $(\neg \mathbf{X}p)@clk \equiv \neg((\mathbf{X}p)@clk)$, and
- $(\mathbf{X}(p \wedge \mathbf{X}(q))@clk1)@clk \equiv ((\mathbf{X}p@clk1)@clk) \wedge (\mathbf{X}(\mathbf{X}q)@clk1)clk.$

(8.1)

```
Synopsis:  Create property circuit node at depth i
Input:     prop_tree_node f, depth i, clock clk
Output:    uckt_node f^i
Procedure: Prop_node@

if (Is_ckt_node_valid(f→clk)) clk = f→clk;
j = Get_clk_tick_depth(clk,i);

switch(f→type) {
  case LEAF://ckt_node
    return Unroll(f,j);//uckt_node at j
  case AND://f = f1 ∧ f2
    return And(Prop_node@(f1,j),
               Prop_node@(f2,j));
  case NOT: //f = !f1
    return Invert(Prop_node@(f1,j));
  case X: //f = X(f1)
    j = Get_clk_tick_depth(clk,j+1);
    return Prop_node@(f1,j);
}
```

Fig. 8.6: Unrolling of clocked property tree node *p*

```
Synopsis:  Customized BMC solve for F(p)
Input:     prop_tree_node p, bound N, initial state I
Output:    TRUE/FALSE/ABORT
Procedure: BMC_solve_F@

C = I; k=0;
while (k<N) {  //L1 is active always
 i= Get_clk_tick_depth(clk,k);
 p^i = Prop_node@(p,i,clk);
 if (Is_sat(C ∧ p^i)) return TRUE; //wit found
 C = C ∧ ¬p^i; //L3 Learning
 Merge(p^i, const_0); //Simplify
 for (j=i;j>=0;j--) {//loop-free check
   if (!Is_clock_enable(clk,j)) continue;
   if (!Is_clock_state_equal(i,j)) continue;
   C = C ∧ ¬_jL_i;
   if (!Is_sat(C)) return FALSE; //no wit exists
 }
 k = i+1; //increment depth
}
ABORT("Bound Reached"); //wit not found
```

Fig. 8.7: BMC Customization for clocked Property ***F****(p))@clk)*

One can easily see that the procedure *BMC_solve_F*$^{@}$ shown in Figure 8.7 is identical to *BMC_solve_F*, except for the highlighted lines that we described below. For clocked properties, we perform additional pruning of loop checks using the procedure *Is_clock_state_equal(i,j)* which returns

TRUE if and only if *((i-j) mod R) = 0*, where *R* is the *recurrence length.* In other words, loop-checks between *i* and *j* depths are performed by SAT solver only if the global clock repeat at those depths (Section 8.3.3).

Example 8.1

For the clocked LTL formula $(\boldsymbol{F}(a \wedge \neg \boldsymbol{X}(b \wedge \boldsymbol{X}(c))@clk1)@clk$, when we apply the procedure $Prop\_node^{@}(p,i,clk)$ at depth *i* on the sub-expression *p* with *clk* enabled at *i*, *i+2*, *i+4*, and *clk1* enabled at *i+1*, *i+3*, *i+5*, the procedure returns $uckt\_node\ p^{i} = a^{i} \wedge \neg (b^{i+3} \wedge c^{i+5})$. To compare, the general clock translation [120] to an equivalent unclocked LTL would give

$$p=(\neg clk\ \boldsymbol{U}\ clk \wedge a) \wedge \neg(\neg clk\ \boldsymbol{U}\ (clk \wedge \boldsymbol{X}(\neg clk\ \boldsymbol{U}\ (clk \wedge (\neg clk1\ \boldsymbol{U}\ clk1 \wedge b) \wedge (\neg clk1\ \boldsymbol{U}\ (clk1 \wedge \boldsymbol{X}(\neg clk1\ \boldsymbol{U}\ (clk1 \wedge c))))))))).$$

Our approach of translating clocked sub-formulas *directly* into property circuit nodes (such as f^{i}) overcomes the problem of devising customized translations for deeply nested equivalent un-clocked formulas. This also allows us to take advantage of sharing, partitioning, and SAT-based incremental learning, as we did in the unclocked BMC translations (see Chapter 5). We can similarly extend our translation approach to handle other commonly occurring clocked specifications such as $(\boldsymbol{F}(f \wedge \boldsymbol{G}(g)))^{@}$ by adding the highlighted lines appropriately to the Figures 5.13 and 5.14.

8.7 Experiments

In the previous sections, we have described how we can seamlessly combine various solutions for handling multi-clock systems with recent advancements in BMC. We have implemented these ideas in an integrated system, called *BMC*$^{@}$, using SAT-based model-checking framework (*Verisol*). For evaluating the effectiveness of such an integrated solution, especially, the customization of clocked specification, we compare with *BestBMC* obtained by disabling *only* the customization for clocked properties in *BMC*$^{@}$, but keeping all other improvements [45, 153, 154, 184, 188]. Thus, *BestBMC* uses a general BMC formulation [66, 67] on translated [120] clocked properties, while *BMC*$^{@}$ handles clocked properties directly using the BMC customization procedure as shown in Figure 8.7.

We experimented on a workstation with 2.8 GHz Xeon processor with 4GB running Linux 2.4.21-27. We considered two multi-clock systems from *OpenCores* [189] (a suite of publicly available benchmarks): *VGA/LCD Controller* and *Tri-mode Ethernet MAC Controller.* We obtained input clocking characteristics, reset sequences and other constraints from the accompanying test benches. Based on the specification documents, we identified several clocked LTL reachability properties. We used a time limit of 2 hours for each BMC run.

8.7.1 VGA/LCD Controller

The controller core provides VGA capabilities for embedded systems supporting many available CRT and LCD displays with video memory outside the core. It has two positive edge triggered input clocks:

1. *wishbone clock* (*freq=416.66Mhz, T=2.4ns*)
2. *pixel clock* (*freq=33.33Mhz, T=30ns*).

Using this clocking information, we computed automatically the clock scheduling and constraints, and a recurrence length of 55 (repetition period = 83.33ns). The core design has 162 FFs on *pixel clock*, 2340 FFs on *wishbone clock*, 87 primary inputs, and 44K 2-input gates. We identified 13 clocked properties *P1-13*. The properties can be classified as

- *reachability* of control condition/states of the horizontal timing generator (*P1-P6*) and the vertical timing generator *(P7-P11)*, and
- *assertability* of line FIFO request *(P12)*, and line underflow interrupt across clock-domain *(P13)*

The properties have the following forms:

- P1-P12 are of the form $(\mathbf{F}(p \wedge \mathbf{X}(q)))@px\_clk\_r$ and
- P13 is of the form $\mathbf{F}(p@wb\_clk\_r \wedge \mathbf{X}(q)@px\_clk\_r)$.

We present the comparison results in Table 8.2. Columns 1 lists different properties *P1-13*, Column 2 indicates the number of unrollings in a witness (depth #D) if we were to consider every tick of a global clock with LCM frequency; Column 3 reports the number of non-redundant BMC unrollings (#U) based on using our derived clock schedules; Columns 4 and 5 show whether the witness was found (F?), and time taken (in sec) respectively by $BMC^{@}$, and similar statistics for *BestBMC* in Columns 6 and 7. Note, if a time-out occurs (TO), Columns 4 and 6 also present number of time frames (U*) analyzed just before time-out.

BestBMC finds witnesses for only 5 properties in the given time limit while $BMC^{@}$ easily finds witnesses for all 13 properties, outperforming *BestBMC* by 1-2 orders of magnitude. Note, using automatically derived clock schedules, we require far fewer non-redundant unrollings (#U) in comparison to witness depth (#D) for all ticks of a global clock at LCM frequency.

Table 8.2: Comparative evaluation of BMC for clocked properties on VGA_LCD

Prp	WIT #D	#U	BMC@		BestBMC	
			F?	sec	F?(U*)	sec
P1	2	1	Y	<1	Y	<1
P2	50	27	Y	1	Y	19
P3	101	55	Y	3	Y	186
P4	151	82	Y	5	Y	694
P5	351	190	Y	16	N(160)	TO
P6	101	55	Y	3	Y	186
P7	401	217	Y	18	N(161)	TO
P8	600	324	Y	32	N(162)	TO
P9	800	432	Y	52	N(162)	TO
P10	1000	540	Y	78	N(162)	TO
P11	800	432	Y	54	N(162)	TO
P12	850	459	Y	61	N(61)	TO
P13	906	489	Y	2.1k	N(81)	TO

F?: Witness Found (Y/N)?
#U: Number of BMC Unroll
U*: Depth analyzed before TO

8.7.2 Tri-mode Ethernet MAC Controller

This core implements a MAC controller conforming to the IEEE 802.3 specification with support for 10/100/1000 Mbps. It has five external clock inputs:

1. *Clk_125M* (freq=125Mhz)
2. *Clk_user* (freq=100Mz)
3. *Clk_reg* (freq=50Mhz)
4. *Rx_clk* (freq=125/25/2.5Mhz)
5. *Tx_clk* (freq=125/25/2.5Mhz)

where frequencies of *Rx_clk* and *Tx_clk* depend on the input mode selected. In addition, there are five gated clocks derived from these external clocks.

Using the clocking information, we computed automatically the clock schedules and constraints, and a recurrence length of 19. The design has 3961 FFs, with 815 clocked on *Clk_reg*, 835 clocked on *Clk_user*, 764 clocked on *Rx_clk* (and its derivative), 775 on *Tx_clk* (and its derivative), and the rest on the gated clocks. It has 142 primary inputs and 33K 2-input gates. We identified 16 clocked properties *E1-E16* corresponding to receiver and transmitter modules, and input speed modes. These properties can be classified as

- reachability of control states *(E1,E3-8,E10,E12-16)*,

- assertability of a high water mark of a receiving FIFO *(E2,E11)*, and
- update of packet size across clock domains *(E9)*.

Note,
E2, E11 are of the form *(**F**(p))@Clk_user_r*,
E9 is of the form *(**F**(p*__X__(q)@Clk_user_r))@Rx_clk_gated_r*, and
rest are of the form *(**F**(p *__X__(q)))@Clk_user_r*.

We present the results in Table 8.3, with descriptions as in Table 8.2. Again, *BestBMC* finds witnesses for only 5 properties in the given time limit while *BMC*$^{@}$ easily finds for all 16, outperforming *BestBMC* by 1-2 orders of magnitude. As an example, for the *E2* property, *BMC*$^{@}$ takes 16 sec while *BestBMC* takes 4400 sec. Our integrated approach *BMC*$^{@}$ also requires far fewer non-redundant unrollings in BMC.

Table 8.3: Comparative evaluation of BMC for clocked properties on Ethernet MAC

Prp	WIT #D	#U	BMC$^{@}$		BestBMC	
			F?	sec	F ?(U*)	sec
E1	299	149	Y	41	N(143)	TO
E2	269	134	Y	16	Y	4.4k
E3	279	139	Y	14	Y	5.3k
E4	463	232	Y	1.6k	N(145)	TO
E5	289	144	Y	21	144	5.9k
E6	309	154	Y	25	N(148)	TO
E7	299	149	Y	19	Y	6.7k
E8	319	159	Y	48	N(149)	TO
E9	434	216	Y	126	N(127)	TO
E10	299	159	Y	2	Y	3.1k
E11	2110	1235	Y	202	N(224)	TO
E12	2120	1240	Y	261	N(221)	TO
E13	2130	1247	Y	314	N(221)	TO
E14	2150	1259	Y	277	N(213)	TO
E15	2140	1252	Y	240	N(221)	TO
E16	2160	1264	Y	268	N(220)	TO

F?: Witness Found (Y/N)?
#U: Number of BMC Unroll
U*: Depth analyzed before TO

8.8 Related Work

We discuss various approaches that have addressed some of the issues in the verification of multi-clock systems. In approaches [184-187], the goal is to reduce the number of state elements in the model using *phase abstraction* techniques. First, clock-like signals, which exhibit periodicity, are identified (manually or using 3-valued simulation [184]). Based on these clock signals, one identifies non-overlapping latch layers (or phases), and then retains latches in one layer as flip-flops, and replaces latches in the remaining layer by wires or multiplexers. Subsequently, one obtains a verification model by making C (equal to #phases) copies of the transition relation, and simplifying the logic by propagating the phase values of the clock signals. However, the presence of gated clocks and multiple clocks with arbitrary clock frequencies and ratios, can severely restrict the identification of clock-like signals (and various phases) and therefore, limit the size reduction of the verification model. Note that these approaches focus mainly on reducing the number of flip-flops in order to improve the scalability of BDD-based model checking, and not so much on reducing the number of logic gates that is better-suited for SAT-based BMC.

In another approach by Clarke *et al.* [188], given multiple clock frequency constraints, a clock state machine is built based on *event queue semantics*. Each clock state maps to a configuration (≡ a set of events) in an event queue, where each event corresponds to a tick of an *active* clock. They formulate a BMC problem instances by unrolling the design composed with the clock state machine only at clock events, thereby avoiding the redundant unrollings. However, the authors have not proposed any solution to combine their approach with dynamic simplification procedures in the BMC framework, or to handle clocked specifications.

8.9 Summary

We discussed an integrated verification solution for verifying multi-clock synchronous systems with PSL-style clocked specifications. We provide a uniform modeling scheme for various design features such as multiple clocks with arbitrary frequencies (non-integral ratios), multiple phases, gated clocks and latches. Using event queue semantics, we generate automatic scheduling and clocking constraints for BMC unrolling, to avoid computation at every tick of the global clock and to filter loop-checks. Furthermore, we use dynamic simplification to reduce the size of the BMC problem instance by reusing the previous frame's unrolled sub-circuit for the currently inactive clock domain. We also customize translations of clocked specifications for BMC, and show its effectiveness on two large *OpenCores* multi-clock systems. We believe that such an integrated customized solution provides significant improvement in addressing the scalability issues in verification of multi-clock systems.

8.10 Notes

The subject material described in this chapter are based on the authors' previous work [52] © 2007 IEEE.

PART III: PROOF METHODS

In Part III (Chapters 9—10), we discuss various SAT-based proof techniques.

In Chapter 9, PROOF BY INDUCTION, we describe induction-based proof techniques augmented with automatically generated inductive invariants such as reachability invariants. For embedded memory systems, we also describe techniques to model arbitrary initial memory state precisely and thereby, provide inductive proofs using SAT-based BMC for such systems.

In Chapter 10, UNBOUNDED MODEL CHECKING, we describe an efficient circuit-based cofactoring approach for SAT-based quantifier elimination that significantly improves the performance of pre-image and fixed-point computation in SAT-based UMC. We also describe customized formulations for determining completeness bounds for safety and liveness using SAT-based UMC, rather than using loop-free path analysis. These formulations, comprising greatest fixed-point and least fixed-point computations, handle nested properties efficiently using SAT-based quantification approaches.

9 PROOF BY INDUCTION

9.1 Introduction

Although BMC can find bugs in larger designs than BDD-based methods, the correctness of a property is guaranteed only for the analysis bound. However, one can augment BMC for performing proofs by induction [66, 67]. A *completeness* bound has been proposed [66, 67], to provide an inductive proof of correctness for safety properties based on the longest loop-free path between states. Induction with increasing depth k, and restriction to loop-free paths, consists of the following two steps (shown in Figure 9.1):

- *Base*: to prove that the property holds on every k-length path starting from the initial state.
- *k–step Induction*: to prove that if the property holds on a k-length path starting from an arbitrary state, then it also holds on all its extensions to a (k+1)-length path.

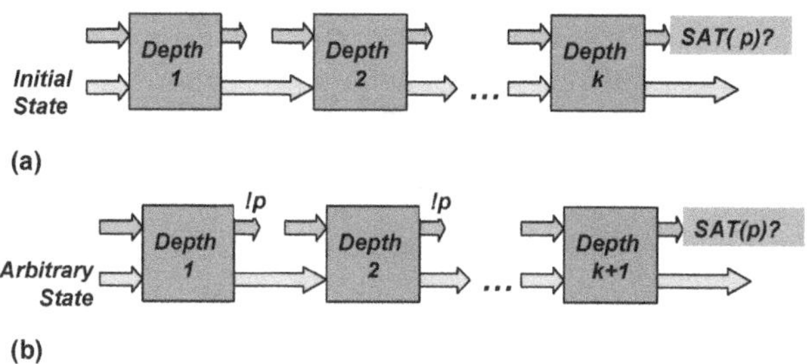

Fig. 9.1: BMC-based inductive proof steps (a) Base (b) Induction on $\neg F(p)$

The restriction to loop-free paths imposes the additional constraints that no two states in the paths are identical. Note that the base case includes use of the initial state constraint, but the inductive step does not. Therefore, the inductive step may include unreachable states also. In practice, this may not allow the induction proof to go through without the use of additional constraints, i.e., stronger induction invariants than the property itself. To shorten the proof length, one can use any circuit constraints known by the designers as inductive invariants. In particular, we discuss use of *reachability constraints* [73] to obtain proofs earlier than the longest loop-free path. We also discuss how this method can be combined effectively with EMM techniques (discussed in a previous chapter) for providing proofs for properties on design with embedded memories [50, 51].

9.2 BMC Procedure for Proof by Induction

We describe a k-step inductive procedure in the augmented BMC procedure *BMC_solve_F_induction*, highlighted in Figure 9.2.

```
Synopsis:  Customized BMC solve with Induction
Input:     prop_tree_node p, bound N,
           Initial state I, Inductive invariant invar
Output:    TRUE/FALSE/ABORT
Procedure: BMC_solve_F_induction

C = I; k=0; IC=invar;
while (k<N) {  //L1 is active always
  p^k = Prop_node(p,k);
  IC = IC ∧ EMM_constraints(k); //for EMM
  if (!Is_sat(IC ∧ p^k)) return FALSE;
  C = C ∧ EMM_constraints(k); //for EMM
  if (Is_sat(C ∧ p^k)) return TRUE; //wit found
  C = C ∧ ¬p^k; //L3 Learning
  Merge(p^k, const_0); //Simplify for falsification
  for (j=k;j>=0;j--) {//fwd loop-free check
    C = C ∧ ¬_jL_k;
    if (!Is_sat(C)) return FALSE; //no wit exists
  }
  IC = IC ∧ ¬p^k;
  for (j=k;j>=0;j--) {//bwd loop-free check
    IC = IC ∧ ¬_jL_k;
    if (!Is_sat(IC)) return FALSE; //no wit exists
  }
  k = k+1; //increment depth
}
ABORT("Bound Reached"); //wit not found
```

Fig. 9.2: Customized BMC with Induction Proof

Inductive proof is obtained when the procedure returns *FALSE* at lines 11, 18, and 23, indicating non-existence of witness to property ***F***(p). In the procedure, *invar* denotes the inductive invariants, *IC* denotes the clause data base for doing inductive checks. Recall, *C* and *I* denote property clauses and initial state constraints which will be used for base checks. For EMM, constraints are added (line 10, 12) using the procedure *EMM_constraints* (see Figure 7.6). We discuss the EMM extension for inductive checks more in the Section 9.4.

The first SAT check (line 11) is performed to see if any state satisfies *invar* and the property node. If no such state exists, clearly the negated property p can not be falsified – this provides an early termination case without even starting a proof by induction. Lines 12–19 correspond to a k-step base check for increasing k, and lines 20–24 correspond to inductive steps with a backward loop-free path check. Note that inductive checks use the constraints *IC* without the initial state constraints *I*, while the base checks use the constraints *C* with the initial state constraints *I*. If the constraints in the base check are satisfiable (line 13), no further induction check is required as the property is falsified.

If the forward loop check (as described in Figure 5.12), does not return false (line 18), we perform the backward induction check by adding pair-wise constraints to *IC* in order to ensure that there is no loop from current state s_k to any previous state s_j on the path with no initial state constraints. If the current path segment cannot be extended to remain loop-free and satisfy the inductive hypothesis (line 23), the procedure returns *FALSE*, indicating that we have found a proof by induction. Note, like the forward loop-free check, the pair-wise loop constraints are added incrementally in order to obtain early termination. When $(k \geq N)$, the procedure aborts (line 26), denoting non-conclusive result.

9.3 Inductive Invariants: *Reachability Constraints*

A set of over-approximate reachable states of the design can be regarded as providing reachability constraints. These can be used as inductive invariants to strengthen a proof by induction [73]. In principle, any technique can be used to obtain such over-approximate reachability constraints, including information known by the designer. Here, we describe a framework using BDD-based symbolic traversal on a conservative model (abstract model) to obtain an over-approximation of the set of reachable states. Note that the over-approximation is a trade-off between strength of the resulting inductive invariant and scalability of BDD-based symbolic traversal. As BDD manipulation tends to be resource intensive (both in time and memory), we try to obtain a conservative abstraction of the design by abstracting away some latches as pseudo-primary inputs. One can obtain such abstraction statically, i.e., by analyzing the circuit structure, or dynamically by using more sophisticated proof-based abstraction techniques (discussed in Part IV, Chapter 9). Good candidates for static abstraction [86] are peripheral latches that do not contribute to the sequential core i.e., *cone-of-influence reduction*, or latches that are farther in the dependency closure of the property signals, i.e., *localization reduction*. Essentially, the abstract design should contain a superset of the paths in the real design.

BDD-based symbolic traversal is used on the abstract design to obtain an over-approximate set of reachable states. For simple safety properties, we store the union of the state sets computed iteratively at each step i by the pre-image operation, denoted as B_i, backwards from the bad states until convergence, as shown in Figure 9.3(a). We also store the union of the state sets in each step i, denoted as F_i, of the forward reachability analysis, starting from the initial state until convergence, as shown in Figure 9(b). Instead of exact traversal on the abstract design, approximate traversal techniques can also be used [190, 191]. In addition, over-approximation techniques for BDDs can be used to further reduce the size of the final BDDs [33].

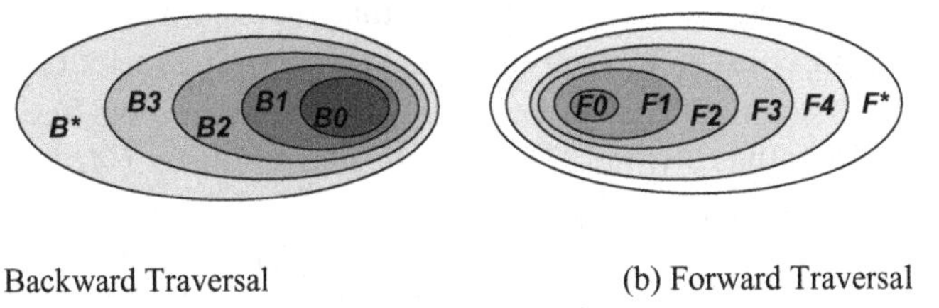

(a) Backward Traversal (b) Forward Traversal

Fig. 9.3: BDD traversal: (a) backward from the bad states B_0, (b) forward from Initial states F_0

Once the forward and backward over-approximate sets of reachable states, i.e., F^* and B^* respectively, are obtained, each as a BDD (or a set of BDDs), the task is to convert the BDDs to a form suitable for the BMC engine. We convert a BDD to a circuit form, where each internal BDD node f is represented as a multiplexer, *ITE(v,hi,lo)*, which is controlled by the BDD variable v (in this case, a "state" variable) as shown in Figure 9.4. The size of constraint is linear in the size of the BDD. We use reordering heuristics [24-27] as well as over-approximation methods to keep down the BDD sizes. Note, we use a hybrid SAT solver (discussed in Chapter 3) that allows circuit representation of these constraints quite efficiently. One can choose to represent the constraints explicitly by enumerating all BDD paths to zero [192]. In our experience, such explicit enumeration has a high overhead and is detrimental to SAT solver.

The derived circuits are added as additional constraints in the BMC engine, as shown in Figure 9.5. The forward reachability constraint F^* serves as an inductive invariant *invar* that constrains the arbitrary state at the start of the $k+1$ cycle of the inductive step, as shown in Figure 9.5(b). In many cases, this strengthening is enough to make the inductive proof succeed.

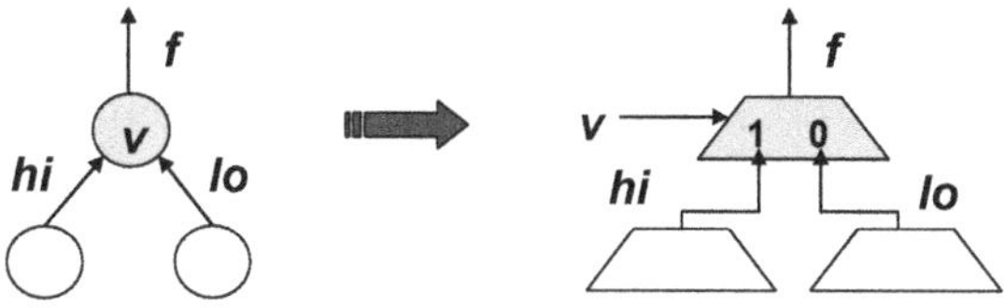

Fig. 9.4: BDD to circuit structure

Similarly, one can use the backward reachability constraint B^* in the base step to check for satisfiability after the k-th step as shown in Figure 9.5(a). If the SAT problem $C\wedge(B^*)^k$ is unsatisfiable, we obtain a proof that the negated property cannot be falsified. It is interesting to note that the backward set B^* complements the forward reasoning of the base step, while the forward set F^* complements the backward reasoning of the inductive step.

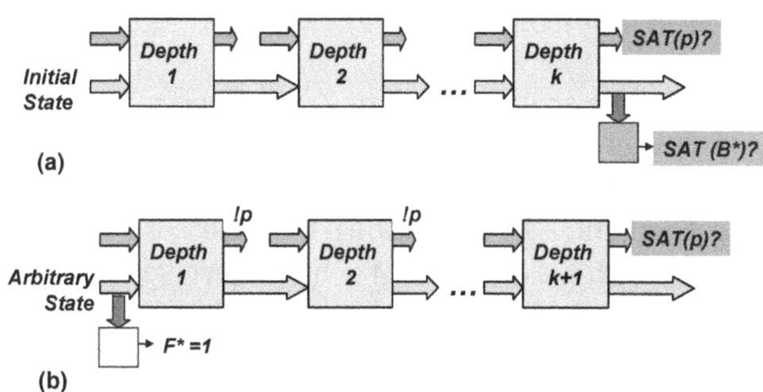

Fig. 9.5: Induction Proof using reachability constraints

9.4 Proof of Induction with EMM

We describe extensions of our EMM method (see in Chapter 7) to provide correctness proofs for embedded memory systems. To derive induction proofs for BMC with EMM, we model the initial state using procedure *EMM_arb_init_constraints* (see Section 7.4.5) to capture the arbitrary initial states. Recall, to model a memory with an arbitrary initial state, we introduce new symbolic variables at every time frame. Simply introducing new variables introduces additional behaviors in the verification model. Therefore, we need to identify a sufficient set of constraints that models the arbitrary initial state of the memory correctly. This is done by identifying pairs of read operations from the same addresses, which should forward the same data, provided no other write operations have been performed on the same address between the two reads. We add these constraints using a hybrid representation in the sub-procedure *EMM_arb_init_constraints,* within the procedure *EMM_constraints.* We augument the inductive constraints *IC* (line 10, Figure 9.2) with the new set of arbitrary initial memory state constraints (Eq 7.8) during the proof steps of BMC with EMM constraints.

9.5 Experiments

In the first set of experiments, we discuss the effectiveness of reachability constraints in strengthening the induction proofs. Further, we also show that BDD-based symbolic traversal can be used effectively to generate these invariants. In the second set, we experimented on designs with multi-port embedded memories and discuss how an induction proof with EMM can be used effectively.

9.5.1 Use of Reachability Invaraints

We implemented these techniques in *VeriSol* and experimented on some large industrial designs, with more than 2k flip-flops, and more than 10k gates in the cone of influence of the safety properties. The experiments were conducted on a Dual Intel Xeon 1700 MHz processor, with 4GB memory, running Linux 7.2, with a time limit of 3 hours. We searched for more than 100 time frames for each design, but could not find a counterexample. We also attempted proofs by induction of depth more than 40, but the induction proofs did not succeed, despite use of about 20 universal constraints (enforced in every state) provided by the designers.

Next, we tried BMC with *reachability inductive invariants* derived using BDD analysis in *VeriSol*. These experiments were conducted on a 440 MHz. Sun workstation, with 1GB memory, running Solaris 5.7. The results are shown in Table 9.1, where Columns 2 – 5 report the results for BDD-based analysis on the abstract model, while Columns 6 – 9 report the results for a BMC-based proof by induction on the concrete design, with use of the BDD constraints. We obtained the abstract models automatically from the unconstrained designs, by abstracting away latches farther in the dependency closure of the property variables. For these experiments, we performed only the symbolic traversal on the abstract model, since our BDD-based symbolic model checker did not allow checking on universally constrained paths. Columns 2 – 5 report the size of the abstract model (number of flip-flops #FF, number of gates #G), the CPU time taken for traversal, the number of forward iterations, and the final size of the BDD F^*, respectively. Columns 6 – 9 report the size of the concrete design, the verification status, and the time and memory used by the BMC engine, respectively.

Note that due to the small size of the abstract models, we could keep the resource requirements for BDDs fairly low. The important point is that despite gross approximations in the abstract model analysis, the BDD reachability invariants were strong enough to let the induction proof go through successfully with BMC in each case. Though neither the BDD-based engine, nor the BMC engine, could individually prove these safety properties, their combination allowed the proof to be completed very easily (in less than a minute).

Table 9.1: Results of proof by induction using BDD-based reachability invariant

D	BDD-based Abstract Model Analysis				Induction Proof with BDD Constraints on Concrete Design			
	#FF / #G	Time(s)	Depth	Size of F*	#FF / #G	Status	Time(s)	Mem (MB)
D1-p1	41 / 462	1.6	7	131	2198 / 14702	TRUE	0.07	2.72
D2-p2	115 / 1005	15.3	12	677	2265 / 16079	TRUE	0.11	2.84
D3-p3	63 / 1001	18.8	18	766	2204 / 16215	TRUE	0.1	2.85

9.5.2 Case Study: Use of Induction proof with EMM

This case study makes use of EMM for memory with multiple ports, and for finding invariants that can aid proofs by induction.

The design has 2400 latches (excluding the memory registers), 103 inputs and ~46K 2-input gates. It has one memory module with AW=12 and DW=32. The memory module has 1 write port and 3 read ports, with memory state initialized to 0. There are 8 reachability properties.

When we abstracted out the memory completely, we found spurious witnesses at depth 7 for all properties,. Thus, we need to include the memory module in our verification. Using EMM, we were not able to find any witnesses for these properties up to depths of 200, in about 10 seconds. Next, we tried obtaining a proof of unreachability for all depths. Using EMM with proof-based abstraction (discussed in Chapter 11), we were able to reduce the model to about 100 latches requiring 4-5 minutes. However, the model was not small enough for either our BDD-based model checker, or SAT-based BMC, to provide a proof.

During the forward search (upto depth 200), we observed that the *WE* (write enable) control signal stayed inactive throughout. Therefore, we hypothesized that the memory state does not get updated, i.e., it remains in its initial state always. This can be expressed using the following LTL property:

$$AG(WE=0 \vee WD=0)$$

i.e., always, either the write enable is inactive or the write data (*WD*) is 0. Using augmented BMC (Figure 9.2), we were able to prove the above hypothesis using backward induction at depth 2 in less than 1s. Explicit Modeling using BMC (without line 10 in Figure 9.2) takes 78s to prove the same.

The above invariant implies that the data read is always 0 (could potentially be a design bug). Next we abstracted out the memory, but applied this constraint to the input read data signals. We used proof-based abstraction (discussed in Part IV, Chapter 11) to further reduce the design to only 20-30 latches for each property (taking about a minute). We then proved each property unreachable on the reduced model using forward induction proof in BMC in less than 1s. As a side note, BDD-based model checker was unable to build even the transition relation for these abstract models.

9.6 Summary

A BMC engine can be easily augmented to provide correctness proofs by using *k*-step induction. The technique discussed here constrains the inductive checks on loop-free paths. Though such checks can be used to prune out cases with short proof diameter, it is not effective in general. One can, however, strengthen the induction proofs using inductive invariants. Often, the inductive invariants based on circuit structure are not adequate. We discussed a lightweight and effective use of reachability constraints as inductive invariants. We also discussed how to generate these constraints effectively using BDD-based reachability analysis on an abstraction of the given design, and how to integrate them in our hybrid SAT-based BMC framework. Further, we discussed how to extend induction proofs with EMM to derive proofs of correctness for systems with embedded memories.

9.7 Notes

The subject material described in this chapter are based on the authors' previous works [73], used with kind permission of Springer Science and Business Media, and [51] © 2005 IEEE.

10 UNBOUNDED MODEL CHECKING

10.1 Introduction

SAT-based Bounded Model Checking (BMC) [45, 66, 109, 178] has been shown to be more robust and scalable compared to symbolic model checking methods based on Binary Decision Diagrams (BDDs) [12, 17]. Unlike BDD-based methods, BMC focuses on finding bugs of bounded length, successively increasing the bound to search for longer traces. Although BMC can find bugs in larger designs than BDD-based methods, the correctness of a property is guaranteed only for the analysis bound. A *completeness bound* has been proposed [66, 67, 72], to provide a proof of correctness for safety properties based on the longest loop-free path between states. Unfortunately, the longest loop-free path can be exponentially longer than the reachable diameter of the state space (for example the longest loop-free path for an n-bit counter is 2^n while the reachable diameter is 1). One can use inductive invariants like *reachability constraints* [73] to obtain proofs earlier than the longest loop-free path. Alternatively, one can obtain a shorter *completeness bound*, i.e., the *longest shortest backward diameter*, for BMC using SAT-based fixed-point computations [67], described in the procedure *Fixed_point_EF* (see Figure 2.10). Such an approach uses cube-wise enumeration of SAT solutions, as described in the procedure *All_sat* (see Figure 2.9), for computing exisitential quantifications.

In this chapter, we describe our efficient and scalable approach [47], based on *circuit cofactoring* for SAT-based quantifier elimination that dramatically reduces the number of required enumeration steps, thereby, significantly improving the performance of pre-image and fixed-point computation in SAT-based UMC. The circuit cofactoring method uses *Reduced AIG* representation for the state sets as compared to BDDs [68], zBDDs [71], and CNF representations [69]. The novelty of the method is in the use of circuit cofactoring to capture a large set of states, i.e., several state cubes in each SAT enumeration step, and in the use of circuit graph simplification based on functional hashing to represent the captured states in a compact manner.

The approach is guaranteed to contain the set of new states captured in each enumeration step by the cube-based enumeration approaches. Therefore, our approach is also guaranteed to require a smaller number of enumeration steps. Moreover, as compared to other SAT-based UMC algorithms, our approach does not require a redrawing of the implication graph [71] or enlarging the enumerated cube as a post-processing step [121].

Using our SAT-based UMC analysis, we discuss obtaining completeness bounds for safety and liveness properties. Previous bounded model checking (BMC) approaches have relied on either converting such properties to safety checking, or finding proofs by deriving termination criteria using loop-free path analysis. In our previous discussion (see Chapter 5), we have seen that customized translations improve the BMC performance significantly in comparison to standard monolithic LTL translations for both safety and non-safety properties. Here we discuss how customized SAT-based BMC can be augmented with unbounded SAT-based analysis to obtain the *longest shortest diameter* as the completeness bounds for these properties, instead of using loop-free path analysis. One can use inductive invariants such as reachability constraints [73] to shorten the completeness bound further. In particular, these formulations comprise greatest fixed-point and least fixed-point computations to efficiently handle nested properties using SAT-based quantification approaches such as circuit cofactoring.

We use hybrid SAT-solver (see Chapter 4) that works seamlessly on *Reduced AIG* representation. We have implemented our techniques in a SAT-based UMC framework where we show the effectiveness of SAT-based existential quantification on public benchmarks and on a number of large industry designs that were hard to model check using other methods. We show several orders of improvement in time and space using our approach over other CNF-based approaches [71]. We also present several heuristics to further enlarge the set of state cubes captured per SAT enumeration and controlled experiments to demonstrate their effectiveness. Importantly, we were able to prove using our method the correctness of a safety property in an industry design that could not be proved otherwise. We also show the effectiveness of our overall approach for checking liveness on public benchmarks and several industry designs.

Outline

In Section 10.2 we give the basic idea with a motivating example; in Section 10.3 we present the theoretical background of the circuit cofactoring approach; in Section 10.4 we discuss state set representation using circuit graphs; in Section 10.5 we discuss enumeration using hybrid SAT and present various heuristics to further enhance circuit cofactoring; in Section 10.6 we present SAT-based UMC algorithms for safety and liveness properties; in Sections 10.7 and 10.8 we describe our experiments using the circuit cofactoring approach on safety and liveness properties respectively; and in Section 10.9, we discuss related work in UMC; and summarize our discussion in Section 10.10.

10.2 Motivation

Quantifier elimination is one of the key tasks in symbolic model checking. In previous work based purely on SAT, the focus has been either on a better representation of the enumerated set [69], or on enlargement of the enumerated cube [71]. Here, we focus on improving the SAT-based existential quantification procedure with a circuit cofactoring [47] technique that captures a larger solution set at each enumeration step than that captured by cube-based methods, and an efficient representation of the enumerated set. In this section, we give an overview of the approach with a motivating example shown in Figure 10.1.

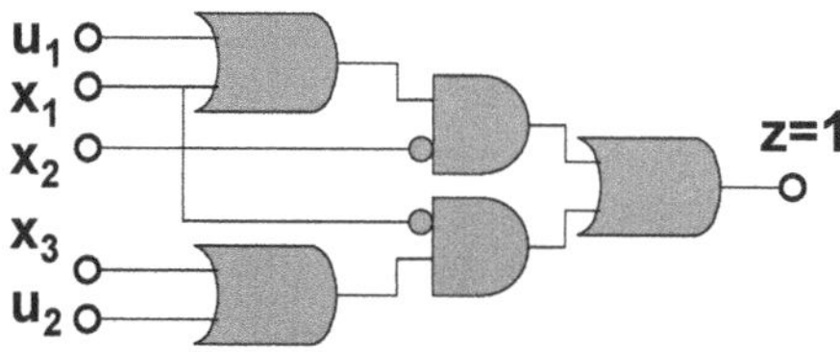

Fig. 10.1: An example circuit

Example 10.1

Let the output node z represent the function $f = \neg x_1 \wedge (x_3 \vee u_2) \vee \neg x_2 \wedge (x_1 \vee u_1)$ where u_1 and u_2 are primary input variables, and x_1, x_2, and x_3 are state variables. Our task is to compute $\exists u_1 u_2\, f$ (which equals $\neg x_1 \vee \neg x_2$) using SAT-based quantifier elimination. Suppose, we get a partial satisfying assignment $\{u_1=1, x_2=0\}$ in the first enumeration. Previous approaches [69, 71] would add the blocking constraint $\Omega = x_2$ as an additional clause to f. As a result, the solution $x_2=0$ will never be enumerated again. In the next enumeration, suppose the solver produces a satisfying assignment $\{x_1=0, x_2=1, u_2=1\}$. With the new solution cube $\neg x_1 \wedge x_2$, the updated blocking constraint becomes $\Omega = x_2 \wedge (x_1 \vee \neg x_2) = x_2 \wedge x_1$. After the addition of this constraint, no more solutions exist, and the procedure terminates. The negation of the blocking constraint $\neg\Omega$ represents the desired set $\exists u_1 u_2 f$.

We address the issues of enumeration time and representation size by answering the following questions in this chapter:

- Can we capture several cubes, hence more new solutions per SAT solver enumeration, than that captured by cube-wise enumeration approaches, and thus, reduce the number of enumerations required?
- Can we efficiently represent the captured solution sets so as to mitigate the space-out problem inherent in the previous approaches and also obtain faster SAT search?

For this example, we show in Section 10.3 that the circuit cofactor can method discover the blocking constraint $\Omega=x_2 \wedge x_1$ in one enumeration step.

10.3 Circuit Cofactoring Approach

In this section, we describe the key insights and theoretical background for the SAT-based existential quantification approach. Later, we discuss how the circuit cofactoring approach is used in SAT-based UMC algorithm for improved performance.

10.3.1 Basic Idea

Consider the example in Figure 10.1 after we obtain the first satisfying assignment $\{u_1=1, x_2=0\}$. We observe that:

- If all the unassigned inputs (u_2 in this example) were given some assignment (say, $u_2=1$), the resulting assignment would be still satisfying.
- The blocking constraint ($\Omega=x_2$ in this example), i.e., the negation of the enumerated cube, would remain unchanged even after the application of the previous step.

Based on these two observations, we present a theorem that forms the basis of our SAT-based quantifier elimination strategy. First, we present a known lemma.

Lemma 10.1: For a cube c and function f, $c \subseteq f \Leftrightarrow f_c=1$.
Proof:
($\Rightarrow$) Given $f_c = 1$. As $c \wedge f_c \subseteq f$, clearly $c \subseteq f$.
($\Leftarrow$) Given $c \subseteq f$. Assume $f_c \neq 1$, i.e., there exists a minterm u over $supp(f_c)$ such that $f_c(u) = 0$. Note, u is independent of the literals of c. Now

we construct a minterm $m = u \wedge c$. Clearly, $m \in f$. However, as $c(m) = 1$ and $f_c(u) = 0, f(m) = c \wedge f_c \vee \neg c \wedge f_{\neg c} = 0$, i.e., $m \notin f$, a contradiction. Therefore, $f_c = 1$. □

Given a set of input variables U and a set of state variables X, we say an assignment $\alpha: V_\alpha \rightarrow B$ for a function f is complete with respect to the input set if $U \subseteq V_\alpha$. For such an assignment, we call the satisfying input cube an input minterm. Note that by assigning an arbitrary Boolean value on all unassigned input variables $u \in U \backslash V_\alpha$, we can make a satisfying assignment complete with respect to the input set.

Theorem 10.1: For a satisfying assignment for f, let s be the satisfying state cube, and u be the satisfying input cube. If a minterm $m \in u$, then $s \subseteq f_{\mathrm{m}}$, i.e., f_{m} captures more satisfying minterms than s.

Proof: As the cube $m \wedge s$ is a satisfying cube, i.e., $m \wedge s \subseteq f$, using Lemma 10.1, we get $f_{m \wedge s} = 1$. Again, using Lemma 10.1 $f_{m \wedge s} = 1 \Rightarrow s \subseteq f_{\mathrm{m}}$, giving us the proof. □

10.3.2 The Procedure

Based on Theorem 10.1, we present the steps of the procedure *Circuit_cofactor* to capture a larger set of satisfying states from a given satisfying assignment, in the following:

- Step 1: Pick a satisfying input minterm by choosing an assignment on the unassigned input variables in the satisfying cube assignment.
- Step 2: Cofactor the function f with respect to the satisfying input minterm obtained in step 1.
- Step 3: Use Reduced AIG to represent the cofactor obtained in step 2, capturing the set of satisfying states.

Example 10.1 (contd):

Following the above three steps, we can show that we can get a larger set of satisfying states than the satisfying cube $\neg x_2$ for the example in Figure 10.1. We first make the satisfying assignment (obtained in the first enumeration) complete with respect to input set by choosing $u_2 = 1$. When we cofactor $f = \neg x_1 \wedge (x_3 \vee u_2) \vee \neg x_2 \wedge (x_1 \vee u_1)$ with respect to the satisfying input minterm $m = u_1 \wedge u_2$, we get $f_{\mathrm{m}} = \neg x_1 \vee \neg x_2$ as shown in Figure 10.2(a). Clearly, f_{m} represents a larger set than the satisfying cube. Importantly, we required just one enumeration to capture the *entire* solution set. Though, in general, we may require more than one enumeration, we observed in our experiments that the number of enumerations required is several orders of magnitude less compared to cube-wise enumeration approaches.

We also observed that the total number of enumerations required depends, though less severely, on the choice of values used for the unassigned input variables in the satisfying cube assignment. For example, if we had chosen $u_2 = 0$, then with input minterm $m = u_1 \wedge u_2$, we would get $f_m = (\neg x_1 \wedge x_3) \vee \neg x_2$. Although f_m represents more states than the satisfying cube, we will require one more enumeration to capture all the satisfying states.

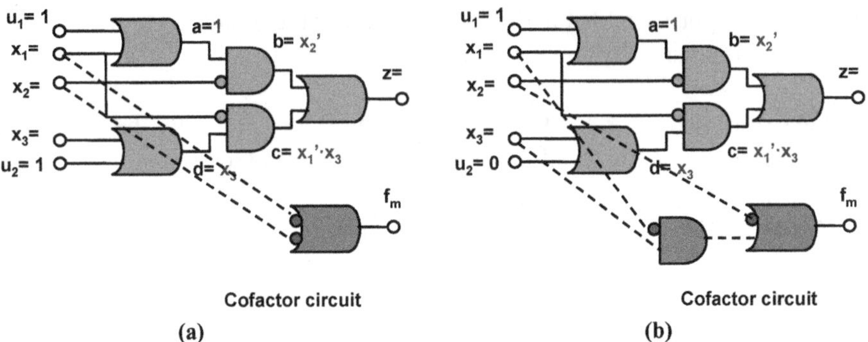

Fig. 10.2: Cofactor-based quantification with minterm (a) $\{u_1=1, u_2=1\}$ (b) $\{u_1=1, u_2=0\}$

10.3.3 Comparing *circuit cofactoring* with *cube-wise enumeration*

Cube enlargement is applied as post-processing process in cube-wise enumeration approach. We show that such post-processing is not needed in circuit cofactoring approach. Specifically, we show that even if we enlarge the satisfying assignment cube (i.e., add new solutions), and hence the state cube, the state set captured by *Circuit_cofactor* method remains a superset of the states represented by the enlarged state cube. This demonstrates that our approach is inherently superior to cube enlargement approaches.

Theorem 10.2: Let there be two satisfying assignments $\alpha: V_\alpha \rightarrow B$ and $\beta: V_\beta \rightarrow B$ such that $V_\beta \subseteq V_\alpha$ and $\forall v \in V_\beta\, \beta(v)=\alpha(v)$. We denote s_α and s_β as the satisfying state cubes, and u_α and u_β as the satisfying input cubes for the assignments α and β, respectively. If an input minterm $m \in u_\alpha$, then $s_\beta \subseteq f_m$.

Proof: As $V_\beta \subseteq V_{\alpha,}$, clearly $u_\alpha \subseteq u_\beta$ and $s_\alpha \subseteq s_\beta$. As $m \in u_\alpha$, $m \in u_\beta$. Using Theorem 10.1, we get $s_\beta \subseteq f_m$. □

One can, therefore, argue that the approaches to enlarge a state cube—by redrawing the implication graph [71], or by identifying a necessary assignment on state variables [121] — are redundant, when solution states are captured by *Circuit_cofactor* method. Note that the above theorem applies to cube enlargement that involves addition of new solutions only.

Arguably, one can also enlarge the state cube by subsuming previously enumerated solutions. Though such an enlargement yields a more efficient representation, it does not capture additional new solutions. Hence, such enlargement *per se* does not reduce the total number of enumerations required. Based on the above, we claim that the union of solution sets captured by our method is guaranteed to subsume the union of solution sets captured by a blocking clause approach [71] at any enumeration step.

In a later section, we discuss several heuristics to choose an input satisfying minterm (Step 1 of *Circuit_cofactor*) that enlarges the set of new solutions represented by the cofactor f_m. Before that, we discuss our method of efficient representation of the state set, and use of a hybrid-SAT solver that has clear advantages over a CNF-based SAT solver for analysis over this representation.

10.4 Cofactor Representation

We use *Reduced AIG* circuit representation as discussed in Chapter 3 for representing both the transition relation and for representing the set of enumerated states. This is similar to some previous approaches [159, 178]. A circuit-based characteristic function, unlike a BDD, is non-canonical and is less sensitive to the order in which it is built. However, all operations cannot be done efficiently. In particular, existential quantification $\exists_B f(A,B)$ can be regarded as a disjunction of formulas obtained by cofactoring f with all minterms over variables in B. Previous approaches [159, 178] perform this disjunction explicitly for each variable in B, leading to an exponential increase in the formula size. On the other hand, in our circuit cofactoring approach, we use a SAT solver to identify satisfying assignments to $f(A, B)$, propagate the B values of the satisfying minterm m, through the circuit graph making it much simpler, and add this simplified circuit graph, i.e., *cofactor* f_m, to the representation of the solution states on-the-fly. Note that,

- Each addition to the representation of the solution set is a simple and reduced circuit graph
- Cofactors share those nodes with the original circuit f which do not have support of B variables.
- Constraint f_m=0 acts as a blocking constraint to prune the subsequent search for solutions

As a result, the larger the number of solution minterms captured in an added circuit graph, the smaller the number of future SAT-based enumeration steps, and the smaller the number of additional solutions that need to be explored. As noted previously in Chapters 3-4, representing the solution set using circuit graphs has two main advantages [39, 42]:

1. compact representation is achieved due to circuit simplification, and
2. faster Boolean reasoning is achieved on a reduced and simplified formula, leading to faster SAT searches in comparison to the unreduced formula.

Example 10.2

Consider the example shown in Figure 10.3, with z node in the circuit graph representing the function f. We apply a hybrid SAT solver iteratively on the problem with constraint {z=1}.

Suppose after the first iteration, we obtain a partial satisfying solution $\{x_1=1, x_3=0, u_2=0\}$. Next, we pick $\{u_1=0\}$ for the unassigned input variable, and cofactor f with the input minterm $\{u_1=0, u_2=0\}$. We obtain the cofactor $c_1 = \neg x_3 \wedge (x_1 \vee x_2)$. Note that the cofactor circuit c_1 captures an additional cube $\neg x_3 \wedge x_2$, other than the enumerated cube $\neg x_3 \wedge x_{1c}$. We add the constraint $c_1=0$ as a blocking constraint, and continue SAT enumeration. Note that the blocking constraint prevents subsequent enumerations of SAT solutions with cubes contained in c_1. Suppose, in the next enumeration, we obtain a partial satisfying solution $\{x_3=1, u_1=1, x_4=1\}$. Next, we pick $\{u_2=1\}$ and cofactor f with the minterm $\{u_1=1, u_2=1\}$. We obtain the cofactor $c_2 = (\neg x_1 \wedge \neg x_2) \vee x_4$. Note the reuse of of the circuit sub-structure representing $x_1 \vee x_2$. When we add the cofactor c_2 as a blocking constraint, i.e., $c_2 = 0$ and continue SAT search, we obtain the result unsatisfiable and stop further enumerations. Note that a disjunction of the cofactor circuits, i.e., $c_1 \vee c_2$ represents the result $\exists u_1 u_2 f$.

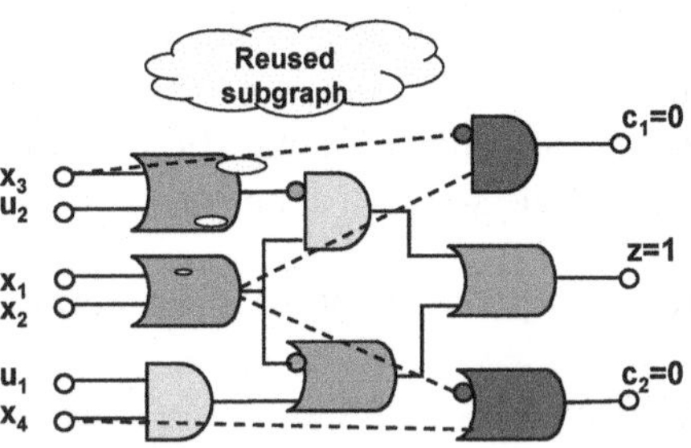

Fig. 10.3: Example of circuit-graph reuse in circuit-based cofactor quantification

10.5 Enumeration using Hybrid SAT

We use a hybrid SAT solver (see Chapter 4) that operates directly on the simplified circuit graph representation and uses CNF representation for learned clauses. We prefer to use hybrid SAT solver for solution enumeration due to following two main advantages over using CNF solvers:

- We do not have to convert the circuit graph to a CNF problem, saving both time and memory.
- A hybrid SAT solver allows the use of circuit-based decision heuristics, for example, a justification frontier [144] to find a *partial satisfying solution* that is highly desirable for generating a large satisfying cube. Using this heuristic, a SAT solver can declare a partial assignment as satisfying when there are no more frontiers to justify, generally much earlier than the stopping criteria used in a CNF-based SAT solver.

Due to the above reasons, we prefer the use of circuit-based justification heuristics within the SAT-solver, rather than the use of post-processing methods (after every SAT result) that either call a separate justification procedure [121] or redraw the implication graph [71]. We also prefer it to the use of dynamic detection and removal of inactive clauses before each decision [147] in a CNF-based SAT solver.

Now, we discuss several low overhead circuit heuristics for further enlarging the sets captured by the cofactors.

10.5.1 Heuristics to Enlarge the Satisfying State Set

The choice of values used to make a satisfying cube assignment complete with respect to the input variables has some impact on the formula size and the size of the representation of satisfying states. We present some heuristics to address this issue. These heuristics pick values on the unassigned input variables so that the latch support of the resulting circuit cofactor is small, hence resulting in a capture of larger solution sets.

In the sequel, we use *circuit* to refer to a *Reduced AIG* graph and *node* to refer to 2-input AND gate, a primary input, or a latch. The two immediate fanins of an AND gate are referred to as the *left* and *right* children. The node can be inverted or non-inverted. Here are some more definitions:

Definition 10.1: A *support variable* of a node is a primary input or latch in the transitive fanin cone of the node.

Definition 10.2: A *latch frontier* is a node that has only latches in its support, and at least one of its fanouts has one or more primary input variables in the support.

Definition 10.3: A *justification-frontier (jft)* is a node that can justify the value 0 on the immediate fanout node. Note that when a node has value 0, then potentially the left or/and right children can be justification-frontiers by taking value 0. A SAT-solver, typically, selects one of the justification frontiers as a future decision variable. For a satisfying assignment, the selected justification-frontier variable is required to have an assignment [41].

Definition 10.4: A *positive* (negative) *fanout score* of node is the number of its fanouts where the node appears as a non-inverted (inverted) child. The higher of the positive or negative fanout scores of a node is referred to as the *score* of a node.

We explain various heuristics on the example shown in Figure 10.1. For the problem constraint $\{z=1\}$ in Example 10.1, we assume that the SAT solver has produced the following partial satisfying solution: $\{u_1=1, x_2=0\}$. We now explain various heuristics to pick the unassigned values on input variables, i.e., on u_2. We classify the criteria for various assignments as score-based (S), decision-based (D), or implication-based (I) in the following.

Heuristic 1 (H1): All score-based

Given a satisfying assignment $\alpha{:}V_\alpha \rightarrow B$ and the satisfying input cube u, we choose a satisfying input minterm $m = u \wedge w$, where literals in w are chosen based on the score of corresponding variable. If the positive (negative) score of the primary input variable is more than the negative (positive) score, we choose the positive (negative) literal. The basic idea is that the controlling values of primary input variables can provide an alternate path for justification that may have been performed through assignments on the state variables.

For the example, as shown in Figure 10.4, we choose $u_2 = 1$ since the positive score of u_2 is 1, and its negative score is 0. Cofactor of f with respect to minterm $\{u_1=1, u_2=1\}$ is $(\neg x_1 \vee \neg x_2)$.

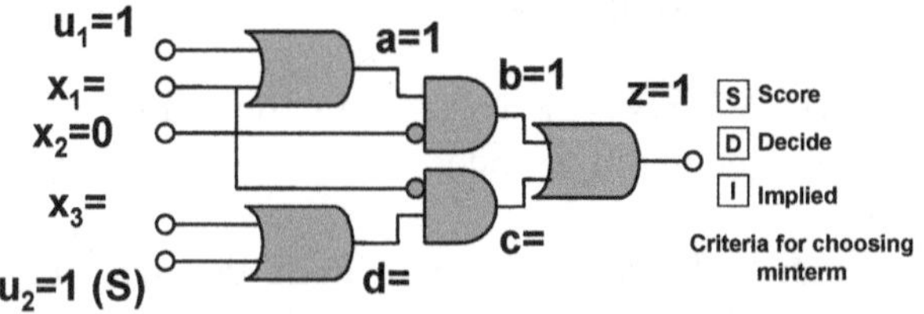

Fig. 10.4: Circuit-based cofactor quantification using heuristic *H1*

Heuristic 2 (H2): All jft-decision based

When using the justification-frontier decision heuristic, our SAT-solver terminates with a satisfying assignment when there is no remaining justification-frontier. In this heuristic, we use the SAT solver to find further assignments on the unassigned primary input variables even after the first satisfying assignment has been found. Once the first satisfying assignment is obtained, we use the procedure *Get_more_frontiers,* as shown in Figure 10.5.

```
1.  Synopsis:   Obtain more justification frontiers
2.  Input:      satisfying assignment α, domain of α i.e., Vα,
3.              current set frontiers F, node n (AIG),
4.              current heuristics H2-H5
5.  Output:     final set of frontiers F
6.  Procedure: Get_more_frontiers
7.
8.  if (Is_visited(n)) return;
9.  mark_visited(n);
10. l = Left_child(n); r = Right_child(n);
11. if (α(n)=1)  {
12.   Get_more_frontiers(l,α,F);
13.   Get_more_frontiers(r,α,F);
14. }else {// α(n)=0
15.   if (r ∉ Vα) { // r is unassigned
16.     if (H2 || H3)  F=F∪{r};
17.     if ((H4 || H5) && (!Is_latch_frontier(r)) F=F∪{r};
18.   }
19.   if (l ∉ Vα) { // l is unassigned
20.     if (H2 || H3)  F=F∪{l};
21.     if ((H4 || H5) && (!Is_latch_frontier(l)) F=F∪{l};
22.   }
23.   if (α(l)=0) Get_more_frontiers(l,α,F);
24.   if (α(r)=0) Get_more_frontiers(r,α,F);
25. }
```

Fig. 10.5: Procedure *Get_more_frontiers*

The procedure traverses the graph from the objective towards the inputs and collects unassigned nodes (in lines 16 and 20) that are potentially justification-frontiers. These nodes are then added to SAT solver as additional justification-frontiers. The SAT solver, then, resumes the process of finding a satisfying assignment that includes more input variables. This procedure is called repeatedly after every satisfying assignment until there are no more additional frontiers to add. Note that as this heuristic does not change the SAT decision procedure, it does not affect the SAT time for getting the first satisfying assignment. The subsequent satisfying assignments are much easier to find, as the problem size gets smaller at each step.

The procedure *Get_more_frontiers* using heuristic *H2* with assignment set $\{u_1=1, x_2=0\}$ produces the frontier set $\{c=1, x_1=1\}$ as shown in Figure 10.6. Continuing SAT with the frontier set, and assuming SAT decides (D) on $c = 1$, we get a set of implied assignments (I) $\{d=1, x_1=0\}$. Next, let SAT decide $x_3=1$. Invoking the procedure *Get_more_frontiers* again with the updated assignments $\{c=1, d=1, x_1=0, x_3=1\}$, we obtain a new frontier set $\{u_2=1\}$. Continuing SAT leads to the complete assignment of the input variables $\{u_1=1, u_2=1\}$. As seen before, cofactor of f with respect to minterm $\{u_1=1, u_2=1\}$ is $(\neg x_1 \vee \neg x_2)$.

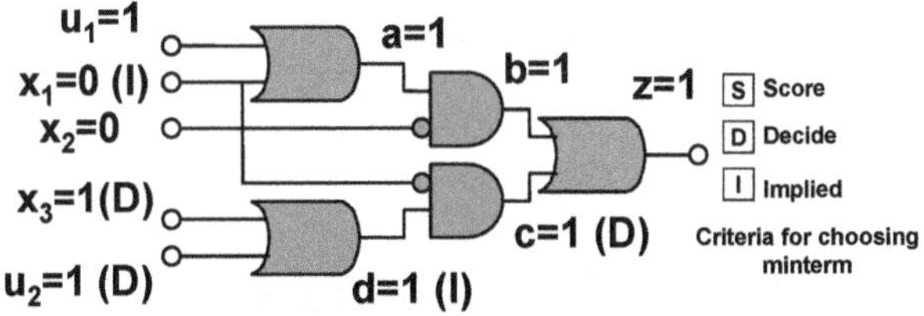

Fig. 10.6: Circuit-based cofactor quantification using heuristic *H2*

Heuristic 3 (H3): First jft decision-based and rest score-based

This heuristic is similar to heuristic H2, with the difference that the procedure *Get_more_frontiers* is called only once. The remaining unassigned input variables are then assigned based on scores (like *H1*).

For the running example, the first call to *Get_more_frontiers* using *H2* and the subsequent SAT call gives assignment set $\{c=1, d=1, x_1=0, x_3=1\}$. The assignment on u_2 is chosen as 1 using heuristic *H1*, i.e., score-based as shown in Figure 10.7.

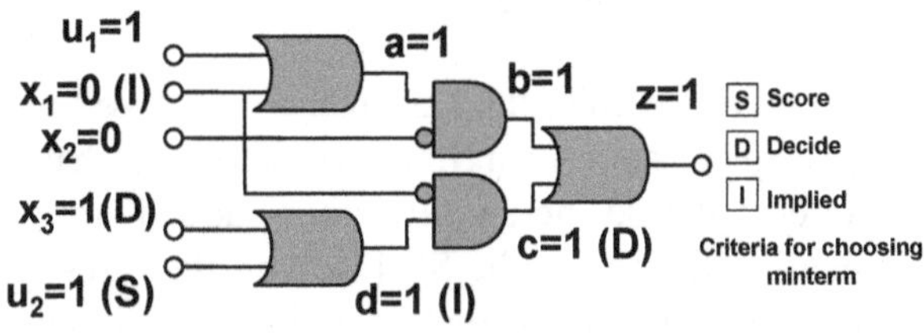

Fig. 10.7: Circuit-based cofactor quantification using heuristic *H3*

Heuristic 4 (H4): All non-latch jft-decision based

Like heuristic *H2*, this heuristic also invokes the procedure *Get_More_Frontiers,* with the difference that the frontiers added to the list *F* are not latch frontiers as shown in lines 17 and 21 (Figure 10.5). Since latch frontiers, by definition, do not have input variables in their support, any assignment to justify them will not increase the number of assigned input variables. This heuristic, unlike heuristic *H2*, reduces the subsequent justification problem size even further in the repeated calls to the SAT solver.

Applying *Get_more_frontiers* using heuristic *H4* with assignment set $\{u_1=1, x_2=0\}$ produces the frontier set $\{c=1\}$. Continuing SAT with the frontier set, and decision (D) on $c=1$, we get implied assignments (I) set $\{d=1, x_1=0\}$. Let SAT decide $x_3=1$. Invoking the procedure *Get_more_ frontiers* again with updated assignments $\{c=1, d=1, x_1=0, x_3=1\}$, we obtain a new frontier set $\{u_2=1\}$. Continuing SAT leads to the complete assignment of the input variables $\{u_1=1, u_2=1\}$. Figure 10.6 also illustrates this heuristic.

Heuristic 5 (H5): First non-latch jft decision-based and rest score-based

This heuristic is similar to *H4* with the difference that the procedure *Get_More_Frontiers* is called only once. The remaining unassigned input variables are assigned based on scores i.e. like *H1*.

For the running example, the first call to *Get_more_frontiers* using *H4* and the subsequent SAT call gives the assignment set $\{c=1, d=1, x_1=0, x_3=1\}$. The assignment on u_2 is chosen as 1 using heuristic *H1*, i.e., score-based. Figure 10.7 also illustrates this heuristic.

In our experience, *H5* gives more robust performance compared to the rest. We study the effect of various heuristics H1-H5 in our experimental sections.

10.6 SAT-based UMC

In the following subsections, we discuss various algorithms for SAT-based unbounded model checking using the circuit cofactoring approach discussed in the previous sections.

10.6.1 SAT-based Existential Quantification using Circuit Cofactor

We present an improved SAT-based existential quantification procedure *All_sat_circuit_cofactor* based on circuit cofactoring, as shown in Figure 10.8. The procedure is similar to procedure *All_sat* in Figure 2.9. In line 10 (Figure 10.8), we obtain a satisfying input minterm *m* over the *B* variables from the recently found satisfying assignment using one of the heuristics H1-H5 discussed in the previous section. The function *f* is cofactored with the input minterm *m* using the procedure *Cofactor* in line 11 to obtain a set of satisfying states (a superset of the set of states obtained in line 10 in Figure 2.3). Note that the derived circuit-based cofactors are represented compactly using *Reduced AIG* representation as discussed in Section 10.4. Especially, the nodes that do not have support of *B* variables get reused in such a representation.

```
1.  Synopsis:   Quantification using circuit-cofactor
2.  Input:      function f, input keep set A,
3.              input quantifying set B
4.  Output:     ∃B f(A,B)
5.  Procedure:  All_sat_circuit_cofactor
6.
7.  C=∅; //initialize constraint
8.  while (Is_sat(f=1∧C=0)=SAT) {
9.     α=Get_assignment_cube();
10.    m=Get_satisfying_input_minterm(α,B);
11.    fm=Cofactor(f, m);//Circuit_cofactor: lines 10-11
12.    C= C ∨ fm;
13. }
14. return C;   // return when no more solution
```

Fig. 10.8: SAT-based existential quantification using circuit cofactor

10.6.2 SAT-based UMC for *F(p)*

We present the algorithm *Fixed_point_EF_circuit_cofactor* for computing a least fixed-point for checking an ***F*** property using our circuit cofactoring approach, as shown in Figure 10.9. The algorithm uses the *All_sat_circuit_cofactor* procedure for pre-image computation. Note that, unlike the standard algorithm (see Figure 2.10), where a pre-image computed in the previous step is used to compute the next pre-image, we use the unrolled f^i (f at i^{th} unroll depth) to represent the target states at the ith step, as shown in Figure 10.10. This change does not modify the overall algorithm for obtaining the fixed-point, and gives better performance in practice. This is likely due to the fact that the representation of f^i is more compact in practice than that of $\delta \wedge T(<X \leftarrow Y>)$. Moreover, one can use the *reachability constraint* $C(X)$ (over-approximated reachable states from initial state) [73], as a *care set* in line 12 (i.e., *All_sat_circuit_cofactor* is called with $f^i \wedge \neg R \wedge C$) to potentially get a backward fixed-point in less number of pre-image steps than the backward diameter [67].

For robustness, one can alternately compute under-approximated pre-images in line 12, with a resource bound such as number of SAT enumerations in *All_sat_circuit_cofactor.* These state sets can be used to obtain a proof at a depth larger than the backward diameter, but potentially smaller than the longest loop-free path. The idea is similarly used in approximated BDD-reachability computation [33].

```
1.  Synopsis:   Least fixed-point using circuit-cofactor
2.  Input:      property node f(X,U)
3.  Output:     set of states satisfying EF(f)
4.  Procedure:  Fixed_point_EF_circuit_cofactor
5.
6.  i=-1; R(X)=T(x)=∅; //initialize
7.  do {
8.    R(X)=R(X)∨ T(X); //add i^th pre-image of f
9.    i = i+1; //increase the pre-image depth
10.   f ^i(X,U)  = Prop_node(f, i); //get f at i^th depth
11.   //Compute states for f ^i but not in R
12.   T(X)=All_sat_circuit_cofactor(f ^i ∧¬R(X), X, U);
13. } while (T(X) ≠ ∅) // fixed-point reached?
14. return {R(X),i}; // returns states satisfying EF(f)
```

Fig. 10.9: Least fixed-point computation using circuit cofactor

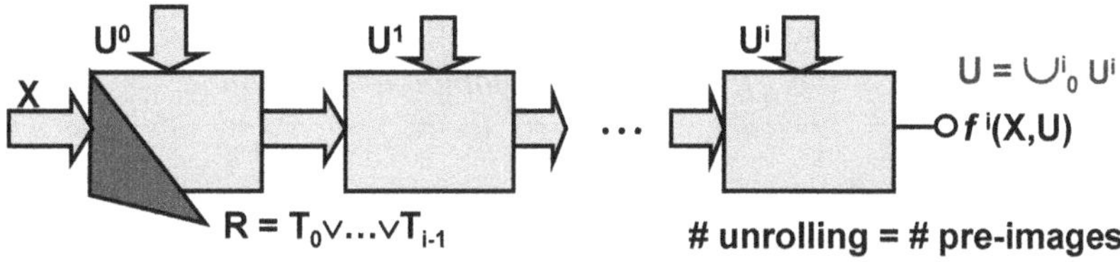

Fig. 10.10: Least fixed-point computation using unrolling

10.6.3 SAT-based UMC for *G(q)*

A *completeness bound* for BMC on liveness property can be determined by forward *longest loop-free path* length analysis [67, 119]. If there is no witness to the property $\boldsymbol{G}(q)$ of length less than the forward longest loop-free path length, the property provably cannot have a witness. As explained earlier, such a bound can be exponentially larger than the longest shortest forward diameter, and hence, not very practical. Here, we discuss a procedure to determine a *completeness bound* for BMC on liveness property such that the bound is much shorter than the loop-free bound. Such a bound computation is based on cofactor-based greatest fixed-point computation, similar to the least fixed point computation for safety property. Such a bound can also be used to determine the existence of a witness to the (negated) liveness property without actually doing loop-checks. Note, for the liveness property $\boldsymbol{AF}(\neg q)$, the negated counter-part is $\boldsymbol{EG}(q)$. We drop the existential $\boldsymbol{E}$ path quantifier for LTL and implicitly check for a witness. Proving no witness to $\boldsymbol{G}(q)$ is same as proving correctness for $\boldsymbol{AF}(\neg q)$.

Let $\boldsymbol{G}^k(q)$ denote the k-bounded $\boldsymbol{G}(q)$ property. It is defined as:

$$\boldsymbol{G}^k(q) = q \wedge \boldsymbol{X}(q) \wedge \ldots \wedge \boldsymbol{X}^k(q) \tag{10.1}$$

Note, $\boldsymbol{X}$ is a next state LTL operator and $\boldsymbol{X}^k$ is shorthand for $\boldsymbol{X}\ldots\boldsymbol{X}$ for k-times. Let H^k denote the set of states that satisfy $\boldsymbol{G}^k(q)$, i.e., the set of states from which there is a path $s_0 \ldots s_k$ such that q is true in all $k+1$ states. Note that, $H^{k+1} \subseteq H^k$ as shown in Figure 10.11. For non-empty $\boldsymbol{G}(q)$, there exists some $k = m$ such that $H^{m+1} = H^m$. We call m the greatest fixed-point bound *gfp bound*. We make the following observations:

- If there is a finite path of length m+1 where q = *true* in all the m+1 states, then this finite segment is also a prefix to an infinite path where q is globally true (a witness loop).
- In other words, if there is no path from initial states $I(s)$ of length m+1 where q is true in all the m+1 states, then the property $\boldsymbol{G}(q)$ is proved to have no witness. Note, this is equivalent to satisfiability checking of $(I(s) \wedge H^m)$. Thus the *gfp bound* is a *completeness bound* for BMC on $\boldsymbol{G}(q)$.

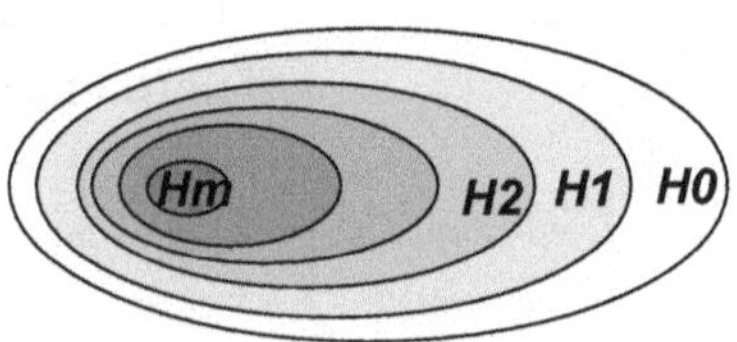

Fig. 10.11: Pre-image computation using unrolling

Using the *gfp bound m* one can determine if a witness *exists* by checking satisfiability of q for m time frames, without checking for any witness loops, which are quite expensive in practice. Though the *gfp bound* does not provide any bound on the length of a witness loop, yet it may not be worth the effort to find such a bound. If an m-length witness prefix exists, it is a matter of expending enough resources to find the witness loop. Note, *gfp bound* could be much smaller than the witness loop length as determined by longest path analysis [119]. We have seen examples where it is much easier to find m-length witness prefixes confirming the existence of a witness, than finding the witness loop.

We present the procedure *Fixed_point_EG_circuit_cofactor* for computing the greatest fixed-point algorithm for ***G(q)***, as shown in Figure 10.12. It uses the *All_sat_circuit_cofactor* algorithm (line 10) for pre-image computation.

```
1.  Synopsis:   Greatest fixed-point using circuit-cofactor
2.  Input:      property node q(X,U)
3.  Output:     set of states satisfying EG(q(X))
4.  Procedure:  Fixed_point_EG_circuit_cofactor
5.
6.  i=-1; T(X)=1;
7.  do {
8.     H(X) = T(X); i = i+1;
9.     q^i(X,U)=Prop_node(q, i); //get q at i^th depth
10.    //Compute states satisfying q for depths ≤i
11.    T(X) = All_sat_circuit_cofactor(q^i ∧H(X), X, U);
12.    if (T(X)==0) return (NULL,0); //gfp is zero
13. } while (SAT_Solve(H(X)∧¬T(X))==SAT)); //gfp?
14. return {T(X),i)}; //states satisfying G(q)
```

Fig. 10.12: Greatest fixed-point computation using circuit cofactoring

Again, note that unlike a standard symbolic algorithm where a pre-image computed in the previous step is used to compute the next pre-image, we use the conjunction of unrolled q^i (q at i^{th} unroll depth) with *H(X)* to represent the target states (satisfying $\boldsymbol{G^i(q)}$) as shown in Figure 10.13. This allows efficient handling of a nested ***G(q)*** without modifying the overall algorithm (discussed in the next section). The greatest fixed-point check is done by satisfiability check of *H(X)\T(X)* (line 13). Note that if no state satisfies ***G(q)***, then it will be discovered at line 12 (an empty *T(X)* corresponds to zero enumeration steps in *All_sat_circuit_cofactor*). For a top-level (i.e. unnested) property ***G(q)***, an additional check for containment of the initial state(s) in the resulting set *T(X)* is required.

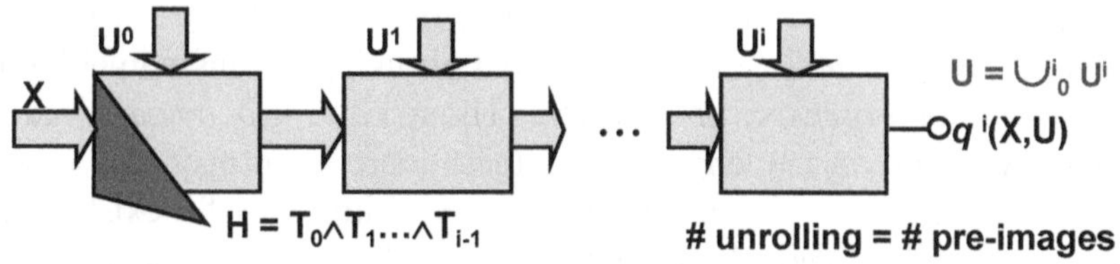

Fig. 10.13: Greatest fixed-point computation using unrolling

10.6.4 SAT-based UMC for *F(p∧G(q))*

The customized translation of the property ***F****(p∧****G****(q))* (in Section 5.5.3) uses two bounds: *N* for finding a state where *p=true*, and *M* for finding a witness loop where *q=true* in each state (and where the fairness constraints are also satisfied) as shown in Figure 10.14. In the previous section, we provided the completeness bound for ***G****(q)*, i.e., *M* as the *gfp bound m*. Here, we provide the completeness bound *n* (for *N*), referred as the *lfp bound,* using a least fixed-point computation. As *m* is the *gfp bound* for the nested ***G****(q)*, states satisfying $\boldsymbol{G}^m(q)$ are equivalent to states satisfying ***G****(q)*. Note, that we do not represent $\boldsymbol{G}^m(q)$ directly as a set of states. Instead, we unroll the transition relation *m* times and impose the constraint *q=true* on each time frame. In our experience, SAT-based UMC does not work well with state set representations for nested properties, but works better on unrolled transition relations.

For obtaining the *lfp bound n*, we can simply use the algorithm *Fixed_point_EF_circuit_cofactor* in Figure 10.9 on $p \wedge \boldsymbol{G}^m(q)$ (instead of on *p*). If the least-fixed point computation completes, then we can check for containment of the initial state(s) in the resulting state set. Alternately, the *lfp bound n* and the *gfp bound m* together give the completeness bounds for BMC on the property ***F****(p∧****G****(q))*. In other words, for our customized formulation of property ***F****(p∧****G****(q)),* i.e., in the procedure *BMC_solve_FG* (see Figure 5.14), we first set *N=n*. This searches for a path prefix (called *stem*) of length at most *n* to a state where *p=true*, and from which there exists a finite path (essentially, a loop) of length *m* where *q=true* in each state, as shown in Figure 10.14. The search for such a path is done efficiently in the procedure *BMC_solve_G* (Figure 5.13) using the following steps 1-3:

1. Set *M=m*,
2. Skip the expensive loop checks (lines 14-33),
3. Interpret *ABORT* result (line 35) as *true.*

If no such *stem* of length $\leq n$ exists, ***F****(p∧****G****(q))* is proved to have no witness.

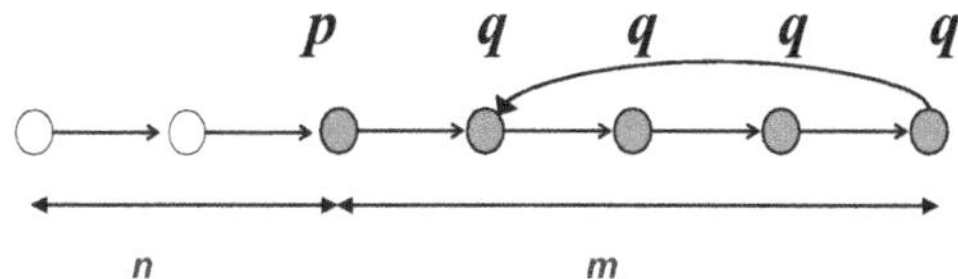

Fig. 10.14: Customized translation for $F(p \wedge G(q))$

10.7 Experiments for Safety Properties

We experimented on a workstation with 2.8 GHz Xeon Processors with 4GB running Red Hat Linux 7.2. We report our experiences on several large industry designs and publicly available benchmarks. We compare Circuit Cofactoring approach (CC) (*Fixed_point_EF_circuit_cofactor* procedure using various heuristics H1-H5) with a Blocking Clauses (BC) approach [71] (*Fixed_point_EF* in Figure 2.10 with redrawing of implication graph), BDD-based model checking (VIS) [166], and a SAT-based BMC (using induction [66, 67]). In the sequel, we use H1 to denote CC approach with heuristic H1 (similarly for other heuristics).

10.7.1 Industry Benchmarks

For our experiments, we chose five industry designs, each with a safety property that was very hard to verify. These designs were abstracted to a comparatively smaller model using iterative abstraction techniques [48], (discussed later in Chapter 11). Even after abstraction, we were not able to get a proof or a counter-example using VIS (version: VIS-2.0) [166] and SAT-based BMC. We conducted our experiments on these abstract models D1-D5 with a three hours time limit imposed for each UMC run.

In the first set of experiments, we compared, as showin in Table 10.1, the performance figures (number of enumerations, time and memory used) at all pre-image depths of our approach H1 against BC and VIS.

Table 10.1: Performance summary of SAT-based UMC methods

Design	#FF	#PI	#G	#pre-images		
				VIS	BC	H1
D1	166	168	2462	4	2	11
D2	294	2000	9592	NA	8	13
D3	1119	2199	16015	NA	0	5
D4	1773	1993	15849	0*	1	8
D5	1748	1870	14630	0*	1	8

Columns 2-4 show number of latches (FF), number of primary inputs (PI), and number of 2-input gates (G), respectively. For designs D2 and D3, we have additional external constraints that restrict the state space for checking the safety property. Columns 5-7 show the number of pre-images computed using approaches VIS, BC, and H1 in a 3 hours time limit. We could not use VIS on D2 and D3 as it did not have the capability to handle external constraints. For designs D4 and D5, VIS could not finish even a single pre-image. Compared to BC, our approach is able to compute more pre-images in a given time limit on all examples, showing the effectiveness and potential of such an approach.

In Table 10.2, we compare in detail the performance of approaches BC and H1 for four pre-image computations. Column 1 shows the pre-image depth (d). Column 2 itemizes the different comparison parameters for a given depth: number of enumerations (#E), time taken (in seconds), and memory used (in MB) for the corresponding row. The remaining columns show the results for the different methods, on designs D1-D5 respectively, where the values shown with "*" correspond to those just before time-out. Note that as the properties for designs D1-D3 are state properties we require only one enumeration at depth 0. As we see, the approach BC times out in all designs except D2 during or before the fourth pre-image. Further, the number of enumerations required by BC is several orders of magnitude more than that required by H1. For example, on design D4 at depth 1, only one enumeration is required by H1 as compared to 870 by BC. The time taken and memory used by H1 are also far lower. We believe that other cube-wise enumeration approaches will suffer similarly.

In the second set of experiments, we compare the performance of several heuristics, including H1-H5 discussed earlier, on design D2. We also compare these with a *random-based heuristic Hr*, which picks a satisfying input minterm by assigning random values to the unassigned primary inputs after the first satisfying assignment is found by the SAT solver. The results are shown in Columns 1-9 of Table 10.3.

Column 1 shows the pre-image depth *d* (shown starting from depth 5 for better readability). Column 2 itemizes the different comparison parameters: number of enumerations (#E), time taken (in seconds), and memory used (in MB) for the corresponding row. Columns 3-9 show the results for BC, Hr, and H1-H5 respectively. Compared to BC, which could complete only 8 pre-images, CC method using any heuristic is able to complete at least 11 pre-images in the 3 hours time limit. These heuristics help further in enlarging the states. For example, with H5, we were able to compute 14 pre-images requiring only 282 MB compared to at most 13 pre-images by the rest.

Table 10.2: Performance comparison of UMC methods: CC vs/ BC

		D1		D2		D3		D4		D5	
d	P	BC	H1	BC	H1	BC	H1	BC	H1	BC	H1
0	#E	1	1	1	1	1	1	1.1k	1	932	1
	sec	0	0	0	0	0	0	49	0	36	0
	MB	3	2	3	3	4	4	6	4	6	4
1	#E	10	1	6	1	54.4k*	1	870	1	981	1
	sec	0	0	0	0	>3hr	0	116	0	105	0
	MB	3	3	4	4	30*	5	9	5	8	5
2	#E	24	1	582	4	-	1	27.5k*	1	36.6k*	1
	sec	0	0	12	0	-	0	>3hr	0	>3hr	0
	MB	3	3	5	5	-	7	50*	6	39*	6
3	#E	86.1k*	1	37.9k	7	-	10	-	2	-	2
	sec	>3hr	0	2268	0	-	3	-	1	-	1
	MB	19*	3	9	5	-	12	-	8	-	8
4	#E	-	92	7.7k	19	-	69	-	4	-	3
	sec	-	0	3080	0	-	73	-	3	-	3
	MB	-	3	11	7	-	48	-	10	-	10

In the third set of experiments, we used additional reachability constraints [73] to get a fixed-point for design D2 by using the following 3 steps:

- We abstracted the design by removing the external constraints. The design with no constraints has 66 FFs, 92 PIs and 1434 gates.
- Using BDD-based forward reachability on this abstract design we get a set of over-approximated reachable states R^+(in less than 4 sec).
- We use R^+, represented as a circuit, as a care set (discussed in Section 10.6.2) in line 12 of Figure 10.9 (or line 10 in Figure 2.10) to obtain the pre-image set.

We show the results for fixed-point computation in Table 10.3, where Columns 10 and 11 show the results using R^+ as a care set with BC and H5, respectively. CC approach reaches the fixed-point in 13 pre-images (note, #E=0 at 14th step) in less than a minute, while BC times out at the 9th step. Since the initial state was not contained in any of the pre-images up to 14 steps (checked using a SAT-based BMC in less than a second), this guarantees the correctness of the property.

Table 10.3: Comparison of heuristics in CC on D2 and proof

d	P	BC	Hr	H1	H2	H3	H4	H5	BC+R$^+$	H5+R$^+$
	#E	3.0k	14	20	14	22	17	21	3.0k	1
5	sec	3459	2	2	3	2	4	3	2348	0
	MB	13	8	9	8	9	9	8	9	4
	#E	758	32	30	28	24	27	24	758	1
6	sec	3571	10	9	11	9	12	10	2432	0
	MB	14	12	12	12	12	11	12	10	4
	#E	831	40	41	24	30	29	33	831	1
7	sec	3721	50	37	37	33	42	38	2549	0
	MB	17	24	21	17	19	18	19	12	4
	#E	9.2k	82	60	58	70	57	56	9.2k	1
8	sec	5853	280	138	159	173	175	140	4267	2
	MB	24	51	35	36	36	37	32	18	5
	#E	13.1k*	156	111	91	64	85	63	18.7k*	1
9	sec	>3hr	1610	621	727	533	662	550	>3hr	4
	MB	34*	107	68	80	56	82	76	30*	7
	#E	-	113	95	88	114	98	81	-	4
10	sec	-	4085	1637	1858	1771	1708	1349	-	8
	MB	-	306	107	121	111	116	108	-	9
	#E	-	58	71	44	47	51	43	-	3
11	sec	-	8873	3731	3946	3690	3194	2890	-	15
	MB	-	319	261	231	191	181	220	-	12
	#E	-	17*	31	27	47	48	29	-	1
12	sec	-	>3hr	5752	6664	6023	5192	4542	-	25
	MB	-	372*	157	414	266	227	221	-	15
	#E	-	-	48	34*	34*	23	32	-	3
13	sec	-	-	10140	>3hr	>3hr	7520	6060	-	36
	MB	-	-	370	265*	263*	375	236	-	18
	#E	-	-	20*	-	-	32*	38	-	0
14	sec	-	-	>3hr	-	-	>3hr	8963	-	58
	MB	-	-	166*	-	-	370*	282	-	27
	#E	-	-	-	-	-	-	13*	-	-
15	sec	-	-	-	-	-	-	>3hr	-	-
	MB	-	-	-	-	-	-	511*	-	-

10.7.2 Public Verification Benchmarks

We experimented on 102 verification problems for checking safety properties on various designs from the VIS verification benchmark suite (VVB) [166], with a time limit of 1000s for each problem. We present the comparison of our approach CC (using H5) with BC, BDD (VIS), and BMC as scatter plots in Figure 10.15. As shown in the scatter plot of CC/BDD in Figure 10.15(a), our approach performed better in 68 cases, while BDD performed better in 16 cases. We also observe the complementary strengths of the BDD-based and SAT-based approaches. As shown in the scatter plots of CC/BC and CC/BMC, in Figures 10.15(b) and (c), respectively, our approach is almost always superior to BC and SAT-based BMC approaches.

We also experimented on a design *swap* [71] that non-deterministically swaps consecutive elements in an array of length *n*. The correctness criterion is that all elements of the array are distinct. We compared our approach CC (using H5) with BC and BDD (VIS) for varying *n*, each with a time limit of 1000s. The performance results are shown in Figure 10.15(d). Note that the BDD approach times out for *n*>8 (also noted in [71]). With our approach CC, we go up to *n*=24, while with BC we could go up to *n*=16. Note that our approach is about an order of magnitude faster than the BC approach on this example.

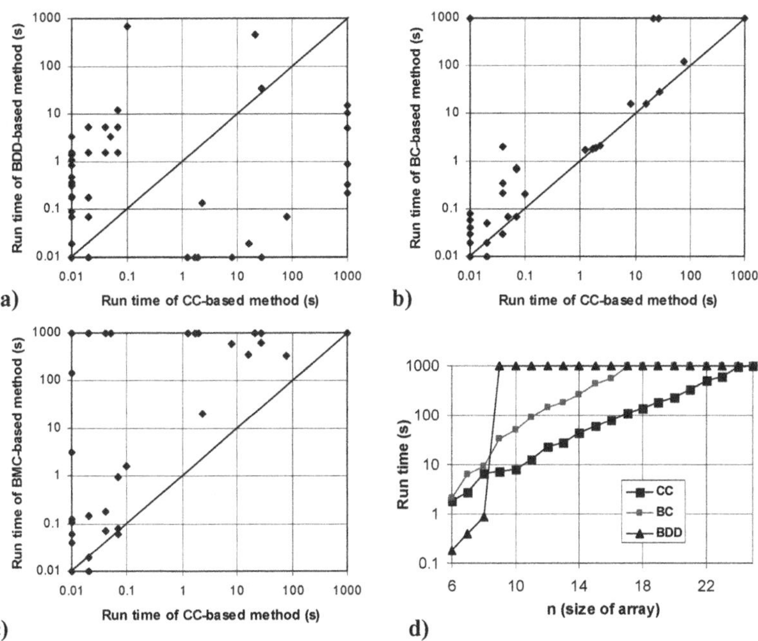

Fig. 10.15: a) CC vs BDD on VVB, b) CC vs BC on VVB, c) CC vs BMC on VVB, d) CC vs BC vs BDD on swap

10.8 Experiments for Liveness Properties

We used SAT-based UMC techniques using circuit cofactoring to find completeness bounds – greatest fixed-point and least fixed-point – for checking liveness properties of the form $F(p \wedge G(q))$. We compare our results with bounds obtained using longest loop-free path analysis [119] for the sub-formula ***G(q)***. We used a time limit of 1800s for each run. The results are shown in Table 10.4 (Note, 'TO' ≡ timeout, 'NA' ≡ Not Applicable.) Column 1 shows the designs and the number in the bracket denotes the property; Columns 2 and 3 show the completeness bound obtained for the sub-formula ***G(q)*** using the longest loop-free path analysis with standard monolithic translations and the corresponding time taken, respectively. Bound *k* with '*' indicates that the analysis timed out after that depth. The 0 value denotes that there is no state from which there is an infinite path where *q* holds always, where the number in the bracket indicates the number of unrollings required to discover it. The remaining columns show the results for deriving bounds related to customized BMC translations. Columns 4 and 5 show the *gfp bound* as obtained by the greatest fixed-point algorithm described in Figure 10.12 and the time taken, respectively. The 0 value has a similar meaning as in Column 2. Columns 6 and 7 show the *lfp bound* for ***F****(p∧**G**(q))* computed using the algorithm described in Section 10.6.4 and the time taken, respectively. We do not compute *lfp bound* when the corresponding *gfp bound* is 0 (indicated by NA). It is not surprising that the loop-free analysis needs to search much deeper than the *gfp bound* analysis. However, note that deriving completeness bounds is much more efficient (requiring less resources) for customized partitioned translations for ***F****(p∧**G**(q)* than the corresponding monolithic standard BMC translations.

Table 10.4: Completeness Bounds for ***F****(p∧**G**(q)*

Design	Loop-free ***G***(.)		Fixed-point ***G***(.)		Fixed-point ***F***(.)	
	k	Sec	gfp	sec	Lfp	sec
Feistal	0(8)	<1	0(8)	<1	NA	NA
Miim(1)	172*	TO	2	<1	5	<1
Miim(2)	169*	TO	2	<1	5	<1
Coherence(1)	56*	TO	11	56	10	1169
Coherence(2)	55*	TO	13	146	11	1128
Palu (1)	95*	TO	3	<1	5	<1
Palu (2)	93*	TO	3	<1	5	<1
Pathfinder	9	< 1	2	<1	9	<1
s1269 (1)	0(7)	<1	0 (7)	<1	NA	NA
Smult	134*	TO	1	<1	3	<1
Unidec	0(12)	<1	0(12)	6	NA	NA
VsaR	112*	TO	79	262	7	34

Next, we used the *gfp* and *lfp bounds* as completeness bounds to find violations or proofs for the properties using BMC. We omitted those properties for which the *gfp bound* is 0, since that corresponds to a proof of correctness. We compare the customized translation *BMC_solve_FG* (Figure 5.14) with the standard monolithic translation. For our customized translation, we set *N* to the *lfp bound* and *M* to the *gfp bound,* and for the monolithic translation we set *k* to *lfp+gfp bound.*

We present the results in Table 10.5. Column 1 shows the designs (with non-zero *gfp bound*). Columns 2-5 show the results using our customized translation for BMC; specifically, Columns 2-3 indicate the existence (*Y*) of counter-example (*CEX*) or proof (*N*) and the corresponding time taken, respectively; Columns 4-5 show the *CEX* length (if any) and the corresponding time taken to find the *CEX* loop, respectively. Columns 6-7 show the results for standard translation for BMC in VIS; specifically, Column 6 shows the *CEX* length or proof (*N*) and Column 7 shows the corresponding time taken. Note that, in the standard translation, computation of existence of *CEX* is not separated from finding the actual *CEX.* (Note, '?' denotes result is undetermined.) Clearly, our translation is able to use completeness bound much more efficiently to find violation/proofs than the standard translation. Also, as shown in the *Miim* examples, it is easier to detect the existence of a loop without actually finding one.

Table 10.5: BMC with Completeness Bounds

Design	Custom (N=lfp, M=gfp)				Standard (k=lfp+gfp)	
	CEX?	sec	CEX	sec	CEX	sec
Miim (1)	Y	<1	?	TO	?	TO
Miim (2)	Y	<1	?	TO	?	TO
Coherence (1)	N	5	NA	NA	N	288
Coherence (2)	N	10	NA	NA	N	479
Palu (1)	Y	<1	1	<1	1	<1
Palu (1)	Y	<1	1	<1	1	<1
Pathfinder	Y	<1	2	<1	2	<1
Smult	Y	<1	4	<1	4	<1
VsaR	N	7	NA	NA	N	665

10.9 Related Work

Symbolic model checking based on combining non-canonical decision diagrams like Reduced Boolean Circuits (RBCs) [178] and Boolean Expression Diagrams (BEDs) [159] with SAT-solvers were proposed to alleviate the problems seen in pure BDD-based approaches. However, the use of circuit-based existential quantification in pre-image/image computation results in a formula size that grows exponentially with the number of quantified variables, in spite of all heuristic attempts to mitigate the growth. This approach is, therefore, limited to designs with a very small number of inputs.

The use of SAT-based quantifier elimination through enumeration of SAT solutions has been the focus of symbolic model checking algorithms proposed in [68-71].In these approaches, the transition relation is maintained in a conjunctive normal form (CNF) and a SAT procedure is used to enumerate all state cube solutions. Similar to traditional model checking, the number of pre-image computations required is bounded by the diameter of the state space, and the method provides a guarantee of correctness when the property is true.

In [69, 71], a blocking clause representing the negation of the enumerated state cube is added at each step in order to prevent repeating the solutions. In [71] a redrawing of the implication graph is carried out to enlarge the state cube. In [69], a two-level minimizer is used to compact the CNF formula after addition of new blocking clauses. Note that in both approaches, at any enumeration step only a single state cube is captured. Since the number of required enumerations is bounded below by the size of a two-level prime and irredundant cover of the entire solution state set, quantifier elimination based on cube-by-cube enumeration tends to be expensive.

In a slightly different approach [70], SAT-based quantifier elimination is achieved using a PODEM-based ATPG solver, which performs decisions only on input variables. The approach uses a *satisfying cut-set* to prune the search space for new solutions and uses a BDD representation for the enumerated solutions. Though this approach reduces the number of back-tracks due to efficient pruning, it still has to enumerate all state cubes and is prone to the BDD explosion problem. In yet another approach [121], an ATPG solver is used as the search engine, and state cube enlargement is achieved using a separate justification procedure once a satisfying result is found. This approach is also limited by its cube-wise enumeration strategy.

In [68] the transition relation is expressed as CNF while the enumerated states are represented as one or more BDDs. The image computation is performed by invoking BDD-based quantification on the CNF formula at intermediate points in the SAT decision tree. Bounding against the already enumerated BDDs provides additional pruning of the SAT search as well as a means to detect the fixed-point. While this method does try to enumerate more than one cube in every SAT enumeration, it is based on BDDs, while our method is not.

A different model checking approach based on use of SAT techniques and *Craig interpolants* has been proposed by McMillan [112]. Given an unsatisfiable Boolean problem, and a proof of unsatisfiability derived by a SAT solver, a *Craig interpolant* can be efficiently computed to characterize the interface between two partitions of the Boolean problem. In particular, when no counterexample exists for depth k, i.e., the SAT problem in BMC for depth k is found to be unsatisfiable, a Craig interpolant is used to obtain an over-approximation of the set of states reachable from the initial state in 1 step (or any fixed number of steps). This provides an approximate image operator, which can be used iteratively to compute an over-approximation of the set of reachable states, i.e., till a fixpoint is obtained. If at any point, the over-approximate reachable set is found to violate the given property, then the depth k is increased for BMC, till either a true counterexample is found, or the over-approximation converges without violating the property. The main advantage of the interpolant-based method is that it does not require an enumeration of satisfying assignments by the SAT solver. Indeed, the proof of unsatisfiability is used to efficiently compute the interpolant, which serves directly as an over-approximated state set. However, this approach can computes approximate reachable states, which are sufficient for proving safety properties, but are harder to apply for exact model checking.

10.10 Summary

BDD-based model checking tools do not scale well with design complexity and size. SAT-based BMC tools provide faster counter-example checking, but a proof may require unrolling of the transition relation up to the longest loop-free path. Previous approaches to SAT-based unbounded model checking suffer from large time and space requirements for solution enumeration, and hence are not viable. We discussed SAT-based quantifier elimination using a circuit cofactoring approach that captures a larger set of solution states compared to a cube-wise enumeration approach. Circuit cofactoring method typically requires several orders of magnitude fewer SAT solver enumerations compared to a cube-wise enumeration. Moreover, using an efficient circuit representation for states and a hybrid SAT-solver, our approach is less dependent on variable ordering and scales well with design size and complexity. Using our method, we were able to show orders of magnitude performance improvement over the cube enumeration methods. On a public benchmark suite, our method performs better in more number of cases than BDD-based and blocking clause-based approaches. Moreover, we were able to prove the correctness of a safety property on an industry design with our approach, when none of the other approaches were able to do so.

We discussed how customized translations allow efficient derivation and use of completeness bounds. We also discussed formulations for determining completeness bounds using SAT-based UMC for checking liveness properties, rather than using loop-free path analysis. These comprise greatest fixed-point and least fixed-point computations in a way that allows efficient handling of nested properties. We showed the effectiveness of our approach on several public benchmarks and industry designs.

10.11 Notes

The subject material described in this chapter are based on the authors' previous works [47] © 2004 IEEE, and [46] © ACM 2005.

PART IV: ABSTRACTION/REFINEMENT

In Part IV (Chapter 11), we discuss various techniques for generating an abstract model that preserve the property correctness, including techniques specialized for design with embedded memories.

In Chapter 11, PROOF-BASED ITERATIVE ABSTRACTION (PBIA, $PBIA^+$, $PBIA^+$ with EMM)), we describe an iterative resolution-based proof technique using SAT-based BMC to generate an abstract model that preserves the property up to a bounded depth. We also describe use of lazy constraints and other SAT heuristics ($PBIA^+$) for further reducing the size of the abstract model and for improving verification of the abstract model. This technique when applied iteratively on the abstract model obtained in the previous step further reduces the abstract model size. Later, we describe techniques ($PBIA^+$ with EMM) to combine proof-based iterative abstraction with EMM techniques to identify irrelevant memory and ports.

11 PROOF-BASED ITERATIVE ABSTRACTION

11.1 Introduction

As defined in Wikipedia [193], *"Abstraction is the process of reducing the information content of a concept, or of an observation of an empirical phenomenon, typically in order to retain only information which is relevant for a particular purpose. Abstraction typically results in complexity reduction leading to a simpler conceptualization of a domain in order to facilitate processing or understanding of many specific scenarios in a generic way."*

Obtaining appropriate abstract models that are small and suitable for applying proof methods have been the subject of research [75, 86] for quite some time. Here we discuss proof-based techniques for generating abstract models that preserve the property validity for a bounded depth, including techniques specialized for design with embedded memories.

A proof-based abstraction technique [48, 49, 77, 78] for SAT-based BMC works as follows: BMC is performed for increasing depths on the concrete design (or a previously abstracted model). When there is no counterexample of (or up to) a given depth to the given property, an *unsatisfiable core* is identified for this depth, and a gate-based [78] or a latch-based [48] abstraction is used to generate an abstract model. This abstract model is *conservative*, i.e., it has more behaviors than the concrete design. Furthermore, due to the sufficiency property of the *unsatisfiable core*, the abstract model — also referred to as a *sufficient abstract model* — is guaranteed to preserve validity of the property up to the analyzed depth. In many practical cases, for such an abstract mode, the property can also be proved *valid for all depths*, using proof techniques discussed in Part III. Since the abstract model has more behaviors than the concrete design, this also implies validity of the property on the concrete design. Note that the abstract model obtained using proof-based abstraction technique is property-specific, i.e., each property may lead to a different abstraction.

Here we describe a latch-based abstraction [48] and several heuristics to reduce the size of ensuing abstract model that are guaranteed to *not have* a counterexample of (or up to) that depth. We describe how our resolution-based proof analysis techniques can be adapted to work with the hybrid SAT solver used in our BMC framework (see Chapters 3-5). Unlike the techniques that are based on CNF-based SAT solvers [77, 78], we describe a technique that is based on hybrid SAT solver and uses circuit information to generate "good" abstract models. Recall, a hybrid SAT solver uses a hybrid representations of Boolean constraints, e.g. where a circuit netlist is used to represent the original circuit problem, and CNF is used to represent the learned constraints.

We also describe the idea of using *lazy constraints* [49], in order to delay propagating the effect of values implied by such constraints. In a standard DPLL-based SAT solver [35], the BCP (Boolean Constraint Propagation) procedure treats all unit-literal clauses as *eager* constraints, i.e., implications due to these constraints are performed as soon as possible (modulo some ordering). Rather than modify the SAT solver to handle such constraints, we modify the constraint clauses in CNF into another equivalent CNF form to achieve the desired lazy effect. This allows us to directly use the latest improvements in SAT solver technology, without any modification to SAT search process. We also describe how lazy constraints can be used automatically in SAT-based BMC for handling certain kinds of constraints. In particular, we use them for handling initial state constraints for latches, and for handling environmental constraints provided by the designer. Intuitively, lazy initial state constraints provide a way to get away from the "irrelevant" initial state values, potentially leading to a smaller invariant abstract model. Similarly, lazy constraints provide a way to delay enforcing "irrelevant" environmental constraints, potentially leading to a smaller set being actually used in the proof of unsatisfiability. In this sense, use of lazy constraints can be regarded as a heuristic targeted at deriving a smaller abstract model for performing unbounded model checking. Similarly, we identify a *sufficient* set of environmental constraints for a given property, in order to generate smaller abstract models.

We also discuss how we combine BMC with Efficient Memory Modeling (EMM) technique (see Chapters 5, 8) with proof-based abstraction (PBA) technique to identify irrelevant memories and ports, in order to obtain a smaller abstract model.

Our overall verification procedure comprises a top-down *iterative abstraction* framework. Starting from the concrete design, we apply SAT-based BMC with proof analysis on a *seed (abstract) model* obtained in the previous iteration to further reduce the abstract model size, preserving the validity of the property. By iterating SAT checking on the *unsatisfiable core*, the number of original clauses/variables needed for unsatisfiability can be reduced significantly [104]. Also, the constraints constituting the unsatisfiable core are related to the particular conflict clauses learned by a SAT solver. These, in turn, depend upon other heuristics in the SAT solver, e.g. decision heuristics, heuristics for choosing a cutset during conflict analysis [103]. We vary these heuristics in different runs of the SAT solver, in order to obtain a potentially smaller unsatisfiable core. Subsequent iterative steps have shown to be effective and faster due to smaller input model at each step as compared to the previous step. In practice, we stop the iteration when size of the seed model converges, i.e., when there is no further reduction. Under certain practical conditions, we allow a refinement step, which can potentially increase the size of the seed model.

Alternatively, one can obtain an *insufficient core* by removing clauses/variables heuristically from the *unsatisfiable core* so as to keep the size of the core small. Note that, in general, the removal of clauses/variables may not preserve the unsatisfiability of the core. The abstract model obtained from such insufficient core is not guaranteed to preserve the validity of the property for the bounded depth. We refer to the abstract model so obtained as *insufficient model.* Such *insufficient models*, due to their small size, are suitable for applying proof-methods. Note, a proof of correctness on any of these models guarantees correctness on the concrete design, while a counterexample may require a refinement, or going back to a previous iteration in our iterative flow.

The main advantage of the iterative framework is that it is targeted at reducing the size of the seed models across successive iterations. The potential benefit is that for properties that are false, BMC search for deeper counterexamples is performed on successively smaller models, thereby increasing the likelihood of finding them. For properties that are true, the successive iterations help to reduce the size of the abstract models, thereby increasing the likelihood of completing the proof by unbounded verification methods.

We have implemented these techniques in our verification framework, *VeriSol* [53], and report our experiences on some large industry designs. For some of the examples, we obtained proofs which had been elusive so far. For other cases, we could perform deeper BMC searches for counterexamples. We observed that iterative abstraction typically gives an order of magnitude reduction in the final model sizes. For many examples, this reduction was crucial in enabling the successful application of the unbounded verification methods.

Outline

The rest of the chapter is organized as follows: We give a brief overview of proof-based abstraction in Section 11.2; present latch-based abstraction algorithm in Section 11.3; discuss improvements in latch-based abstraction in Section 11.4, discuss generation of abstract models in Section 11.5; describe techniques to further improve the abstract model using lazy constraints in Section 11.6; present the iterative abstraction framework in Section 11.7; discuss various applications of proof-based abstraction in Section 11.8; discuss combining proof-based abstraction with embedded memory modeling techniques in Section 11.9; present various experimental results in Sections 11.10-11.12; discuss briefly the related work in Section 11.13 and summarize in Section 11.14.

11.2 Proof-Based Abstraction (PBA): Overview

We describe our proof-based abstraction (PBA) technique for SAT-based BMC in the procedure *BMC_Solve_F_PBA*, as shown in Figure 11.1 (highlighted lines 11-13). When the SAT problem at line 10 is unsatisfiable, i.e., there is no counterexample for the safety property at a given depth k, the *unsatisfiable core R(k)* (reasons for depth k) is obtained using the procedure *SAT_get_refutation* in line 11. This procedure simply retraces the resolution-based proof tree used by the hybrid SAT solver and identifies a subset formula that is sufficient for unsatisfiability (see Chapter 4, Section 4.3.1). We use latch-based abstraction technique to obtain an abstract model from the *R(k)* that results in smaller size abstract model. Rather than optimize at the level of each gate in the original design, our *latch-based abstraction* technique is targeted to minimize the set of latch reasons LR^k at depth k, using the procedure *Get_latch_reasons* (line 12), while still retaining the useful property that there is no counterexample of depth k. An abstract model is then generated for depth k by converting those latches in the given design that are *not in the set* LR^k to pseudo-primary inputs[2]. Due to the sufficiency property of *R(k)*, the resulting abstract model is guaranteed to preserve correctness of the property up to depth k. Depending on locality of the property, the set LR^k can be significantly smaller than the total latches in the given design. Using some heuristic in the procedure *Is_model_stable* (line 13), we choose to abort the abstraction steps at some suitable depth d.

[2] Pseudo-primary inputs refer to output signals of those latches whose next state logic is removed and the signals behaves as primary inputs.

Specifically, a depth d ($<N$) can be chosen such that the size of the set LR^d does not increase over a certain number of consecutive depths, called *stability depth*. In many cases, the property can be proved correct on the abstract model generated at depth d and hence, for the given design. We apply PBA techniques iteratively, called *iterative abstraction* [48], to further reduce the set LR^d and hence, obtain a smaller abstract model.

```
Synopsis:     Customized BMC with PBA for F(p)
Input:        prop_tree_node p, bound N, Initial state I
Output:       TRUE/FALSE/ABORT
Procedure:    BMC_solve_F_PBA

C = I; k=0;
while (k<N) {  //L1 is active always
  p^k = Prop_node(p,k);
  C = C ∧ EMM_Constraints(k);//update constraints
  if (Is_sat(C ∧ p^k)) return TRUE; //wit found
  R(k) = SAT_get_refutation(); //get proof of UNSAT
  LR^k = LR^(k-1) ∪ Get_latch_reasons(R(k));
  if (Is_model_stable(LR^k)) ABORT("Model Stable");
  C = C ∧ ¬p^k; //L3 Learning
  Merge(p^k, const_0); //Simplify
  for (j=k;j>=0;j--) {//loop-free check
    C = C ∧ ¬_jL_k;
    if (!Is_sat(C)) return FALSE; //no wit exists
  }
  k++; //increment depth
}
ABORT("Bound Reached"); //wit not found
```

Fig. 11.1: Customized BMC with PBA for ***F(p)***

11.3 Latch-based Abstraction

We use the following notation to describe our latch-based abstraction, which we also refer to as *latch interface abstraction*. For a node v in the unrolled design, let *F(v)* denote the node corresponding to it in the transition relation of the design. (Note, we use the term *transition relation* to denote the entire combinational logic of the design, including next-state logic for the latches, as well as output logic for the external constraints, either due to the property, or enforced by the designers.) For a given node e in the unrolled design, let *Ext(e)* denote the (possibly empty) set of external constraints imposed on node e. For a given latch L in the transition relation of the design, let *IF(k,L)* denote the set of its *latch interface constraints* in the unrolled design up to depth *k*. The set *IF(k,L)* consists of constraints corresponding to equality of the latch output at time frame *i*, with the latch input at time frame *i-1*, for $1 \le i \le k$. Note that this set includes the initial state constraint. Note also that any *k*-depth unrolling of the design would necessarily include these constraints for each latch, either explicitly, or implicitly, in the problem representation. The abstraction procedure *Get_latch_reasons()* (line 12, Figure 11.1) consists of following steps:

1. Given *unsatisfiable core R(k)*, we first mark a node *v* in the unrolled design, if variable *v* appears in some constraint in *R(k)*. For each such *v*, we say that node *v* is *marked*.

2. For each marked node *e*, such that some constraint in *Ext(e)* belongs to *R(k)*, we perform a backward circuit traversal starting from *e*, through only marked nodes. Note that the recursive traversal on marked nodes is terminated at any unmarked node, but is otherwise continued through the fanin nodes. Any marked node which is visited during such a traversal is called *visited*.

3. For each latch *L*, we say that *L* is *visited* if any of the nodes denoting its output at time frame *i*, $0 \le i \le k$, is visited.

4. We extract the combinational fanin cones of all latches that are visited. This set of latch reasons at depth *k* are denoted as LR^k in the Figure 11.1. We also extract combinational fanin cones of all nodes *F(e)* such that *Ext(e)* is not empty. These fanin cones represent the transition relation of our abstract model. In particular, all latches that are not visited are abstracted away as pseudo-primary inputs (PPI).

The resulting abstract model, as shown in Figure 11.2, is called a *sufficient model* for depth *k*, denoted *SM(k)*. Since it is generated by abstracting away some latches as pseudo-primary inputs, it is known to be conservative for LTL properties [12, 86, 194], i.e., truth of a property on the abstract model guarantees its truth on the given design. It has an additional useful property, stated in the following theorem.

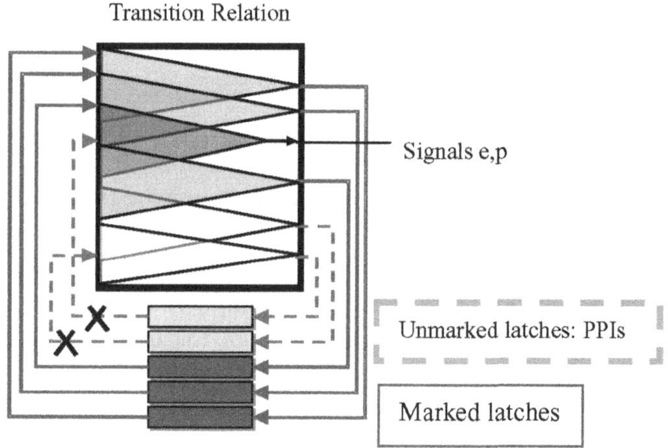

Fig. 11.2: Latch-based abstraction

Theorem 11.1: The sufficient model *SM*(*k*) generated using the latch interface abstraction does not have any counterexample of depth *k*.

Proof: Recall that constraints in *R(k)* are sufficient to generate implications (without taking any decisions), which lead to a conflict at some node in the unrolled design, corresponding to the final conflict graph. The latch interface abstraction uses circuit connectivity information to prune away those constraints in *R(k)* which are not needed to obtain the final conflict. Consider a marked node that does not have a transitive fanout path, through other marked nodes, to a marked node with an external constraint in *R(k). We claim that such a node is not needed to obtain the final conflict.* This is because implications only on inputs of a circuit node cannot cause a final conflict on the output of that node. The same reasoning can be used to show that there can be no final conflict on any of its marked but unvisited transitive fanouts. Therefore, at the end of Step 2, the set of visited nodes and their associated constraints are guaranteed to lead to a final conflict.

Due to the structure of the unrolled design, all transitive paths connecting visited nodes in different time frames have to go through latch interfaces between those time frames. Therefore, each visited node is contained in the combinational fanin cone of some visited latch output node, or some visited node on which an external constraint is in *R(k)*. The abstract model *SM(k)* includes all such combinational fanin cones in its transition relation (Step 4). Therefore, all visited nodes and their constraints are included in a *k*-depth unrolling of the abstract model. Therefore, we are guaranteed to get a final conflict without any decisions, thereby proving that the abstract model *SM(k)* cannot have any counterexample of depth *k*. □

We also use an *alternative abstraction*, where we skip the backward circuit traversal in Step 2 (in the procedure *Get_latch_reasons*) altogether. We consider a latch *L* to be marked, if any of its output nodes in any time frame is marked (not visited), i.e., if any constraint in *IF(k, L)* belongs to *R(k)*. In this case, the abstract model consists of the combinational fanin cones of all marked latches, and all external constraint nodes. The reasoning in our proof works also for this cheaper (to compute) abstraction. Indeed, the proof of sufficiency works, while pruning *R(k)*, for any subset *S* of latches, such that $\{L/\ visited(L)\} \subseteq S \subseteq \{L/\ marked(L)\}$. In the remainder of this paper, we denote the abstract model corresponding to any such set *S* as *SM(k),* since it is guaranteed to not have a counterexample of depth *k*.

11.4 Pruning in Latch Interface Abstraction

The unsatisfiable core *R(k)* can include a node in the unrolled design, on which constraints are not needed to generate the final conflict. The pruning obtained by the latch interface abstraction in Step 2 is geared at throwing away these nodes, without losing sufficiency for the unsatisfiability.

A small example of how this can happen is shown in Figure 11.3. Part (a) of this figure shows the implication graph at the time of learning the conflict clause *C1: (a'+b)*, and the associated antecedents. In Part (b), we show a final conflict graph, where implications from an external constraint on node *e* imply variable *a* to *1*, which leads to use of the conflict clause to imply *b* to *1*. This further leads to *c* and *d* being implied to *1*, with the implication on *d* leading to a final conflict on node *v*. Given this final conflict graph, the recursive marking procedure for *R(k)* starts by including antecedents for implications from the circuit clauses (from *e* to *a*, from *b* to *c*, … etc.). Furthermore, it substitutes the antecedents of the conflict clause, leading to *R(k)* as shown in the figure.

Since there is no transitive fanout path from *b* to *e,* as noted in the proof of Theorem 11.1, our abstraction prunes away the constraints associated with *b*. Note that implication values existing only on inputs of node *b* cannot lead to a conflict on *b* and therefore, it does not matter what value *b* takes. Observe from the circuit shown at the top of Part (b), that when the gate corresponding to variable *b* is removed, the implication from *a* to *d* can still be used to obtain the final conflict. Indeed, if the SAT solver had used this implication directly, the conflict clause *C1* may never have been used at all. However, in general, we cannot rely on the SAT solver to use the implication from *a* to *d*, instead of the transitive implications from *a* to *b*, and *b* to *d*. In case *d* is far away from *a*, say through a chain of buffers, it may actually be faster for the SAT solver to use the learned clause *C1*, than not to use it (which is how conflict clauses help to improve SAT solver performance). The reason that constraints involving *b* even appear in *R(k)* is that the required values on *c* and *d*, needed for the final conflict, are used to imply a consistent value on their fanout variable *b*.

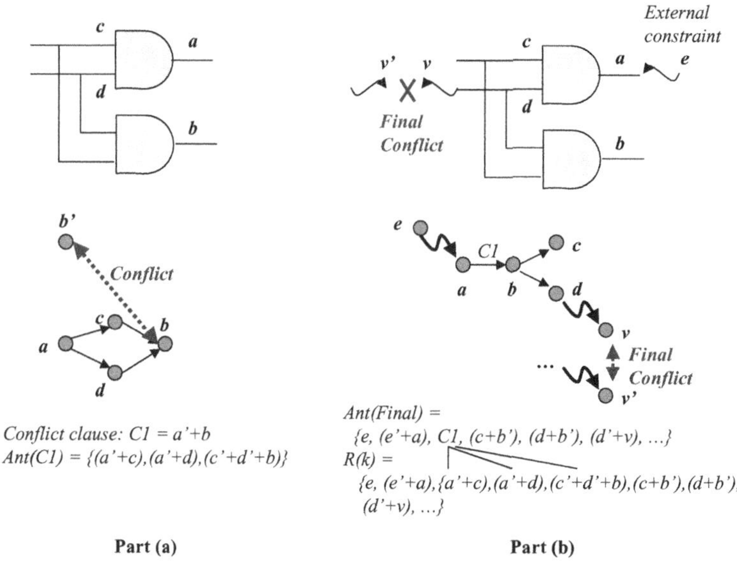

Fig. 11.3: Example for pruning the set *R(k)*

11.4.1 Environmental Constraints

In industry practice, a design is verified under environmental constraints, typically provided by a designer or a verification expert. Typically, these constraints are specified *en masse* at the level of each block, with little guidance about which constraints matter for which property, thereby resulting in large over-constrained models when verifying a single property. This is an important practical issue, especially with the growing popularity of assertion-based design methodology [85].

We address this problem by using the unsatisfiable core *R(k)* to identify a set of *sufficient* environmental constraints for a given property. Technically, this is a straight-forward application of proof-based abstraction. Recall from Step 2 of *Get_latch_reasons*, that we mark all signals *e* on which environmental constraints *Env(e)* are imposed, and extract their combinational fanin cones to add to the abstract model. In cases where environmental constraints are imposed directly on latch outputs, we also include those latches in the abstract model. However, if we modify Step 2 to mark only those signals *e* on which some environmental constraint *Env(e)* belongs to *R(k)*, then we can obtain a potentially smaller abstract model. Note that verification for the resulting abstract model is *conservative*, i.e. correctness on the abstract model with respect to the set of *sufficient* environmental constraints implies correctness for the original model with respect to the entire set of environmental constraints. Using the same principle, one can modify Step 2 to mark only those signals *p* on which some property constraint belongs to *R(k)*.

11.4.2 Latch Interface Propagation Constraints

We discuss some lightweight and effective SAT heuristics that target reducing the involvement of latch interface propagation constraints in implications during SAT search, in order to potentially reduce the number of latches for which interface constraints belong to *R(k)*, thereby leading to smaller abstract models.

We first define *PI logic* as gates in the transitive fanout of the primary inputs (PI), exclusively driven by the primary inputs. As shown in the Figure 11.4, the gate *x*, input of gate *a,* is a part of the PI logic, while gate *a* is not. Based on this, we discuss the following two heuristics:

PI Decision Heuristic

In this heuristic, SAT search is guided towards the gates at the boundary of the *PI logic,* such as gate *a* in Figure 11.4, and then reducing the relative scores of their inputs (gate *y*) that are not driven by PI logic. Recall, the SAT decision engine picks next decision variable based on their relative scores.

PI Implication Heuristic

During BCP, the deduction engine processes the implication queue using either LIFO (Last-In-First-Out) or FIFO (First-In-First-Out) order. In this heuristic, we use a priority queue where we give high priority to the implications on variables in the *PI logic.*

Note that both heuristics are targeted at giving a marginal preference to the *PI Logic*, rather than an overwhelming preference, since it can crucially affect SAT solver performance.

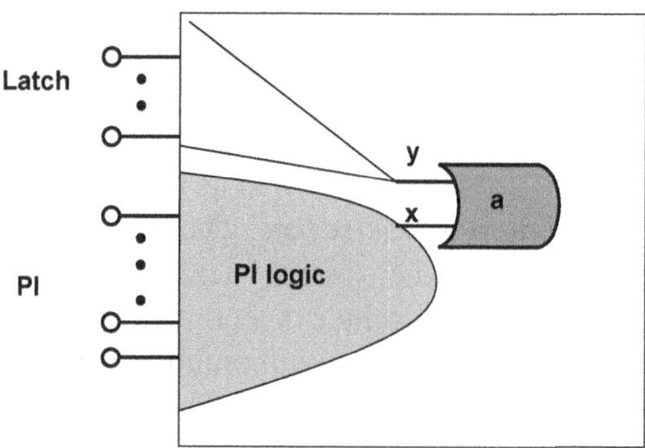

Fig. 11.4: Example of PI logic

11.5 Abstract Models

Using the latch-based abstraction, we obtain the following two types of abstract models that will be useful in various verification tasks: *accumulated sufficient* and *insufficient* abstract models.

Accumulated Sufficient Abstract Models

When using BMC on increasing depth *k*, it is useful to identify the *accumulated* unsatisfiable core for depth *k*, denoted *AR(k)*. It can also be identified from a single unsatisfiable BMC problem, which checks the existence of a counterexample of any depth *up to k*.

We use an abstraction similar to that defined in the previous section, where the accumulated unsatisfiable core *AR(k)* is used in place of *R(k)*. The resulting model is called an *accumulated sufficient model* for depth *k*, denoted *ASM(k)*. Following a similar reasoning as in the proof of Theorem 11.1, it can be shown that the model *ASM(k)* does not have any counterexample of depth less than or equal to *k*.

Insufficient Abstract Models

The main purpose of generating abstract models is to enable use of proof methods. Typically, such proof methods do not work well on large models. Therefore, if the abstract models resulting from the entire *unsatisfiable core* are too large, we may not be able to apply these methods. This is typically the case for many industry designs in our experiments, especially when *k* gets large.

The latch interface abstraction already includes some pruning of the set *R(k)* (or *AR(k)*), which is guaranteed to retain the unsatisfiability at (or up to) depth *k*. It is also possible to arbitrarily pick any subset of visited latches required by the latch interface abstraction. The choice can be dictated by heuristic criteria such as — at what depth was its output node visited, at how many depths was its output node visited, etc. The abstract model derived by retaining some, but not all, of the visited latches is called an *insufficient model.* It is not guaranteed to exclude a counterexample of any length. However, it can potentially exclude many in practice. The important point is that it is still conservative for verification of LTL properties. In comparison to models derived from *localization reduction* [86], which is based on a static cone of influence analysis, an insufficient model based on a semantic proof analysis may better capture the needed invariant for all depths.

11.6 Improving Abstraction using Lazy Constraints

For an *unsatisfiable core R(k)*, consider the following partition of the set of latches in the given design:

- *Propagation latches*: for which at least one interface propagation constraint belongs to *R(k)*
- *Initial value latches:* for which only an initial state constraint belongs to *R(k)*
- *PPI latches:* for which neither the initial state constraint, nor any of the interface propagation constraints belongs to *R(k)*.

Clearly, the set of *PPI latches* can be abstracted away, since they are not used at all in the proof of unsatisfiability – this was done in Step 4 of the procedure *Get_latch_reasons*. On the other hand, a *propagation latch* needs to be retained in the abstract model, since it was used to propagate a latch constraint across time frames for the derived proof. The more interesting case is presented by an *initial value latch.* It is quite possible that an *initial value latch* is not really needed to derive unsatisfiability – its initial state constraint may just *happen* to be used by the SAT solver. We observed empirically that on large designs, a significant fraction (as high as 20% in some examples) of the marked latches are *initial value latches.* Rather than add these latches to an abstract model, our strategy is to guide the SAT solver to find a proof that would not use their initial state values *unless needed.* This is done by use of *lazy constraints* as described in the following.

11.6.1 Making Eager Constraints Lazy

Recall that *BMC(M,f,k)* includes initial state constrains *I*, which are typically added to the Boolean formula as 1-literal clauses on variables representing the symbolic initial state. In a standard DPLL-based SAT solver [35], implications due to 1-literal clauses are performed *eagerly* during pre-processing, i.e. as soon as possible, before any decisions are taken in the SAT search. Furthermore, these implications are recorded, and potentially used by the SAT solver for conflict analysis later, which may lead to these constraints belonging to the *unsatisfiable core R(k)*. Our effort is targeted at depriving the SAT solver of immediate implications from some (or all) of the 1-literal clauses representing initial state constraints.

A naïve way of delaying implications due to initial state constraints is to mark the associated variables, and delay BCP on these marked variables during pre-processing. However, this would involve an overhead of checking for such variables during BCP. Rather than change the standard BCP procedure, we achieve the desired effect by changing the CNF representation of these constraints. This allows us to continue to exploit the latest improvements in SAT solver technology without modifying the SAT solvers.

Definition 11.1: A 1-literal eager constraint *(x)* is converted to a *lazy constraint* by replacing it with *(x+y)(x+ -y)*, where *y* is a new Boolean variable not appearing in the remaining formula.

Note that converting an eager constraint to a lazy constraint according to Definition 11.1 does not affect the satisfiability of the formula. This is because a value *1* implied on *x* satisfies both replacement clauses, while a value *0* implied on *x* due to other constraints causes an immediate conflict. The use of a new variable *y* for each lazy constraint typically avoids *y* getting chosen as a decision variable. The general idea of delaying implications by adding auxiliary variables can be applied to multi-literal clauses as well.

In general, delaying implications by using lazy constraints may result in a performance penalty, since the efficiency of a SAT solver depends upon performing more implications in order to avoid search. However, note that the initial state constraints typically form a tiny fraction (<1%) of the entire BMC SAT problem. Furthermore, we are willing to pay some performance penalty, *provided* there is significant reduction in the size of the derived abstract models. In this sense, we regard the use of lazy constraints as a *heuristic* for deriving smaller abstract models. Its effectiveness in practice is shown by our experimental results, discussed in Section 11.11.

11.7 Iterative Abstraction Framework

Our verification methodology is centered on an iterative abstraction framework, based on the use of BMC with Proof Analysis and the related abstractions in the inner loop. The overall flow is shown in Figure 11.5.

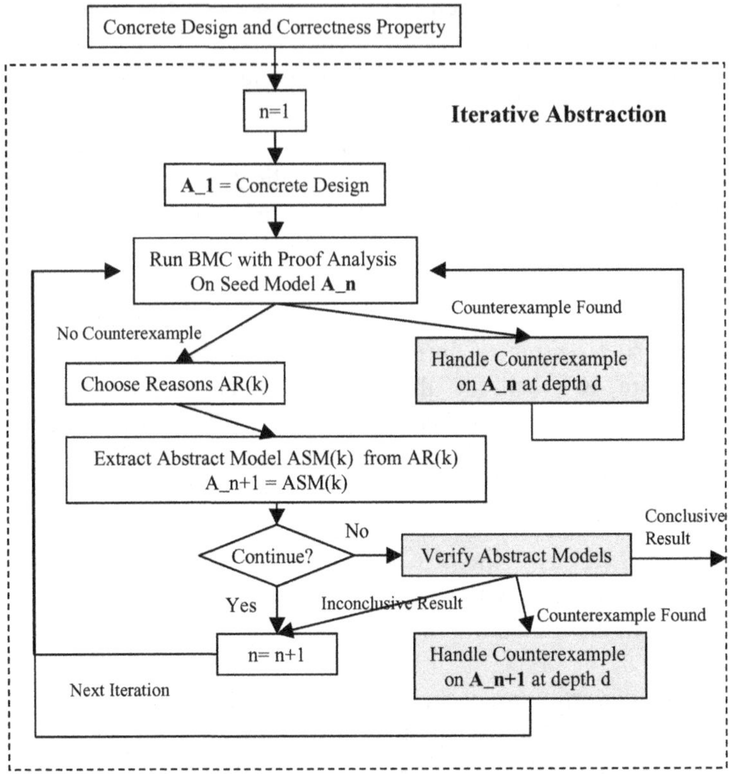

Fig. 11.5: Iterative abstraction framework

11.7.1 Inner Loop of the Framework

Each iteration of our framework, indexed by *n* as shown in the Figure 11.5, consists of applying SAT-based BMC with proof Analysis on a given *seed* model A_n. The seed model for the initial iteration ($n=1$) is the concrete design. At each iteration, we run BMC with proof analysis (Figure 11.1) up to some fixed depth (potentially different for each iteration). The proof analysis technique is used to identify the *unsatisfiable cores* for each depth *k* when there is no counterexample. If a counterexample is found at some depth *d*, it is handled as described in the next section. The result of such handling is that we may obtain a new seed model A_n' potentially larger than A_n, and we repeat the current iteration.

On the other hand, if no counterexample is found by BMC, we heuristically choose one of the sets *AR(k)* at some depth *k*. For example, we can choose a set that remains unchanged for a certain number of time frames. Such a depth is referred as *stability depth*. Then we use any abstraction technique which is guaranteed to exclude all counterexamples of depth less than or equal to *k*, in order to generate the corresponding accumulated sufficient model *ASM(k)*. The *ASM(k)* model is used as the seed model A_{n+1} for the next iteration. If we decide not to continue, e.g., if the seed model A_{n+1} is small enough, or if model A_{n+1} is unchanged from model A_n, we attempt to verify the abstract models generated in this iteration. The result of such verification is that we can get a counterexample (handled as described next), or a conclusive result (we can stop), or an inconclusive result. In the last case, we try to reduce the size of the seed model by performing another iteration.

11.7.2 Handling Counterexamples

Our scheme for handling counterexamples is shown in Figure 11.6. Given a counterexample on a model A_n at depth *d*, we first check if model A_n is the concrete design. If it is, then we have found a true counterexample. However, for *n>1*, the counterexample could be spurious, since it was obtained not on the concrete design, but on an abstract model.

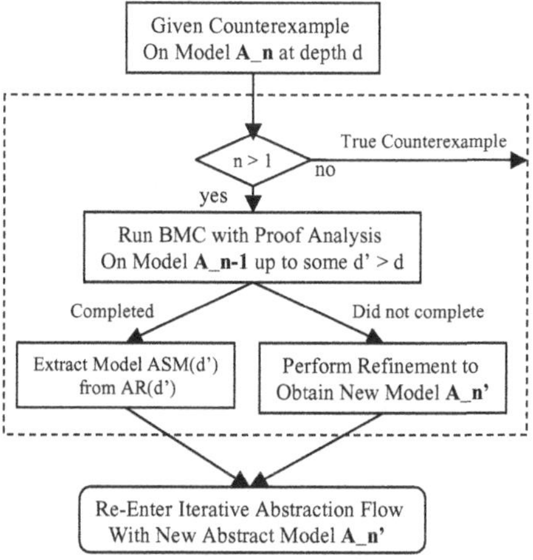

Fig. 11.6: Handling counterexamples

In case the counterexample is spurious, we run a deeper search on seed model A_{n-1} from the previous iteration, in order to choose *AR(d')*, at some depth $d' \geq d$. Note that, in practice, it may not always be possible to run BMC deeper than *d*. If we can, we extract the corresponding model *ASM(d')*. Otherwise, we perform a resolution-based refinement to remove the counterexample in case it is spurious [75]. After refinement, we re-enter the iterative abstraction flow with the new model A_n'. This model is guaranteed to exclude at least the given counterexample, but is potentially larger than the model A_n.

We describe our resolution-based refinement procedure as follows. It is instructive to recall that the set *R(k)* is sufficient, but not necessary, for unsatisfiability. As described earlier, the latch interface abstraction prunes the given sufficient set *R(k)* further, to yield another sufficient set. Rather than choosing all latch nodes at the failure interface marked by *R(k)*, we can use the latch nodes in the pruned sufficient set as refinement candidates for counter-example guided abstraction refinement (CEGAR) [75]. This can potentially reduce the number of candidates.

11.7.3 Lazy Constraints in Iterative Framework

We discuss the following methods for use of *lazy constraints* automatically in the iterative framework. Recall, *lazy constraints* refers to translation of unit-literal constraint clauses into two two-literal clauses, whose resolution gives the unit-literal clause.

Lazy PPI Constraint Approach (LPPIC)

Initial value latches are excluded from the abstract model for the next iteration in iterative abstraction. However, in order to ensure correctness up to depth *k*, lazy constraints on the initial values of the corresponding PPIs are added to the BMC formulas in the next iteration.

Lazy Latch Constraint Approach (LLC)

In this approach, lazy constraints are used for representing initial state values for *all* latches in a given model. We distinguish the first iteration of an iterative abstraction framework (n=1, in Figure 11.5), where BMC is typically applied on a large concrete design, from higher numbered iterations i.e., n>1 where BMC is applied on smaller abstract models. On this basis, we can choose to apply lazy latch constraints in:

a) no iterations,
b) in (n>1) iterations, or

c) in all ($n \geq 1$) iterations.

Lazy Environment Constraint Approach (LEC)

Lazy constraints can also be used to model environmental constraints *Env(e)* that are specified as 1-literal constraints. In practice, we use it mainly for final verification since using a partial constraint set during iterative abstraction may result in spurious counterexamples, requiring refinement steps, as discussed in previous section.

11.8 Application of Proof-based Iterative Abstraction

We describe various ways of using abstract models obtained during the iterative abstraction procedure to improve various verification tasks such as falsification and obtaining proofs of correctness. We perform verification on the seed models (*ASM(d)* models), in order to derive benefits of iterative abstraction. However, in practice, if these models are too large, we use either the *SM(k)* models, or the insufficient abstract models, derived from any depth k checked by BMC.

Falsification: Improved BMC search

In many examples, BMC can complete deeper searches on the smaller abstract models than on the larger concrete design. Due to the sufficiency property of the abstractions, the seed model at each iteration can have counterexamples only at depths strictly greater than the depth d, from which it was generated in the previous iteration. In other words, if there is no counterexample up to depth d on the abstract model, then it is guaranteed that there is no counterexample up to depth d in the concrete design either. Furthermore, the seed model is no bigger than that of the previous iteration, provided there are no refinements (which we use only when we cannot do a deeper search with BMC). The combined effect is that for properties that are false, BMC search for deeper counterexamples is performed on successively smaller models, thereby increasing the likelihood of finding them.

Proof using Unbounded Model Checking

For properties that are true, the successive iterations help to reduce the size of the abstract models, thereby increasing the likelihood of completing the proof by unbounded verification methods. We use proof methods such as BDD-based [12, 17] or SAT-based on the abstract models. Due to the limited capacity of such methods, they are more likely to work on smaller abstract models. Note, if the correctness property is proved true, it is guaranteed to be true on the concrete design as well.

Proof by Induction

If the pruned set of latch interface constraints does not include any constraint due to initial state of a latch, then it represents an inductive invariant. Note that although initial state constraints are enforced in the BMC SAT problem at depth k, if the pruned set of sufficient constraints does not contain any, then this constitutes a proof of unsatisfiability when starting from an arbitrary initial state. Using lazy constraints for *all initial states latches*, one may also quickly find an "invariant" abstract model which does not rely upon the "irrelevant" initial state values. This corresponds to an inductive step in a proof by induction with increasing depth [67].

Generating Reachability Constraints for Proofs

We use BDD-based symbolic traversal techniques to perform a reachability analysis on the abstract model [12, 17]. The computed reachable set corresponds to an over-approximate reachable set for the concrete design. These are used as additional *reachability constraints* during proofs by induction or SAT-based Unbounded Model Checking. In practice, the reachability constraint generated from an abstract model obtained using proof-based methods was found to be a stronger invariant compared to that obtained using a static approach such as *localization reduction* [86].

11.9 EMM with Proof-based Abstraction

As discussed earlier, EMM can significantly reduce the size of the verification model for system designs with multiple memories and multiple read and write ports. However, for checking the correctness of a given safety property, we may not require all the memory modules or the ports. To further reduce the model, we can abstract out *irrelevant* memory modules or ports completely. In this case, we do not need to add the memory modeling constraints for the irrelevant memory modules or ports, thereby further reducing the size of the verification model.

For the purpose of automatically identifying irrelevant memory modules and ports, we combine our EMM approach for BMC with PBA [51] as shown in Figure 11.1 (line 9). This can not only reduce the non-memory logic (from the *Main* module, Figure 7.1) but also identify the memory modules and ports that are not required for proving correctness up to a given bounded depth of BMC analysis.

The dependency of the property on any memory module for a given depth k is determined easily by checking whether a latch corresponding to the control logic for that memory module (the logic driving the memory interface signals) is in the set LR^k. If no such latch exists in the set LR^k, we do not add the EMM modeling constraints for that memory module. In other words, we abstract out that memory module completely. We perform a similar abstraction for each memory port. This reduces the problem size and improves the performance, as observed in our experiments reported in the next section.

11.10 Experimental Results of Latch-based Abstraction

We have implemented the iterative abstraction framework in our SAT-based model checking platform *VeriSol* [53]. For our experiments here, we chose the designs with safety properties, ranging in size up to 416k gates and 12k flip-flops in the static cone of influence of the related property. These properties could not be falsified using BMC search engine, or proved using proof techniques with and without static abstraction such as localization [86]. All experiments were performed on a 2.2 GHz Dual Xeon processor machine, with 4 GB memory, running Linux 7.2.

11.10.1 Results for Iterative Abstraction

The results for use of iterative abstraction are summarized in Table 11.1. The size of the concrete design is listed in Row 2 in terms of number of flip-flops (#FF), and number of gates (#Gates). The results for the different iterations are shown in the remaining rows, where for each iteration, we report the size of the abstract model (#FF, number of flip-flops), the depth at which it was derived (k), and the total CPU time taken by BMC with proof analysis to check up to that depth (T(s), in seconds).

Typically, we used a 3-hour time limit for each iteration consisting of a BMC run on a given seed model. Within each iteration, we used either the last depth completed by BMC, or *stability depth* equal to 10, from which to generate the seed model for the next iteration. For these experiments, we used the cheaper latch interface abstraction, which skips the backward circuit traversal. In addition, we iterated over the inner loop (Figure 11.5) until the size of the seed model converged.

Note that the first iteration was quite successful in generating small abstract models. For most designs, we obtained a magnitude of order reduction, in comparison to the size of the concrete design. Typically, the first iteration was also the most expensive in CPU time. Next, note that for some designs, we can clearly see a reduction across the iterations also. In particular, for the design D1, iterative abstraction allowed the size of the abstract model to be reduced from 1269 (Iteration 1) to 113 (Iteration 7) flip-flops. Though other designs did not exhibit the same level of reduction, we did manage to reduce their sizes as well. This effect is related to that observed by others, i.e. when proof analysis techniques are applied iteratively, the final *unsatisfiable core* can be much smaller than the original problem [104]. Though we did not apply the iterative technique to the *unsatisfiable core* at each depth, one can apply such successive iterations at a given depth.

Table 11.1: Results for iterative abstraction

Concrete Design			D1	D2	D3	D4	D5	D6
		#FF / #Gates	12.7k / 416.1k	4.2k / 37.8k	5.2k / 46.4k	910 / 18k	4.2k / 37.8k	3.6k / 155k
Abstract Models Generated by Iterative Abstraction	Iteration 1	#FF	1269	523	1530	476	330	105
		k	63	47	30	80	43	15
		T(s)	32815	10043	10515	6274	882	1786
	Iteration 2	#FF	541	451	1468	420	303	103
		k	63	56	28	78	34	15
		T(s)	486	2540	2763	3250	27	7
	Iteration 3	#FF	439	445	1434	405		
		k	63	41	29	80		
		T(s)	140	4163	4354	8765		
	Iteration 4	#FF	259	444	1406	397		
		k	63	43	27	85		
		T(s)	75	2293	6812	9588		
	Iteration 5	#FF	212		1356	396		
		k	63		28	78		
		T(s)	34		8650	2660		
	Iteration 6	#FF	118					
		k	61					
		T(s)	20					
	Iteration 7	#FF	113					
		k	60					
		T(s)	5					

11.10.2 Results for Verification of Abstract Models

After we performed iterative abstraction, we applied falsification and proof methods on the generated abstract models. These results are summarized in Table 11.2, and are discussed in more detail in this section.

For each design (Row 1), we ran BMC in falsification mode (procedure *BMC_solve_F*, Figure 5.12) with a 3-hour time limit, and these results are shown in the first set of rows. In Rows 2-4, we report the size of the concrete design (#FF / #Gates), the maximum depth for which BMC search was completed (Depth), and the total CPU time taken for searching all depths up to the maximum (T(s), in seconds). Note that our BMC engine is able to search fairly deep even for large designs. However, we were unable to find a counterexample for any of these designs.

Table 11.2: Results for verification on abstract models

Examples			D1	D2	D3	D4	D5	D6
Concrete Design	Falsify (BMC)	#FF / #Gates	12.7k / 416.1k	4.2k / 37.8k	5.2k / 46.4k	910 / 18k	4.2k / 37.8k	3.6k / 155k
		Depth	96	64	32	89	82	307
		T (s)	10230	7519	8667	9760	3968	3099 **
Abstract Model Proof		#FF / #Gates	113 / 1.5k	451 / 14.5k	1356 / 20.8k	396 / 6k	303 / 12.4k	103 / 17.3k
		Depth	1012	115	30	96	211	3034
		T (s)	10788	7129	7513	10134	10603	2635 **
	Proof	Proof?	Yes	No	No	No	Yes	Yes
		T (s)	40				2 *	2738
		Proof Method	BDD-based MC				IND+ REACH	IND+ REACH

Note:*BDD analysis performed on a different abstract model with 40 FF, derived from depth 4.

**mem-out within 3 hour time limt

In the next set of Rows 5-7, we show results for basic BMC on an abstract seed model generated during iterative abstraction. Again, we report the size of the abstract model (#FF / #Gates), the maximum depth searched by BMC (Depth), and the total CPU time taken to search up to that depth (T(s), in seconds). For all designs except D3, we were able to search deeper on the abstract models than on the concrete designs. For some, there was an increase by an order of magnitude in the maximum depth searched. This is due to an improvement in the SAT checking time on smaller problems, and the ability to unroll the model deeper with bounded memory resources. Since no counterexample were still found, these results constituted at least an increased level of confidence in the correctness.

The last set of rows report the results for complete verification of the abstract models. We report the status of the proof methods. Note that we were able to prove the correctness of 3 of the 6 designs. For successful instances, we also report the time taken (T(s), in seconds) and the verification method used.

For design D1, we were able to prove the property correct on the abstract model in 40 seconds, by using standard BDD-based symbolic model checking. For design D5, we were given external constraints by the designers, which needed to be enforced at every cycle. However, these constraints were not enough to help a proof by induction. Therefore, we performed a BDD-based reachability analysis on a much smaller abstract model derived from depth 4, with 40 flip-flops, which took 1 second. The computed reachable state set was used as a *reachability invariant* by the BMC engine (procedure *BMC_solve_F_induction*, Figure 9.2), to successfully perform a proof by induction on the concrete design, in less than 1 second. Similarly, for design D6 also, we performed a reachability analysis on the shown model with 103 flip-flops, taking 2737 seconds. Again, with the BDD-based *reachability invariant*, our BMC engine was able to prove successfully the property on the concrete design in less than 1 second. However, we were not able to conclusively verify the remaining designs D2, D3, and D4.

11.11 Experimental Results using Lazy Constraints

We also conducted experiments to study the effect of lazy constraints in the iterative abstraction framework. For our experiments, we used several large industry designs B1-B15, ranging in size up to 416 k gates and 12 k flip-flops in the cone of influence of safety properties, for which we had not found any counterexamples or proofs.

11.11.1 Results for Use of Lazy Constraints

We performed experiments to compare the effectiveness of lazy constraints for proof-based abstraction, by considering the following four sets of combinations:

- Set 1: No Lazy Constraints are used – this serves as a baseline for comparison.
- Set 2: Lazy PPI Constraints, but no Lazy Latch Constraints (Method *LPPIC* only)

- Set 3: Lazy PPI Constraints + Lazy Latch Constraints (*LLC*) in n>1 iterations (Method *LPPIC* + Method *LLC*(b))
- Set 4: Lazy PPI Constraints + Lazy Latch Constraints (*LLC*) in all iterations n>0 (Method *LPPIC* + Method *LLC*(c))

The results for the first iteration are shown in Table 11.3, and for subsequent iterations are shown in Table 11.4. The design name (D), along with number of flip-flops (#FF) and gates (#G) in the cone-of-influence of the property in the concrete model are shown in Columns 1-3 (Table 11.3), respectively. (Some examples correspond to different properties specified on the same design.) Columns 4-7 compare the number of latches in the *unsatisfiable core* derived at an identical depth in the first iteration (not necessarily the stable model in the first iteration) with all sets applying BMC on the same concrete model. (Note Set 3 is identical to Set 2 for the first iteration). The best reduction (Best %R) is reported as a percentage of the baseline number in Set 1 (Column 4). Note that Set 4 gives significant reduction overall compared to Set 2 and 3, but can be computationally expensive, as observed for the design B4.

In the remaining columns in Table 11.4, we report data for the *final* abstract model using iterative abstraction. For each set, we report the number of flip-flops in the final abstract model (#FF) including those needed for all environmental constraints, the number of iterations needed to achieve convergence (#I), and the total CPU time for all iterations (T, in seconds), including generation of the abstract models. Again, we report the best reduction (Best %R) as a percentage of the baseline size in Set 1 (Column 4).

Note from the results for the *unsatisfiable core*, that on almost all examples, the use of *lazy constraints* provided significant reduction in the number of latches. The average reduction for these 15 examples is 45%. Though the data are not shown here, for 11 of the 15 examples, this reduction is accompanied by a performance improvement. Of the remaining 4 examples, only 1 shows much degraded performance. Overall, we observe that *lazy constraints* can be quite effectiveness in reducing the *unsatisfiable core*, potentially useful in other applications – abstraction refinement [77] and interpolant-based verification [112].

Since our main interest is in the iterative abstraction application, we focus also on the sizes of the final abstract models. Note first that the abstract model sizes reported here include all environmental constraints, which we use for iterative abstraction. (Therefore, the numbers in #FF columns here are higher than those reported for the *unsatisfiable core* in Columns 4-7, Table 11.3) For iterative abstraction, the use of Lazy PPI Constraints alone (Set 2) dramatically reduced the abstract model size by more than 80% in two examples – B4 and B5. By additional use of Lazy Latch Constraints (Sets 3 or 4), significant reductions in size (14% - 62%) were obtained in another six examples. Note that the remaining seven examples also showed reductions, albeit insignificant (less than 10%).

Table 11.3: Abstraction using lazy constraints (first iteration, *i*=1)

D	Concrete Model		Unsatisfiable Core in First Iteration, i =1			
			Set 1	Set 2	Set 4	Best %R
	#FF	# G	#FF	#FF	#FF	
B1	3378	28384	481	480	322	33%
B2	4367	36471	1190	1190	1146	4%
B3	910	13997	507	437	364	28%
B4	12716	416182	404	330	TO*	18%
B5	2714	77220	187	137	3	98%
B6	1635	26059	116	111	17	85%
B7	1635	26084	110	110	23	79%
B8	1670	26729	30	30	19	37%
B9	1670	26729	115	115	22	81%
B10	1635	26064	38	38	16	58%
B11	1670	26729	30	30	19	37%
B12	1670	26729	104	98	75	28%
B13	1670	26729	62	61	52	16%
B14	1635	26085	74	71	15	79%
B15	1635	26060	27	27	27	0%

Among these, we obtained significant performance improvement in four examples (B2, B10, B12, B15), and not much performance penalty in three examples (B8, B9, B11). These data provide support for our default use of lazy constraints in iterative abstraction – we typically obtain significant reduction in abstract model size or a performance improvement, and when we don't, there is not much size or performance penalty. Indeed, of the 8 examples with significant size reduction, 6 did not suffer any performance penalty. In terms of a preference between the different uses of LLC (Set 3 vs. Set 4), the data are somewhat unclear. Our current strategy is to try LLC in all iterations (Set 4). If it fails to perform well, it does so in the first iteration itself, in which case we switch to Set 3.

Table 11.4: Abstraction using lazy constraints (subsequent iteration, *i*>0)

D	Final Abstract Model Generated by Iterative Abstraction												
	No Lazy			With Lazy PPI Constraints, LPPIC									Best %R
	Set 1			No LLC, Set 2			LLC, i>1, Set 3			LLC, i>0, Set 4			
	#FF	# I	T(s)	#FF	# I	T(s)	#FF	# I	T(s)	#FF	# I	T(s)	
B1	522	9	60476	516	9	50754	294	4	11817	294	4	8993	44%
B2	1223	8	80630	1233	5	39573	1119	9	64361	1136	9	70029	9%
B3	433	5	11156	355	9	32520	166	10	29249	196	6	32291	62%
B4	369	4	1099	71	6	1203	71	6	1310	NA	NA	TO*	81%
B5	187	2	17	3	5	22	3	3	21	3	2	17	98%
B6	228	6	5958	225	4	5324	148	3	4102	146	2	7	36%
B7	244	3	3028	240	2	3039	155	5	2768	146	2	85	40%
B8	149	3	25	149	3	28	148	3	28	148	2	41	1%
B9	162	3	40	162	3	43	147	3	44	149	2	43	9%
B10	159	2	12	158	3	29	146	3	30	145	2	6	9%
B11	149	3	25	149	3	28	148	3	28	148	2	40	1%
B12	183	4	2119	182	4	2316	182	4	2376	180	2	653	2%
B13	180	2	63	179	2	68	154	3	71	174	3	61	14%
B14	190	3	1352	192	3	1515	154	5	1480	142	3	10	25%
B15	153	3	125	153	3	149	153	3	142	151	3	73	1%

11.11.2 Proofs on Final Abstract Models

After using iterative abstraction, we obtained final abstract models on which we applied BDD-based model checking. These results are shown in Table 11.5. For each design, we compare verification for the following sets:

- Set A: Abstract model without use of lazy constraints (Set 1 from Table 11.3)
- Set B: Smallest abstract model with use of lazy constraints and all environmental constraints (smallest final model among Sets 2-4 from Table 11.3)
- Set C: Smallest abstract model derived with additional use of a sufficient set of environmental constraints (as described in Section 11.4.1).

For each set, we report the number of flip-flops (#FF) and the number of environmental constraints (#Env) in the final abstract model, whether a proof was obtained (Prf?), and the CPU time taken (T, in seconds, with "TO" denoting a time-out of 3 hours).

Note that without the use of any lazy constraint techniques (Set A), we could not complete verification for any of the derived abstract models, even though they are much smaller than the concrete designs. With use of lazy constraints (Set B), the reduction in abstract model size was crucial in enabling complete verification for B4 and B5, taking less than 30 seconds to prove correctness. Finally, with additional use of sufficient environmental constraints (Set C), we could successfully complete verification for 10 more examples, each taking less than 100 seconds. Note also that the number of sufficient environmental constraints is typically much smaller than the number of original constraints provided by the designers.

Table 11.5: Verification on final abstraction model using lazy constraints

D	Set A: No Lazy				Set B: With Lazy				Set C: Lazy + Suff. Env			
	# FF	# Env	Prf?	T(s)	# FF	# Env	Prf?	T(s)	# FF	# Env	Prf ?	T(s)
B1	522	142	No	TO	294	142	No	TO	163	11	No	TO
B2	1223	142	No	TO	1119	142	No	TO	994	23	No	TO
B3	433	0	No	TO	166	0	No	TO	166	0	No	TO
B4	369	0	D	TO	71	0	Yes	29	71	0	Yes	29
B5	187	0	No	TO	3	0	Yes	1	3	0	Yes	1
B6	228	264	No	TO	146	264	No	TO	17	87	Yes	18
B7	244	264	No	TO	146	264	No	TO	23	93	Yes	26
B8	149	264	No	TO	148	264	No	TO	18	86	Yes	1
B9	162	264	No	TO	147	264	No	TO	20	89	Yes	21
B10	159	264	No	TO	145	264	No	TO	16	87	Yes	4
B11	149	264	No	TO	148	264	No	TO	18	86	Yes	18
B12	183	264	No	TO	180	264	No	TO	76	112	Yes	70
B13	180	264	No	TO	154	264	No	TO	29	91	Yes	98
B14	190	264	No	TO	142	264	No	TO	14	88	Yes	22
B15	153	264	No	TO	151	264	No	TO	28	93	Yes	22

11.12 Case study: EMM with PBIA

In this experiment, we used PBA to identify irrelevant memory modules using EMM for design with multiple memories and ports. We experimented on a hardware implementation of the *quicksort* algorithm (see Chapter 7, Section 7.6.1). Recall, the *array* is implemented as a memory module with address width AW=10 and data width DW=32, with 1 read and 1 write port; and the *stack* (for recursive function calls) is implemented as a memory module with address width AW=10 and data width DW=24, with 1 read and 1 write port. The design has 200 latches (excluding memory registers), 56 inputs, and ~9K 2-input gates. We checked the property P2, stated as "*after return from a recursive call, the program counter should go next to a recursive call on the right partition or return to the parent on the recursion stack.*" The array is allowed to have arbitrary values to begin with, and is modeled using *EMM_arb_init_constraints* (see Chapter 7, Section 7.4.5).

Note that property P2 depends only on the *stack* and the contents of the *array* should not matter at all. We used the PBA technique to discover this automatically. For property P2, we compared performance of EMM with PBA using BMC, with that of PBA on Explicit Modeling using BMC without line 9, Figure 11.1. We used a *stability depth* of 10 to obtain the stable set *LR*. For different array sizes *N*, we compared the performance of EMM and Explicit Modeling approaches. We present the results in Table 11.6. Column 1 shows different array sizes *N*, Columns 2-5 show performance figures for EMM. Specifically, Column 2 shows the number of latches in the reduced model size using EMM with PBA. The value in bracket shows the original number of latches. Column 3 shows the time taken (in sec) for PBA to generate a stable latch set. Columns 4-5 show the time and memory required for EMM to provide the loop-free path proof using BMC (procedure *BMC_solve_F_induction*, Figure 9.2). Columns 6-9 report these performance numbers for the Explicit Modeling using BMC (procedure *BMC_solve_F_induction*, Figure 9.2, without lines 10 and 12).

Note that by use of PBA, the reduced model in Column 2 did not have any latch from the control logic of the memory module representing the array. Therefore, we were able to automatically abstract out the entire array memory module, while doing BMC analysis on the reduced model using EMM. Note that this results in significant improvement in performance, as clear from a comparison of the performance figures of EMM on property P2 in Columns 4 and 5 of Table 7.7 (see Chapter 7) and Table 11.6. Moreover, we see several orders of magnitude performance improvement over the Explicit Modeling, even on the reduced models. Note, for *N*=5 we could not generate a stable latch model in the given time limit for the Explicit Modeling case.

Table 11.6: Results of abstraction using EMM with PBA

N	EMM +PBA		EMM-Proof on Red. Model		Explicit+PBA		Explicit on Red. Model	
	FF (orig)	Sec	Sec	MB	FF (orig)	Sec	Sec	MB
3	91 (167)	10	5	13	293 (37K)	293	2K	274
4	93 (167)	38	145	40	2858 (37K)	2858	10K	456
5	91 (167)	351	2316	116	- (37K)	>3hr	NA	NA

11.13 Related Work

The abstraction techniques discussed here are broadly related to the many efforts in verification that use abstraction and refinement [75, 86]. Counterexample-guided abstraction refinement (CEGAR) approaches [75] have exploited SAT-based techniques to perform the check for spurious counterexamples and to perform proof-based refinement [48, 49, 77, 78]. Most of these efforts are refinement-based approaches, where starting from a small abstract model of the concrete design, counterexamples found on these models are used to refine them iteratively. In practice, many iterations are needed before converging on a model where the proof succeeds. More frequently, the size of the refined abstract model grows monotonically larger, on which unbounded verification methods fail to complete.

The main reason for the popularity of the refinement-based approaches has been a lack of abstraction-based techniques that could extract relevant information from a relatively large concrete design. This changed with the use of proof analysis techniques for SAT solvers and other theorem-provers [195], and use of interpolants that approximate reachable state sets [112]. We too have employed SAT proof analysis techniques effectively, to obtain abstract models in an iterative abstraction framework based on use of BMC in the inner loop. In our approach, a lack of a counterexample provides an opportunity to perform further abstraction. On the other hand, the presence of a counterexample does not necessarily require a refinement based on that particular counterexample.

In comparison to refinement-based approaches, the abstraction-based approach discussed may need to handle much larger models. However, note that we do not require complete verification on these models for the purpose of abstraction. Instead, the presented abstraction method is based on SAT-based BMC, which is an incomplete method *per se*. Furthermore, in practice, the first iteration of our iterative abstraction framework provides a significant reduction in model sizes, such that all successive iterations need to work on smaller models. Our broader goal is to systematically exploit proof analysis with SAT-based BMC wherever possible. Since it is unlikely that a purely refinement-based or a purely abstraction-based approach will work best in practice, we are currently exploring combinations of the two.

In the context of generating proof of unsatisfiability, there have been some efforts at reducing the size of the unsatisfiability core [48, 49, 104]. These efforts can be broadly classified as, *post-processing* and *online processing*. The *post-processing* efforts refer to techniques that analyze and prune [48] the *unsatisfiable core* to reduce the core size, and another technique that uses SAT solve iteratively [48, 104] on the *unsatisfiable core* to obtain a smaller core, until no further core reduction is obtained. *Online processing* efforts refer to techniques that focus on reducing the size of the *unsatisfiable core* through the SAT search by changing its decision heuristics [49]. However, note that for general SAT applications, the standard "size" metric is the number of clauses/variables. In contrast, we use the number of latches (or flip-flops) in the *unsatisfiable core* as the target metric for reduction, even if it might result in a larger number of clauses/variables (although typically it does not). The ideas for reduction of latches can be potentially applied to reduction of specific variables in other SAT applications as well such as in gate-based abstraction [78].

11.14 Summary

We discussed latch-based abstraction using SAT-based proof analysis in an abstraction-refinement flow using BMC. We also discussed how to apply such abstraction in successive iterations, and integrated it with a hybrid SAT solver and use of lazy constraints to further reduce abstract models. This technique can also be combined with EMM to identify irrelevant memories and ports in designs with embedded memories. The small abstract models enable deeper BMC search for counterexamples. They also enable a conclusive proof of the correctness property using BDD-based or SAT-based proofs methods. We are able to show significant model reductions on industry designs, including a 100x reduction in the number of flip-flops, leading to a conclusive proof of correctness for a large production design.

11.15 Notes

The subject material described in this chapter are based on the authors' previous works [48] © ACM 2003, and [49] © 2005 IEEE.

PART V: VERIFICATION PROCEDURE

In Part V (Chapters 12—13), we describe our SAT-based verification framework, combination of strategies for various verification tasks, and a new paradigm to improve further the scalability of current verification framework.

In Chapter 12, SAT-BASED VERIFICATION FRAMEWORK, we discuss the verification platform *VeriSol* that has matured over 4 years (as of 2006) and is being used extensively in several industry settings. Due to an efficient and flexible infrastructure, it provides a very productive verification environment for research and development. We also analyze the inherent strengths/ weaknesses of various verification tasks, and describe their interplay as applied on several industrial case studies, highlighting their contribution at each step of verification.

In Chapter 13, SYNTHESIS FOR VERIFICATION, we describe a Synthesis-for-Verification paradigm (SFV) to generate "verification friendly" models and properties from the given high-level design and specification. We explore the impact of various high-level synthesis (HLS) parameters on the verification efforts, and discuss how HLS can be guided to synthesize "verification aware" models. We also describe a framework for high-level BMC (HMC) and several techniques to accelerate BMC. Such techniques overcome the inherent limitations of Boolean-level BMC, while allowing integration of state-of-the-art techniques that have been very useful for Boolean-level BMC.

12 SAT-BASED VERIFICATION FRAMEWORK

12.1 Introduction

Verifying modern designs requires robust and scalable approaches in order to meet more demanding time-to-market requirements. We present our SAT-based model checking platform *VeriSol* [53] based on robust and scalable algorithms (as discussed in Parts I-IV) that are tightly integrated for verifying large scale industry designs. We briefly discuss and analyze the strengths and weaknesses of various verification engines in *VeriSol* as each addresses the capacity and performance issues inherent in verifying large designs. Using several industry case studies, we describe the interplay of these engines highlighting their contribution at each step of verification. We also discuss the various modeling issues in handling complex features in real designs.

VeriSol has matured over 4 years (as of 2006), and is currently used in the industry by many designers and verification engineers. This verification framework, as shown in Figure 12.1, provides a tightly integrated environment for various state-of-the-art SAT-based verification engines that are specialized for specific tasks like falsification, abstraction, refinement, and deriving proofs. *VeriSol* has been integrated as "C-level Property Checker" in CyberWorkBench (CWB) [55], a high-level design and synthesis environment that automatically generates RTL design and properties from high-level descriptions. Based on this platform, we share our practical experiences and insights in verifying large industry designs. Because of an efficient and flexible infrastructure, it also provides a very productive environment for research and development.

Outline

We first discuss various modeling issues and styles in generating a verification model in Section 12.2. In Section 12.3 and 12.4, we provide an overview of the various verification engines in *VeriSol* and analyze their strengths, respectively. As each engine is specialized for a particular verification task, we provide various case studies in Section 12.5 that discuss the interplay among these engines to achieve the desired verification goal. We summarize our discussion in Section 12.6.

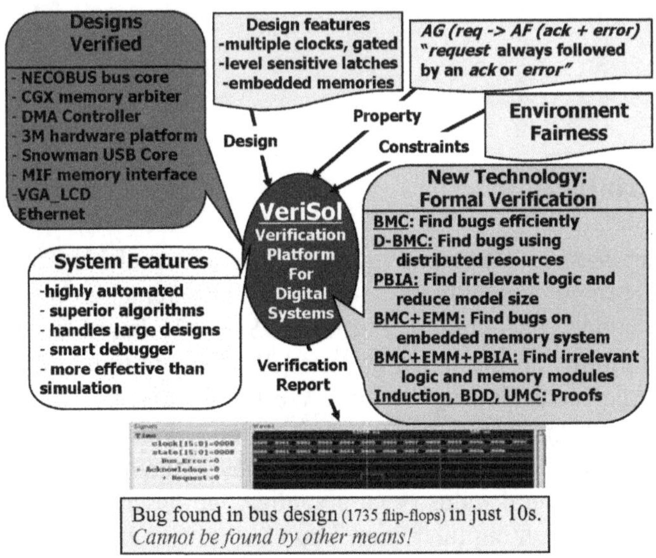

Fig. 12.1: Verification Framework: *VeriSol*

12.2 Verification Model and Properties

We discuss the following three main modeling issues that are handled seamlessly in the *VeriSol* framework: combinational false loops, multiple clocks and phases, and embedded memories. We also briefly discuss the correctness properties needed by the tool.

Combinational false loops

Combinational false loops, in general, are used to allow sharing of hardware resources in area-optimized designs. Since a verification model must not have any combinational loops, we use an algorithm called the *C-Loop*

Breaker that iteratively removes each loop by duplicating logic in the path starting from the inner to the outer most loop, as shown in Figure 12.2. Note that the path $N \to M \to O \to N$ is a false loop as S and $\neg S$ cannot be true at the same time. Though the model size increases, it does not affect SAT-based verification approaches significantly in practice. On the contrary, the resulting disjoint partitioning of the search space is found to be beneficial to SAT solvers in general.

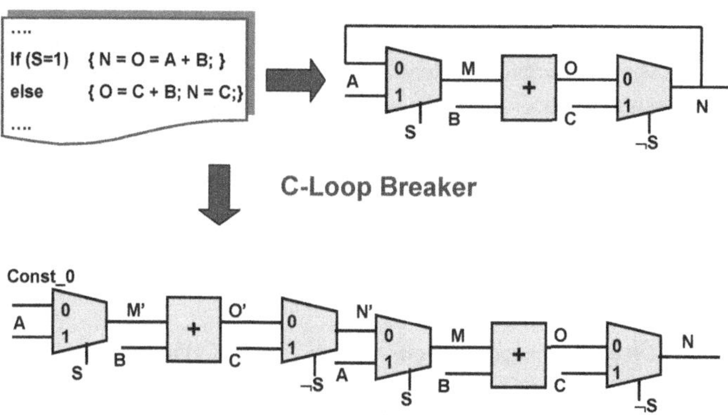

Fig. 12.2: Combinational-Loop Breaker

Modeling Embedded Memory System

For verification of embedded memory designs using EMM, we generate a model without memory modules but with memory-interface signals as input-output signals in the model. These memory interface signal descriptions, as shown in Table 12.1, are provided to the verification tool along with the model.

Table 12.1: Description of memory interface signals

Notation	Description	Type	Values
Mem#	Identify a memory or register file	+<INT>	0, 1,
Addr_Width, Data_Width	Bus width of the signal	N	1, 2,
Read_Port#, Write_Port#	Identify the read (write) port for a given Mem#	+<INT>	0, 1
List of signals	Line separated signals	String	
<init_value>	Init value of memory (uniform over all location)	0,1,x	

Modeling Multi-clock Systems

As discussed in Chapter 8, we adopt a uniform modeling scheme for handling multi-clock systems. Given clock input constraints, we translate a multiple clock system into a single clock system, where update of each state element is guarded by enable signals derived from the input clocks as shown in Figure 12.2. We also generate clocking constraints on the input clocks to avoid redundant checks and to allow dynamic simplification during BMC.

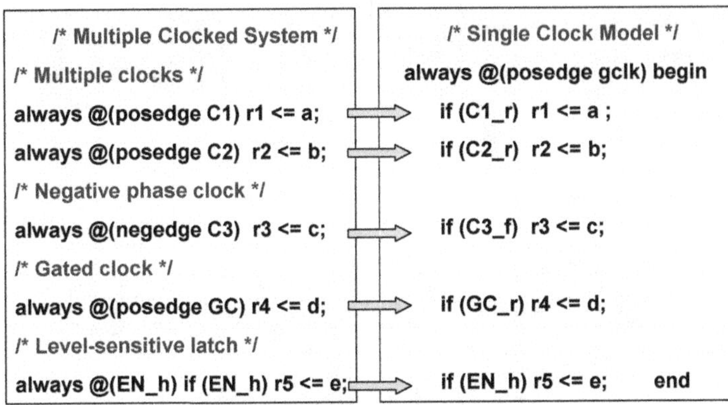

Fig. 12.3: Modeling multi-clock system

Correctness Properties

The checkers or properties required to verify the correctness of the model can be obtained from the following:

- user-specified properties, expressed using formal specification languages like PSL [83]
- common specification primitives, supported by design languages such as System Verilog Assertions [84]
- automatically generated checkers, corresponding to common design errors such as *write before read*, *read before initialization*, *full case*, *parallel case*, *memory access errors*, *memory out-of-bound access*, *unreachable branch*, *synchronization errors* due to clock domain crossing, etc.

12.3 Verification Engines

VeriSol uses *Reduced AIG* representation with on-the-fly simplification algorithms [42, 136], and an incremental hybrid SAT solver [41, 146] that combines the strengths of circuit-based and CNF-based solvers seamlessly.

VeriSol houses the following SAT-based engines, each geared towards verifying large systems: bounded model checking (BMC) [45, 46, 52] and distributed BMC (d-BMC) over a network of workstations (NOW) [152] for falsification, proof-based iterative abstraction (PBIA) for model reduction [48, 49], SAT-based unbounded model checking (UMC) [47] and induction for proofs [67, 73], Efficient Memory Modeling (EMM) [50, 51] and its combination with PBIA in BMC for verifying embedded memory systems with multiple memories (with multiple ports and arbitrary initial state) and to discover irrelevant memories and ports for proving property correctness.

We present the tool as a "wheel of verification engines" in Figure 12.4. All the engines use the common circuit representation and hybrid SAT solver. We use BDDs also, but our applications of BDDs are restricted to lightweight BDD-based learning in SAT solvers (described in Chapter 5), computing inductive invariants such as over-approximate reachable states (described in Chapter 9), obtaining proofs on reduced abstract models generated from proof-based abstractions (described in Chapters 10-11).

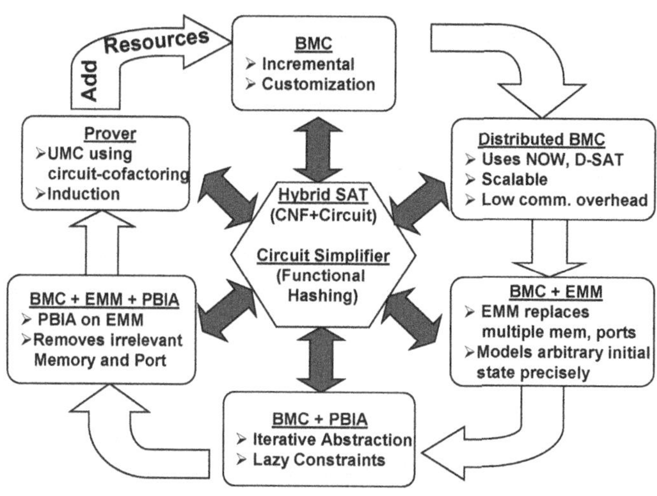

Fig. 12.4: Wheel of verification engine

Efficient Internal Representation

The verification model is represented efficiently as *Reduced AIG*, using an on-the-fly multi-level functional hashing algorithm [42, 136] that detects and removes structural and local redundancies. We use this graph to represent the transition relation, unrolled time frames, and the set of enumerated states. The representation is very effective for Boolean reasoning. (Refer Chapter 3 for more details.)

Hybrid SAT Solver

The hybrid SAT solver consists of a DPLL-based SAT solver that combines the strengths of circuit-based [39] and CNF-based SAT solvers [38] with incremental SAT solving capabilities [154] and BDD learning [146]. The solver uses the deduction and diagnostic engines efficiently on a hybrid Boolean representation, i.e., circuit graph and CNF. The decision engine also benefits from both circuit- and CNF-based heuristics. For satisfiable instances, it can generate a partial satisfying solution, and for unsatisfiable instances, it can generate a refutation proof, targeted for variable reduction (Refer Chapter 4 for further details.)

Customized BMC

Our SAT-based BMC engine uses the simplified circuit graph to represent unrolled time frames and the hybrid SAT solver to falsify the given the property. For commonly occurring properties, both clocked and unclocked, we use *customized translations* of LTL properties that involve partitioning the problem and using incremental model checking [45, 46, 52]. (Refer Chapters 5 and 8 for further details.)

Distributed BMC (d-BMC)

Our d-BMC engine over a network of workstations [152] overcomes the memory limitation of a single server to provide a scalable approach for carrying out deeper search on memory-intensive designs. We achieve:

- *scalability* by not keeping the entire problem data on a single processor, and
- *low communication overhead* by making each process cognizant of the partition topology while communicating,

thereby, reducing the receiving buffer of the process of unwanted information. (Refer Chapter 6 for further details.)

BMC with EMM

Our EMM approach [50, 51] augments BMC to handle embedded memory systems (with multiple read, write ports) without explicitly modeling each memory bit, by capturing the memory data forwarding semantics efficiently using exclusivity constraints. An arbitrary initial state of the memory is modeled precisely using constraints on the new symbolic variables introduced [51]. (Refer Chapter 7 for further details.)

Proof by Induction

We augment our BMC framework to provide induction proofs by adding inductive checks on loop-free paths. We strengthen the proofs using additional reachability constraints as inductive invariants [73], derived from light-weight BDD-based reachability computation on a small abstract model. We also extend induction proofs to BMC augmented with EMM [51]. (Refer Chapter 9 for further details.)

SAT-based Unbounded Model Checking (UMC)

Our UMC approach [46] improves the SAT-based blocking clause approach [71] by several orders of magnitude in performance, by using *circuit-based cofactoring* to capture a larger set of new states per SAT solver enumeration, and representing them efficiently using a simplified circuit graph. The method is combined with inductive invariants, e.g., reachability constraints [73] for faster fixed-point computations. This method is used to derive completeness bounds for both safety and liveness properties, based on fixed-point computations. (Refer Chapter 10 for further details.)

Proof-based Iterative Abstraction (PBIA)

Our PBIA technique [48, 49] generates a *property-preserving* abstract model (up to a certain depth) by:

a) obtaining a set of latch reasons (LR) involved in the unsatisfiability proof of a SAT problem in BMC, and
b) by abstracting away all latches *not in this set* as pseudo-primary inputs.

We further reduce the model size by using the abstraction *iteratively* and using *lazy constraints*. (Refer Chapter 11 for further details.)

EMM with PBIA

We combine the EMM and PBIA techniques [51] to identify fewer memory modules and ports that need to be modeled; thereby reducing the model size, resulting in improved performance. If no latch corresponding to the control logic for a memory module or a port is in the LR set (obtained by PBIA), we do not add the EMM constraints for that memory module or port during BMC. (Refer Chapters 7 and 11 for further details.)

12.4 Verification Engine Analysis

In contrast to typical BDD-based model checking, where various verification tasks are not explicitly instrumented, we use specialized engines in SAT-based model checking for various verification tasks. In the following, we discuss the strengths and weaknesses inherent in these SAT-based engines.

Hybrid SAT Solver: Strengths and Weaknesses

The main advantages are:

- A hybrid SAT solver is efficient at finding a solution with a partial satisfying assignment using *readily available* circuit heuristics.
- A hybrid SAT solver generates resolution proofs targeted for variable reductions.
- The SAT search scales well with problem size as compared to BDDs.
- The SAT search can be made incremental with low overhead, improving the subsequent search significantly. In other words, subsequent SAT searches on overlapping problems can use results from previous SAT search on overlapping problems [154].
- The SAT search performs well on disjoint-partitioned sub-structures.
- SAT technology has seen rapid progress in the last few years, and well-known heuristics exist for CNF-based and circuit-based Solvers. A hybrid SAT combines the strengths of both these solvers.

We list the following inherent weaknesses of SAT solvers:

- SAT solvers are not good at enumerating solutions in general, although there have been some limited successes [47, 68, 71] to address this issue.
- Multiplexors (muxes), in general, are detrimental to DPLL-based SAT solvers [149, 150] as the solvers typically end up explicitly enumerating all possible combinations of the variables corresponding to mux select inputs. Thus, the problems that arise from area-optimized circuits with heavy sharing of registers and functional units can pose serious challenges to SAT solvers.

Bounded Model Checking: Strengths and Weaknesses

The main strengths of BMC are:

- BMC guarantees a shortest counter-example, if there exists one within the bound. Stated differently, the absence of a counter-example for a given bound guarantees that the property is preserved up to that bound.

In some cases, designers can provide a bound with high confidence and use BMC to carry out the exhaustive search up to the bound.

- Successive BMC problem instances overlap, thereby, allowing incremental formulation and exploiting incremental learning in SAT solvers [46, 153, 154].
- The reached states are not stored in SAT-based BMC, unlike in BDD-based model checking [17] where the state explosion problem severely restricts its scalability. In general, SAT-based BMC is less sensitive to the number of state variables compared to BDD-based methods. In practice, BMC has been found to scale much better.
- BMC problem instances provide a natural disjoint partitioning of the overall problem. This can be efficiently exploited to distribute the disjoint partitions of the BMC problem over a network of workstations to overcome memory limitations of a single server [152].

Like other verification techniques, BMC also has some inherent limitations:

- The size of a BMC problem instance grows (at least) linearly with the unrolling bound. For a large bound, we see progressive deterioration in BMC performance. A large design can further impede the performance. However, techniques like on-the-fly circuit simplification [42, 45, 136] during the unrolling of design are used to mitigate this problem.
- The BMC technique is incomplete [66] in practice, as the stopping criteria, also known as *completeness threshold*, is difficult to determine or to achieve in many cases.

Efficient Memory Modeling: Strengths and Weaknesses

The merits of the EMM technique are:

- The EMM technique is sound and complete as it preserves the data forwarding semantics of memory, i.e., data read from a memory location is the same as the last written data at that location. It reduces the verification model size significantly by abstracting away memory state elements, which are exponential in the address width.
- No examination or translation of a design are required to apply the EMM technique with BMC.
- The number of muxes in EMM constraints can be reduced by capturing the exclusivity of a matching read-write pair – this results in improved SAT performance.
- EMM constraints grow lazily with the unroll depth, thereby, avoiding a state explosion due to memory elements. Fewer memory accesses can further reduce the size of EMM constraints.

- It works well with arbitrary as well as uniform memory initialization.

The inherent limitations of the EMM technique are as follows:

- Frequent memory accesses lead to longer search depths as each memory access per port per memory needs one clock cycle in the associated model. Moreover, it also reduces the scope of simplification of EMM constraints.
- Non-uniform memory initialization increases the number of memory cycles and hence, leads to longer search depths.

Proof-based Abstraction: Strengths and Weaknesses

The main strengths of PBA are as follows:

- The PBA technique is able to reduce the model by abstracting out the irrelevant logic, where the reduced model, typically preserves the validity of the property at larger depths than abstraction techniques based on static analysis such as localization reduction [86]. Latch based abstraction approach is specially geared towards reducing model size in terms of latch support. For designs with memories, identification of irrelevant memories and access ports can significantly reduce the complexity of the models.
- The reduced model is often found to have sufficient detail to prove the validity of the property using unbounded model checkers [46-48, 71].
- When the technique is used in an abstraction-refinement framework [196], it reduces the refinement requirement significantly by eliminating all counter-examples up to a bounded depth.
- The reduced abstract model can be used for deeper BMC search.
- The reduced abstract model can also be used for generating reachability invariants, for a proof by induction on the concrete design.

Following are the inherent limitations of PBA:

- As the presence of muxes makes SAT problem difficult i.e., more conflict-learning and backtracks, we hypothesize that the muxes also reduce the effectiveness of the technique by increasing the size of the proof of unsatisfiability and thereby, leading to larger abstract models.
- As PBIA is mainly an abstraction technique, the first iteration of PBA on a concrete model is the most time consuming stage.

12.5 Verification Strategies: Case Studies

Various SAT-based verification tasks have inherent weaknesses as seen previously. It is quite natural to ask how to combine them effectively in order to reap their strengths. We follow two broad interplays of these engines in setting up verification flows for designs with and without embedded memory as shown in Figure 12.5(a-b). For non-embedded memory designs and properties, we first attempt to falsify them using the engines BMC or d-BMC. We use d-BMC typically when we see the memory limitation of a single server causing a bottleneck. If we cannot find a bug, we apply PBIA next, to obtain a small abstract model removing irrelevant logic. On this abstract model, we apply proof techniques such as SAT-based UMC or Induction. If required, we also generate reachability invariants from the smaller abstract models. We follow similar steps for design with embedded memories, by using BMC augmented with EMM.

Using selected case studies from the industry, we demonstrate the role of various engines at each step of the verification flow. Note that without the interplay of the engines we could not have verified any of these designs. The first two case studies use the verification flow shown in Figure 12.5(a) and the next two use that shown in Figure 12.5(b). All experiments were performed on servers with 2.8 GHz Xeon processors with 4GB running Red Hat Linux 7.2. Note, we use solid arrow to denote flow of engines and hollow arrows to denote input to the engines and output result from the engines. Similar shaded blocks indicate engines, while the white blocks denote the input or output to and from the engines.

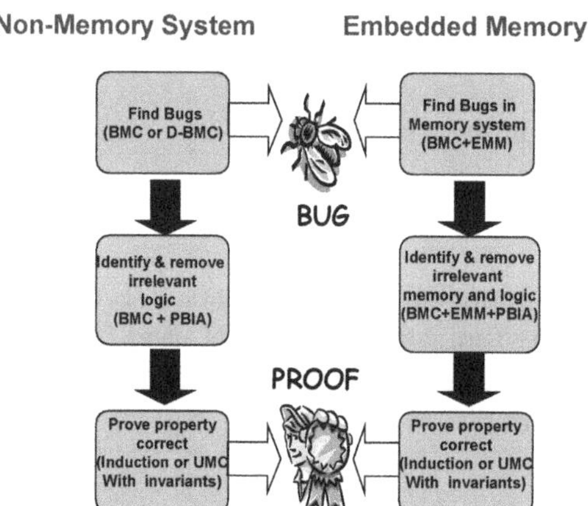

Fig. 12.5: Verification flow without (a) / with (b) embedded memory

Industry Design I

The design has 13K flip-flops (FFs), ~0.5M gates in the cone of influence of a safety property. Verification flow for the design is shown in Figure 12.6. Using BMC, we showed there was no witness up to depth 120 (in 1643s)

before we ran out of memory. As the memory of a single server became a bottleneck for doing deeper BMC, we applied the distributed engine, d-BMC. Even using d-BMC, we could not find witness up to depth 323 (in 8643s) using 5 workstations (configured as 1 Master and 4 Clients and connected with 1Gps Ethernet LAN), with a communication overhead of 30% and scalability factor of 0.1 (i.e., potentially we could do a 10 times deeper analysis than that on a single server.) We hypothesized that the property is correct. We then used the PBIA engine to obtain an abstract model with 71 FFs and ~1K gates in 6 iterations taking ~1200s. With UMC on the abstract model, we proved the property correct in ~30s.

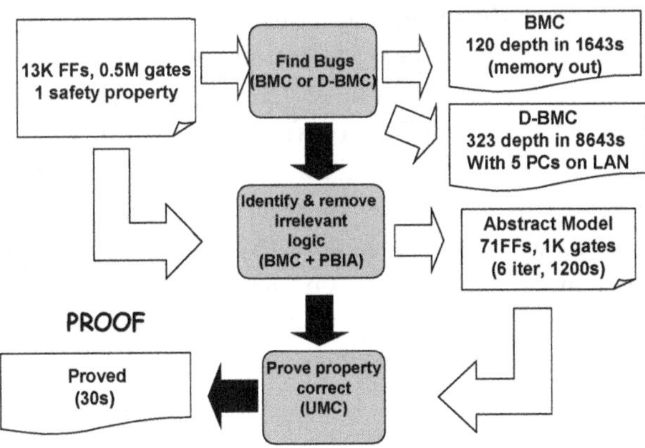

Fig. 12.6: Verification flow for example *Industry I*

Industry Design II

The design II with environmental constraints has 3.3K FFs and ~28K gates for a safety property. Environmental constraints were of the form $\mathbf{G}(p_0)\wedge\ldots\mathbf{G}(p_i)\ldots\wedge\mathbf{G}(p_k)$ where p_i corresponds to an internal node in the design. Verification flow for the design is shown in Figure 12.7.

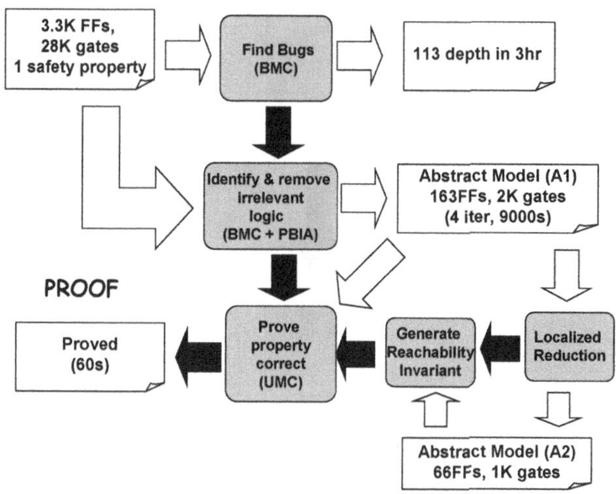

Fig. 12.7: Verification flow on example *Industry II*

Using BMC, we showed there was no witness up to depth 113 (in ~3hr, 720MB). Here, we "heuristically" decided not to use d-BMC; the bottleneck was due to hardness of SAT problem and not due to memory limitation of the server. We hypothesized the correctness of the property. We used PBIA to obtain an abstract model *A1* with 163 FFs and ~2K gates in 4 iterations taking 9000s. We removed the environmental constraints from the model *A1* and obtained (using localized reduction [86]) another abstract model *A2* that has only 66 FFs and ~1K gates. We computed a reachability invariant *reach_invar* [73] on the *A2* model in ~4s. We then used *reach_invar* as invariant with UMC on the *A1* model to obtain a proof in ~60s.

Industry Design III

The design III has 756 FFs (excluding the memory registers), and ~15K gates. It has two memory modules, both with address width AW = 10, and data width DW = 8. Each module has 1 write and 1 read port, with the memory state initialized to 0. There are 216 reachabality properties. The verification flow is shown in Figure 12.8. Using BMC with EMM, we found witnesses for 206 of the 216 properties, taking ~400s and 50Mb. The maximum depth over all witnesses was 51. To compare, using explicit modeling, we required 20540s (~6hrs) and 912Mb to find witnesses for all 206 properties. By using induction with BMC with EMM, we proved the remaining 10 properties in <1s. (It took 25s to get the proof with explicit memory modeling.)

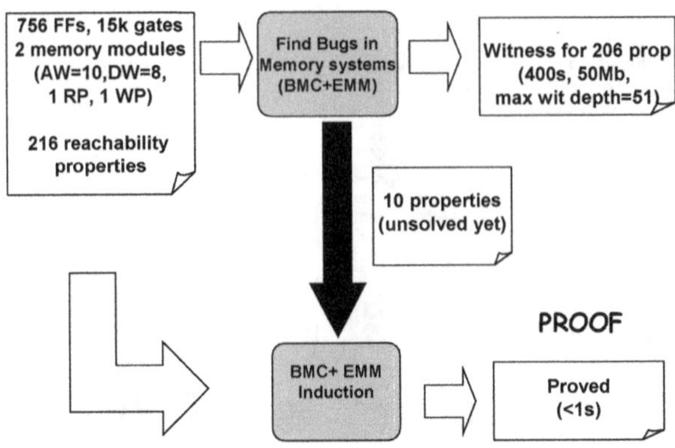

Fig. 12.8: Verification flow on design *Industry III*

Quicksort

We considered an HDL implementation of the *quicksort* algorithm with two memory modules: an un-initialized array (with AW=10, DW=32, 1 read and 1 write port) and an un-initialized stack (for recursive function calls, with AW=10, DW=24, 1 read and 1 write port). The design has 167 FFs (excluding memory registers), and ~9K gates for array size 5. The property checked was "*after return from a recursive call, the program counter should go to a recursive call on the right partition or return to the parent on the recursion stack*". The verification flow is shown in Figure 12.9.

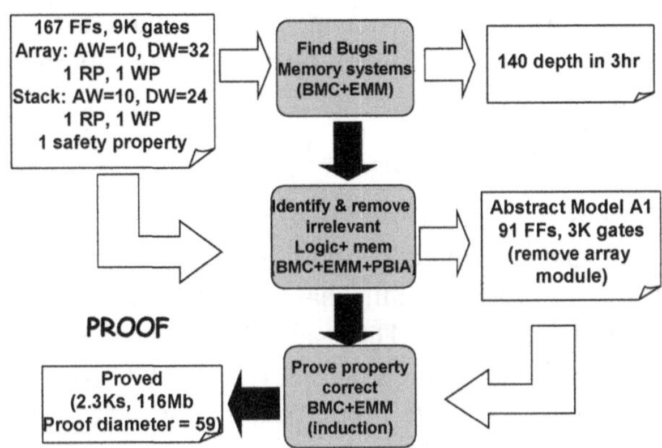

Fig. 12.9: Verification flow on example *Quicksort*

Using BMC+EMM+PBIA, we reduced the model to 91 FFs and ~3K gates, and identified the array module as irrelevant for this property. On this reduced model we proved correctness using forward loop-free path analysis (proof diameter = 59) in 2.3Ks, 116MB. Without the abstraction, the induction proof in BMC+EMM takes ~5Ks, 400MB. For an explicit memory model, however, we could obtain neither a proof nor an abstract model in 3 hours.

12.6 Summary

We describe an overview of our SAT-based verification framework *VeriSol.* We discussed various complex modeling issues that are common in current designs. We analyzed the strengths and weaknesses of various verification engines and presented the unique strengths of our approaches that give us an edge over the previous methods. We also show the effectiveness of our approach by applying various engines in an intertwined manner to achieve the desired goal.

12.7 Notes

The subject material described in this chapter are based on the authors' previous work [53], used with kind permission of Springer Science and Business Media.

13 SYNTHESIS FOR VERIFICATION

13.1 Introduction

To overcome increasing demand for shorter time-to-market and complex designs, significant efforts are being made in high-level synthesis methodologies, design languages and verification methodologies to leverage the expressive power of high-level models and reduce the design cycle time. However, with progression through each stage in the design cycle from abstraction to realization, part of the high-level information gets lost; which can adversely affect the performance and optimality of the verification solution at that stage.

High-level synthesis (HLS) [54, 81], also known as *Behavioral Synthesis*, is a process of generating a structural RTL model from a system-level model (SLM) — algorithmic description written in high-level languages such as C/C++ or SystemC [82]. HLS is usually targeted at meeting three main design constraints: performance, area, and power (PAP). Functional verification, on the other hand, ensures that the designs at various levels of implementation such as SLM, structural RTL, and gate level netlist, meet the functional specifications. Although automatic synthesis tools are typically used in the design process, one cannot rule out functional bugs in these designs due to buggy synthesis tools, or due to erroneous manual modifications in the designs to meet strict design constraints. As verification is increasingly becoming a bottleneck in the design process, significant resources are devoted to bridge the gap between design complexity and verification efforts.

A Design-For-Verification (DFV) methodology [80] has been proposed to bridge the growing verification gap by passing some of the burden to the designers. The underlying principle is to leverage designers' intent and knowledge to strengthen verification efforts. In other words, designers using suitable language primitives should convey their assumptions and design insights to aid the verification process. Specifically, DFV advocates one or all of the following:

- define *interface constraints* and *assumptions* for a sub-system to decompose verification efforts
- instrument the sub-systems with *feature checkpoints* such as assertions or invariants to infer functional-checkers
- add/identify *probes* such as control points in design where interesting behaviors occur to prioritize the bug search efforts
- add/identify *calibration checkpoints* for comparing performance results to infer performance-checkers
- add/identify *measurement checkpoints* for coverage metrics to access quality of verification efforts
- annotate *debugging features* to improve error diagnosis and speedy bug-fixes.

DFV methodology has significantly improved design productivity and shortened the design-cycle time, and thus, is increasingly adopted by the industry.

SAT-based formal verification (FV) techniques such as Bounded Model Checking (BMC) [66, 67] are quite matured and advanced. These verification techniques, applied at a structural-level closer to a realization, are also popular among designers. Part of the popularity is due to a general skepticism among designers, in regards to verification efforts made on an abstract model of the design. Understandably, these structural level verification techniques are likely to remain indispensable due to their ability to find structural bugs including those induced by HLS.

Most of the current research-emphases of FV, however, are on improving the scalability of the model checking techniques and reducing the model size, and not so much on the effect of various modeled design entities that can severely limit the capabilities of the verification techniques. In our work [57, 58, 62], we take one step further by separating the design optimized for PAP from the design optimized for correctness, thereby reducing the verification burden. This separation is imperative because designs optimized for PAP are often not optimized for verification, thereby, adversely affecting the performance and scalability of the verification tools. We have proposed a new paradigm called Synthesis-For-Verification (SFV) that involves synthesizing "verification-aware" designs that are most suitable for functional verification. Further, SFV methodology should be applied along with DFV methodology, to obtain the full benefit of verification efforts. In other words, the design input to HLS should have *verifiability features* as proposed in the DFV methodology. Note, our SFV paradigm requires support only from automated synthesis approaches, such as HLS, and can be easily automated, in contrast to the DFV methodology, which requires designers' reliable insights.

As part of the SFV methodology, we identify the effect of various design entities — muxes (multiplexers), control states, data registers, embedded memories, functional units, and re-converging paths in control flow graph — on the performance of typical model checkers. By guiding the use of such entities in existing behavioral synthesis techniques, we propose to obtain "BMC friendly" models that are relatively easier to model check. We believe such a paradigm of generating "verification aware" models will improve the overall verification effort. We also demonstrate its effectiveness by our experiments using representative industry tools and designs.

Outline

The rest of the chapter is organized as follows: we discuss the current methodology and verification issues in Section 13.2; we present the "Synthesis For Verification" paradigm in Section 13.3; we give short background on verification models and synthesis of such models in Section 13.4; we discuss modeling issues and challenges for the SFV paradigm in Section 13.5; we present techniques to synthesize BMC-friendly models in Section 13.6; we discuss EFSM-based learning and simplification during BMC in Section 13.7; we discuss specific techniques for property-preserving and "verification friendly" EFSM transformations in Section 13.8; we present a high-level BMC framework in Section 13.9; we discuss experimental results in Section 13.10 and summarize and give our future direction in Section 13.11.

13.2 Current Methodology

A combined HLS flow with verification is shown in Figure 13.1, where an RTL design synthesized under area, performance and power constraints is verified using a model checker. The checkers or properties needed for the model checkers are also generated during HLS [55] from specification primitives supported by design and verification languages [197], or from a checker library of common design errors. After all the outstanding bugs exposed by the model checker are fixed, and the model has been verified up to a certain level of confidence, it can be used as the reference model, called the *Golden Model.* This reference model can be used later for checking functional equivalence against automatically generated or manually modified RTL. As HLS can easily keep track of the correspondences between the internal signals of the generated models, specifically at the clock boundaries, the problem of validating the correctness of a synthesized design against the *Golden Model* can be reduced to a combinational or a bounded sequential problem. The state-of-the-art combinational equivalence tools (CEC) [198] use the internal signal correspondence very effectively to prove/disprove the equivalence. In practice, CEC is more scalable compared to model checking (property checking). The latter has to overcome the state explosion problem in proving/disproving the properties, while the former exploits internal structural similarities effectively to prove/disprove the equivalence with significantly less effort. In sequential equivalence checking (SEC) [199], RTL models are compared against reference SLM and its complexity is similar to property checking due to the state explosion problem. However, using structural similarities and clock boundary information (generated by HLS), SEC tools can benefit similarly as CEC tools.

HLS, in general, does not necessarily synthesize models that are easy to model check, i.e., the models are not verification "friendly". On at least the following three counts, we find that verification tools tend to perform poorly on the models synthesized by HLS tools.

Area optimization

Given limited hardware resources, the synthesized design tends to have a large number of multiplexers (muxes) due to increased sharing of the resources. It is quite well-known in the SAT solver community [149, 150] that muxes, like parity logic, are not good for DPLL-style SAT engines and hence, also for SAT-based verification engines.

Sequential scheduling

Because of multi-cycle operators and limited datapath resources, the synthesized design tends to be sequentially deeper, which results in time-consuming deeper searches by verification engines.

Memory optimization

Traditionally, HLS uses register files/arrays for embedded memories, to meet area and performance requirements. Such explicit use of memory elements worsens the state explosion problem further for verification efforts [50, 51, 179, 182].

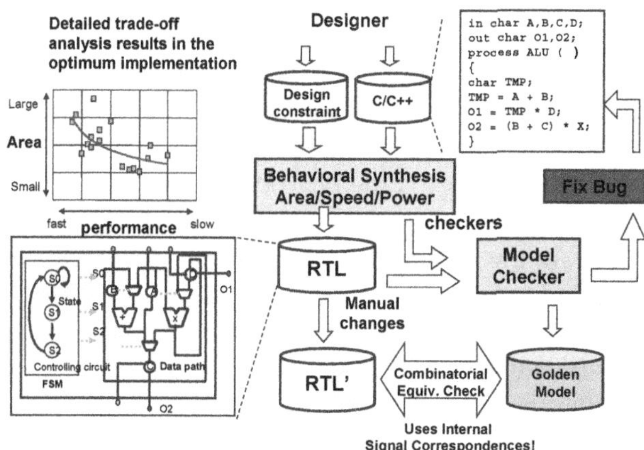

Fig. 13.1: Current Methodology: HLS with Verification

13.3 Synthesis for Verification Paradigm

To leverage off recent advancements in SAT-based and SMT-based BMC we believe that HLS methodologies should also focus on another dimension, i.e., functional verification, as shown in Figure 13.2. Although this dimension is not considered a part of standard design constraints, it helps in reducing the verification costs substantially.

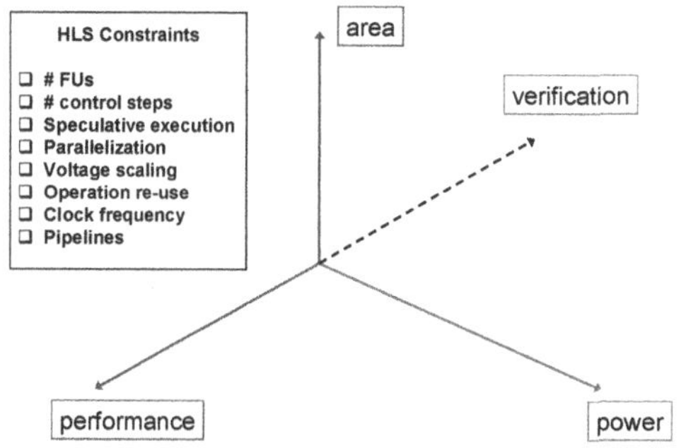

Fig. 13.2: Another dimension to HLS: Verification

As HLS tools, slowly and steadily, become popular among designers, we believe that it would be very beneficial for HLS providers to promote a "synthesis for verification" paradigm. In order to bridge the widening gap between design and verification productivity, we believe that HLS can be guided suitably to generate models that are verification friendly, for application of FV tools. In the following discussion, we also refer to such models as "verification aware" models.

In SFV methodology, we use the existing infrastructure of HLS to first generate verification friendly RTL models and properties from the given SLM design and specification as shown in Figure 13.3. The functional verification (model checking) is then carried out on the friendly model that need not meet any design constraints yet. The SLM and the corresponding verified (with certain level of confidence) RTL model serve as *Golden SLM* and *Golden RTL Models,* respectively, in the next phase where HLS synthesizes RTL that satisfies "design constraints". As HLS generates both the verification-friendly *Golden RTL Model* and the "design constraint satisfying" synthesized model, it is aware of the internal signal correspondences between them. An equivalence checker can very efficiently use this information to validate the functional correctness of synthesized designs, including those that may have undergone manual changes to meet design constraints. We believe that this is also an important step toward removing the skepticism among designers in regards to verification efforts made on a "verification" model, and not on the "synthesized" model. With our experimentation with industry tools and designs, we show that such a paradigm of generating "verification aware" models is practical and will help significantly in improving the overall verification efforts. Here, we explore various high-level synthesis parameters that can be controlled to generate smarter verification models for SAT-based verification, and discuss various challenges and issues with such a paradigm.

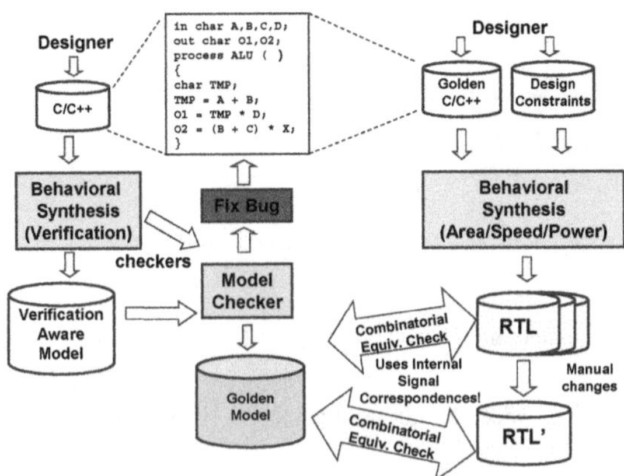

Fig. 13.3: Synthesis for Verification Paradigm

13.4 High-level Verification Models

Here we give short background on high-level synthesis and high-level model representations using extended finite state machines and flow graphs.

13.4.1 High-level Synthesis (HLS)

HLS is defined as a process of realizing a structure from a behavioral description of a system; a structure that implements the behavior of the system and meets the given constraints, typically area, timing, and power [54, 81, 200, 201]. Different design choices in HLS often have conflicting impacts on area, performance, and power requirements. There is a tradeoff associated with various design constraints, and the HLS process has to choose the best RTL, ideally exploring all possible design choices.

The essential steps of HLS are performed in the following order. Given a high-level design, a simplified data-flow graph is generated using compiler-like local transformations. Next, the functional units, storage elements and communication elements are *allocated* to operations, with an objective to minimize the cost associated with hardware resources. The *scheduling* step involves assigning operations to each control step (or control state), typically constrained by either area or performance requirements. The *binding* step refers to the process of assigning modules like adders, ALUs, shifters, storage element, muxes and buses. *Controller synthesis* refers to generation of control signals to drive the assigned modules at each control step.

We differentiate an intermediate *functional RTL* from final *structural RTL* during the synthesis process. The former model corresponds to an Extended Finite State Machine model (described next) soon after the *allocation* and *scheduling* steps, where each datapath operation is associated with a state of the control state-transition graph (STG). The latter model is obtained after *binding* and *controller synthesis,* where circuit implementations of control and data flow in each control state are defined, along with the mux circuits for sharing registers and function blocks assignments. Functional verification can be done on both functional RTL and structural RTL.

13.4.2 Extended Finite State Machine (EFSM) Model

Our method extracts high-level information from an Extended Finite State Machine (EFSM) model of a sequential program/design, with a partitioning of control and datapath. An EFSM has finite logical (control) states and conditionals (guards) on the transitions between the control states. The guards are functions of control states, data-path and input variables.

Formally, an EFSM model is a 6-tuple $<s_0,S,I,O,D,T>$ where, s_0 is an initial state, S is a set of control states (or blocks), I is a set of inputs, O is a set of outputs, D is a set of state (datapath) variables, $T=(T_E, T_U)$ is a transition relation, with T_E denoting an enabling transition relation, $T_E: S \times D \times I \rightarrow S$, and T_U denoting an update transition relation, $T_U: S \times D \times I \rightarrow D \times O$.

An ordered pair $<s,x> \in S \times D$ is called a *configuration* of M. A transition from a configuration $<s,x>$ to $<t,y>$ under an input i, with possible output o comprises of two transitions:

1. an enabling transition, represented as $((s,x,i),(t)) \in T_E$, and
2. an update transition, represented as $((s,x,i),(y,o)) \in T_U$.

For a given enabling transition $s \rightarrow t$, we define an enabling function f such that $f(x,i)=1$ *iff* $((s,x,i),(t)) \in T_E$ and we label the transition as $s \rightarrow_{f(x,i)} t$. For ease of description, we consider deterministic EFSMs where for any two transitions from a state s, *i.e.* $s \rightarrow_{f(x,i)} t_1$ and $s \rightarrow_{g(x,i)} t_2$, $f(x,i) \wedge g(x,i) = FALSE$. We define $out(s) = \{t \mid s \rightarrow_{f(x,i)} t\}$ as a set of outgoing control states of s. Similarly, we define $in(t) = \{s \mid (s,t) \in T_E\}$ as a set of incoming control states of t. We define a NOP state as a control state with no update transition and a single outgoing enabling transition. A NOP state with n incoming transitions can be replaced with n NOP states, each with a single incoming and a single outgoing transition, without changing incoming or outgoing states. In our discussion, a NOP state will have only a single incoming/outgoing transition. We define a SINK state as a control state with no update transition relation and no outgoing transition. We define a transition or state as *contributing* with respect to a variable if it can affect the variable; otherwise, such a transition or state is called *non-contributing*.

Example 13.1

We illustrate an EFSM model M of a FIFO example (implemented using a RAM) using a State Transition Graph (STG) as shown in Figure 13.4, where *S={S0,…,S11}*, *I={ren, wen, fifo_in}*,*O={fifo_out, is_full, is_empty}*, *D={rptr, wptr, is_full, is_empty, RAM[10]}*. The enabling functions are shown in *italics* and update transitions are shown in non-italics in the figure. For example, the transition $S2 \rightarrow_{T23} S3$ has enabling function *T23=(rptr==wptr-1)||(rptr=0 && wptr==9))* and update transition *{RAM[rptr]=in; is_empty=0}*.

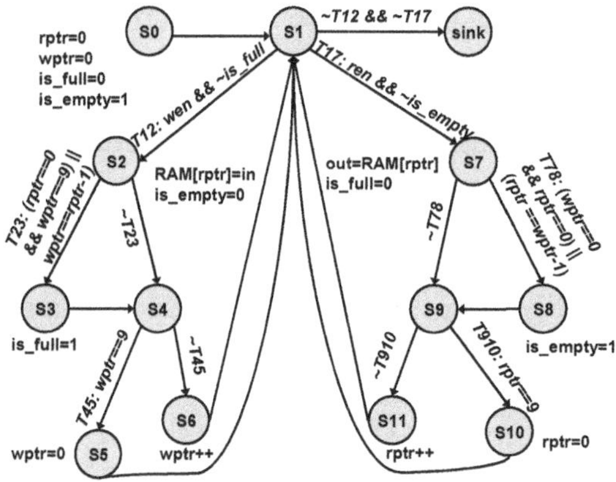

Fig. 13.4: STG of EFSM M

13.4.3 Flow Graphs

A *flow graph* $G(V,E,r)$ is a directed graph with an entry node r. A path from u to v, denoted as $p(u,v)$, is a sequence of nodes $u=s_1,\ldots,s_k=v$ such that $(s_i,s_{i+1})\in E$ for $1\leq i<k$. We denote $Path(u,v)$ as a set of paths between u and v. Length of a path $p\in Path(u,v)$ is the number of edges in the sequence. For $p_x,p_y\in Path(u,v)$, $p_x\neq p_y$ if the sequence of states in p_x is different from p_y; such paths are called *re-converging* paths. A concatenation of paths $p(t,u)$ and $p(u,v)$, denoted as $p(t,v)=p(t,u)\oplus p(u,v)$, represents a path from t to v that goes through u. A node n is said to *dominate* node m, denoted as $dom(m)$ if every path from r to m goes through n. The node r dominates every other node in the graph. A Strongly Connected Component, SCC (V_i,E_i) is a subgraph of G such that for all $u,v\in V_i$, there exists a path $u=s_1,\ldots,s_k=v$ such that $(s_j,s_{j+1})\in E_i$ for $1\leq j<k$. SCC (V_i,E_i,r_i) is a loop with entry r_i if r_i dominates all the nodes in V_i. An edge (u,v) is called a *back-edge* if v dominates u; otherwise it is called a *forward-edge.* Given a back-edge (u,v), a *natural loop* of the edge is defined as the set of nodes (including u) that can reach u without going through v. For a given $G(V,E,r)$, and any pair of natural loops, $L_1=(V_1,E_1,r_1)$ and $L_2=(V_2,E_2,r_2)$ one of the following cases holds:

- they are disjoint i.e., $V_1\cap V_2=\varnothing$,
- they are disjoint but have the same entry node, i.e., $V_1\cap V_2=\{r_1\}=\{r_2\}$, or
- one is completely nested in the other, i.e., $V_1\subseteq V_2$ or $V_2\subseteq V_1$.

A *sink* node is a node with no outgoing edge. A flow graph $G(V,E,r)$ is *reducible* [202] if and only if E can be partitioned into disjoints sets *front-edge set* E^f and *back-edge set* E^b such that $G^d(V,E^f,r)$ forms a direct-acyclic graph (DAG) where each $v \in V$ can be reached from the entry node r. A reducible graph has the property that there is no jump into the middle of the loops from outside and there is only one entry node per loop. Most flow graphs that occur in standard practice fall into the class of reducible flow graphs. Specifically, use of structured control flow statements such as *if-then-else, while-do, continue* and *break* produces programs whose flow graphs are always reducible. Unstructured programs due to the use of *goto* can cause irreducibility of the graphs. Thus focusing on a reducible graph is not a significant restriction of our algorithms and techniques, and indeed accords with practical guidelines.

For a given EFSM $<s_0,S,I,O,D,T>$, let $G=(V,E,r)$ be a flow graph with start vertex r, such that $V=S$, $E=\{(s,t)|\ s \to t\}$, and $r = s_0$. The sets *out(s)* and *in(s)* represent the set of *outgoing* nodes and *incoming* nodes of a node s, respectively. A reachability analysis on a flow graph corresponds to control state reachability analysis of the corresponding EFSM.

13.5 "BMC-friendly" Modeling Issues

SAT-based BMC has been gaining wide acceptance as a scalable solution compared to BDD-based symbolic model checking. With advent of sophisticated SMT solvers built over DPLL-style SAT solvers, SMT-based BMC is also gaining popularity. In the following sections, we discuss "BMC friendly" modeling guidelines based on pros and cons of various BMC techniques. We also describe a simplification and learning scheme [62] — based on high-level control information — for improving both SAT-based and SMT-based BMC.

DPLL-Style SAT Solvers

SAT technology has seen rapid progress in the last few years, and well-known heuristics exist for CNF-based and circuit-based solvers. However, muxes, in general, are detrimental [149, 150] to DPLL-based SAT solvers as the solvers typically end up explicitly enumerating all combinations of the variables corresponding to mux select inputs. SAT solvers are not good at enumerating solutions, although there have been some limited successes [47, 68, 71] to address this issue. Thus, the problems that arise from area-optimized circuits with heavy sharing of registers and functional units can pose serious challenges to SAT solvers.

Embedded Memory

Presence of large embedded memories also adversely affects BMC if modeled explicitly, i.e., as state bits at each memory address. To overcome that, Efficient Memory Modeling approaches [50, 51, 179, 182] have been used to augment BMC to handle embedded memory systems (with multiple read, write ports) without explicitly modeling each memory bit, by capturing the data forwarding semantics efficiently. Sophisticated techniques such as [51] can identify fewer memory modules and ports that need to be modeled; thereby reducing the verification model size.

13.6 Synthesizing "BMC-friendly" Models

Various HLS parameters such as number of functional units, registers, control steps, binding modules, memory modeling styles, parallelization and speculative execution, are used to direct HLS to synthesize a BMC-friendly model, disregarding the design constraints as shown in Figure 13.3. Note, we want HLS to generate model with a small size and sequential depth, and satisfying the following "verification constraints" as well.

Functional Unit and Registers

Although sharing of functional units (FU) and registers reduces the model size, it increases the use of muxes as these datapaths are typically re-used across control steps. Such increased use of muxes adversely affects the performance of a SAT solver, as the partitioning of the search space has to be done dynamically by the SAT solver. Therefore, HLS should use mux-constraint, i.e., number of muxes as a guiding criterion in allocating datapaths.

Control steps

Reducing the number of control steps is useful in general, as it reduces the search depth, but it can increase the sharing of functional units, which in turn can adversely affect SAT performance. In practice, techniques like SAT-based incremental learning [45, 153, 154] and dynamic circuit simplification [45, 137] enable BMC to search deeper effectively. Thus, by *scheduling* under "mux constraints", we can obtain an optimum number of control steps in the model.

Operations

Reuse of operations, as opposed to reuse of datapaths, is helpful as it allows sharing of common sub-expressions without introducing muxes. For example, in the expression $x*y+x*z$, the operation '*' can be shared, i.e., $x*(y+z)$. This also reduces the number of datapaths and hence size of the model.

Speculative Execution and Parallelization

As long as functional units and registers are not re-used, such HLS steps are useful since they reduce the number of control steps.

Memory Modeling

For embedded memories in the design, we direct HLS to use memory modeling instead of explicit register arrays. Efficient memory modeling [50, 51] with BMC effectively reduces the verification model size by abstracting away entire memory arrays *but* retaining the data forwarding semantics.

13.7 EFSM Learning

On-the-fly circuit simplification [42, 43, 136] and SAT-based incremental learning [45, 46, 153, 154] applied on a Boolean-level model have resulted in improved performance of SAT-based BMC. Here we describe our learning and simplification scheme [62], using high-level control-flow information extracted from EFSM models, to improve both SAT-based and SMT-based BMC. Later in Section 13.7, we discuss how to make this learning more effective using EFSM transformations, by adopting a "synthesis for verification" paradigm to obtain "verification-aware" models.

13.7.1 Extraction: Control State Reachability (CSR)

Starting from the initial state $S0$, we compute control state (also called *block*) reachability (CSR) using a breadth first search. A control state S_i is *one-step reachable* from S_j *if and only if* an enabling transition (or edge) exists between them. At a given sequential depth d, let $R(d)$ represent the set of possible states that can be reached (statically, i.e., ignoring the enabling transition function) in one step from the states in $R(d-1)$ with $R(0)=\{S0\}$. We say $s \in R(d)$ is *CSR reachable* at depth d. Note that we are not computing the (fixed-point) diameter. For some d, if $R(d-1)\neq R(d)=R(d+1)$, we say that the reachability *saturates* at depth d and stop; otherwise we compute $R(d)$ and $|R(d)|$ (i.e., size of $R(d)$) up to the BMC bound.

Applying CSR on the FIFO Example 13.1, we obtain the set $R(d)$ as shown in Table 13.1(a). Note, the saturation depth is 15, $|R(15)|=|R(16)|=11$ where $R(15)=\{S1,S2,\ldots,S11\}$.

13.7.2 On-the-Fly Simplification

For n control states $S1 \ldots Sn$, we introduce n Boolean variables $B_{S1} \ldots B_{Sn}$. Let the Boolean variable B_r = *TRUE* iff configuration of M is (r,x) for some $x \in D$. Equivalently, B_r corresponds to a predicate on the control state variable, called the *PC (Program Counter)*, i.e., $B_r \equiv (PC==r)$. Let B_r^d denote the Boolean variable B_r at depth d during BMC unrolling.

At an unrolling depth d of BMC, we apply the following on-the-fly structural and clausal (learning-based) simplification on the corresponding formula. We use a procedure *Simplify (BoolExpr e, Boolean v)* that constrains a Boolean expression *e* to a Boolean value *v*, and reduces the expressions that *use e*. Later, we demonstrate this with an example.

1. Unreachable Block Constraint (UBC)

$$\forall_{r \notin R(d)}\ Simplify(B_r^d, 0)$$

Since the state r is not *CSR reachable* at depth d, the predicate B_r will evaluate to *FALSE* at depth d. Therefore, simplifying the formula by propagating B_r=0 at depth d preserves the behavior of the design. We use the procedure *Merge* (Figure 3.15, Chapter 3 for propagation).

2. Reachable Block Constraint (RBC)

$$Simplify(\vee_{r \in R(d)} B_r^d, 1)$$

At any depth d, at least one state in $R(d)$ is *CSR reachable*.

3. Mutual Exclusion Constraint (MEC)

$$\forall_{r,t \in R(d),\, r \neq t}\ Simplify((B_r^d \Rightarrow \neg B_t^d), 1)$$

At any depth d, at most one state in $R(d)$ is the current state.

4. Forward Reachable Block Constraint (FRBC)

$$\forall_{r \in R(d)} \; Simplify((B_r^{\,d} \Rightarrow \vee_{s \in out(r)} B_s^{\,d+1}),1)$$

At any depth d, if current state is r, i.e., $B_r^{\,d}$=*TRUE*, then the next state must be among the *out(r)* set.

5. Backward Reachable Block Constraint (BRBC)

$$\forall_{r \in R(d)} \; Simplify((B_r^{\,d} \Rightarrow \vee_{s \in in(r)} B_s^{\,d-1}),1)$$

At any depth $d > 0$, if current state is r, i.e., $B_r^{\,d}$=*TRUE*, then the previous state at depth d-1, must be among the *in(r)* set.

6. Block-Specific Invariant (BSI)

$$\forall_{r \in R(d)} \; Simplify((B_r^{\,d} \Rightarrow C_r^{\,d}),1)$$

At any depth d, a given invariant C_r for a given state r is valid only if r is the current state at depth d.

Note, our approach adds only the relevant constraints at each BMC unrolling, thereby reducing the overall formula size. Thus, ideally we would like a smaller set *R(d)* to increase the effectiveness of our simplification. Later, we discuss how we transform EFSM model to reduce the set *R(d)*.

Example 13.1(Contd):

We illustrate simplification constraints at depth, d=4. In particular, we consider the effect of simplification on the unrolled expression for variable *is_full*. The transition relation for the state variable *is_full* is as follows:

$$next(is\_full) = (B_{S0} \,||\, B_{S7})?\, 0 : (B_{S3})\, ?\, 1 : is\_full;$$

The high-level expression for the unrolled variable, corresponding to *next(is_full)* at depth 4, would correspond to:

$$is\_full^5 = (B_{S0}^{\,4} \,||\, B_{S7}^{\,4})?\, 0 : (B_{S3}^{\,4})\, ?\, 1 : is\_full^4;$$

In the following, we explain the application of *Unreachable Block Constraint*. Note, at d=4, only *S4, S5, S6, S9, S10,* and *S11* are reachable (see

Table 13.1(a)). Therefore, we perform the following simplification using *UBC*.

$$\forall_{r \in \{S0,S1,S2,S3,S7,S8\}} \; Merge(B_r^4, 0)$$

Using the above simplification, the expression for *is_full* is mapped to an existing variable, thereby, reducing the additional logic, i.e., $is\_full^5 = is\_full^4$.

At d=4, we also add other constraints such as *RBC*, *FRBC*, and *BRBC* as illustrated below.

RBC: $(B_{S4}^4 + B_{S5}^4 + B_{S6}^4 + B_{S9}^4 + B_{S10}^4 + B_{S11}^4)$
FRBC: $(\neg B_{S4}^4 + B_{S5}^5 + B_{S6}^5), (\neg B_{S9}^4 + B_{S10}^5 + B_{S11}^5), \ldots$
BRBC: $(\neg B_{S4}^4 + B_{S2}^3 + B_{S3}^3), (\neg B_{S9}^4 + B_{S7}^3 + B_{S8}^3), \ldots$

13.7.3 Unreachablility of Control States

CSR provides conservative information, i.e., if a control state is not CSR reachable from the initial state during CSR, it is definitely not reachable in the full EFSM model; however, the other way is not true in general. In the following, we discuss a simple test for proof of unreachability of a control state s. Let n be the total number of control states in the EFSM model. We say s is unreachable if all of the following conditions hold:

- $s \in R(d)$ for some depth d
- $\forall_{d<t<d+n} \; s \notin R(t)$
- state s is not reachable using BMC at any depth k where $0 \leq k \leq d$

As the longest path in a control flow graph (corresponding to the EFSM model) has length n, clearly state s is un-reachable if it is not reachable until depth d.

13.8 EFSM Transformations

Now, we discuss two EFSM transformations that are applied during *allocation* and *scheduling* (of HLS) to reduce EFSM size (especially, datapaths) and to balance re-converging paths in the control flow graph. Both these transformations enhance the effectiveness of high-level learning/simplification during BMC (discussed in Section 13.7).

13.8.1 Property-based EFSM Reduction

Slicing away behaviors (and the elements) unrelated to the specific properties of interest can significantly reduce the model size and thereby, improve the verification efforts. We perform slicing on EFSM [203] with respect to variables of interest as defined by the property, and obtain *contributing* and *non-contributing* states and transitions. Note that such strategies are routinely applied [86] at the Boolean model by performing a static fanin cone analysis of the property under consideration. *We believe that property- preserving slicing at a higher level carried out by HLS can lead to a greater size reduction in the EFSM model and in the corresponding Boolean model.* In the following, we discuss two such reduction schemes, *Cone-of-influence (COI) reduction* and *Collapsing.*

COI reduction

In *COI reduction,* we remove all non-contributing states and their outgoing transitions. Any non-contributing transition $s \rightarrow_{f(x,i)} t$ where s is a contributing state (with t being non-contributing) is replaced by a transition $s \rightarrow_{f(x,i)} SINK$. If we are concerned with reachability of a state $s \in S$ from a start state s_0, we remove the outgoing transition from s since it is non-contributing for the shortest counter-example or proof. For example, the self-loop transition $S1 \rightarrow S1$ (not shown in Figure 13.4) is *non-contributing* and hence, it is replaced by $S1 \rightarrow SINK$ as shown in Figure 13.4.

Collapsing

We define a *collapsing condition* as the condition when all states in *out(s)* are NOP and none of them directly appears in a reachability check. Under such a condition, we collapse all the NOP states and merge them with s. In other words, $\forall t \in out(s)$, (with t being NOP) we remove the transitions $s \rightarrow_{f(x,i)} t$ and $t \rightarrow_{TRUE} q$ and add a new transition $s \rightarrow_{f(x,i)} q$.

13.8.2 Balancing Re-convergence

Recall that the efficiency of on-the-fly simplification during BMC unrolling depends on the size of the set *R(d), i.e.,* $|R(d)|$. A larger $|R(d)|$ reduces the scope of simplification at depth d and hence, the performance of high-level BMC. Re-converging paths of different lengths inside loops is one of the main reasons for the saturation of reachability, and inclusion of all looping control states in the set R. To improve the performance of high-level BMC further, we adopt a strategy called "Balancing Re-convergence" that transforms the original model into a “BMC friendly” model. The transformed model preserves the validity of all properties expressed in LTL\X (Linear Temporal Logic without the neXt-time operator).

Basic idea

For balancing re-convergence and reducing the set *R(d)*, we transform an EFSM by inserting NOP states such that lengths of the re-convergent paths are the same, and control state reachability does not saturate. Reduction in *in R(d)*, in general, improves the scope of on-the-fly simplifications. Note that the additional NOP states have little adverse effect on simplification, although they increase the total number of control states and transitions, and possibly the search depth. As NOP states do not add complexity to the transition relations of any state variables except the program variables encoding the control states, simplification using UBC at depth *d* is practically unaffected by inclusion of such states in *R(d)*. Also, the simplifications due to FRBC and BRBC are not affected, as these additional transitions are single outgoing transition (and hence always enabled) from NOP states. We define $R^{--}(d) = \{s \mid s \in R(d)$ and s is not a NOP state$\}$. We use $\max_d |R(d)|$ and $\max_d |R^{--}(d)|$ to measure the effectiveness of our strategy in improving the scope of simplification of high-level BMC. Note that the above transformation preserves LTL\X properties, as NOP states can only increase the length of traces but not the eventuality and global behavior. As the state of data variables do not change in NOP state, the validity of atomic propositions is not affected.

Example 13.1 (Contd)

For the EFSM model shown in the Figure 13.4, paths $S2\rightarrow S3\rightarrow S4$ and $S2\rightarrow S4$ are re-converging with different lengths. For balancing, we insert a NOP state *S2a* such that transition $S2\rightarrow_{\sim T23}S4$ is replaced by $S2\rightarrow_{\sim T23}S2a\rightarrow_{TRUE}S4$. Similarly, as *paths* $S7\rightarrow S8\rightarrow S9$ and $S7\rightarrow S9$ are re-converging with different lengths, we insert another NOP state *S7a* and replace the transition $S7\rightarrow_{\sim T78}S9$ by $S7\rightarrow_{\sim T78}S7a\rightarrow_{TRUE}S9$. The modified EFSM model M’ is shown in the Figure 13.5. CSR on *M’* is shown in Table 13.1(b). Note that at depth $max_d\ R^{--}(d) = 4$. Also, CSR on *M’* does not saturate.

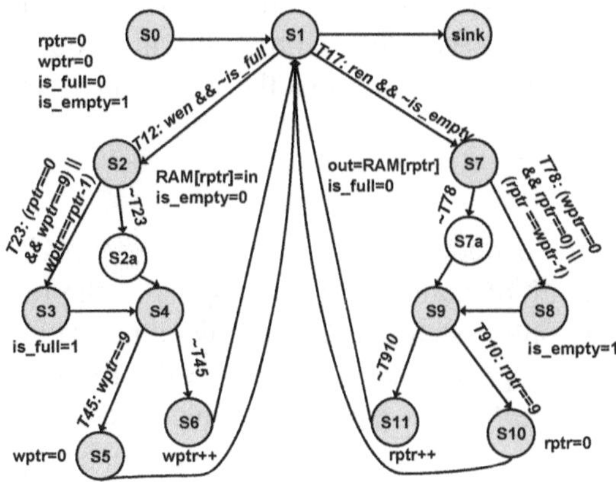

Fig. 13.5: STG of transformed EFSM M'

Table 13.1: Control State Reachability on EFSM (a) M and (b) M'

(a) Model M			(b) Model M'			
D	R(d)	|R(d)|	d	R(d)	|R(d)|	$\|R^-(d)\|$
0	S0	1	0	S0	1	1
1	S1	1	1	S1	1	1
2	S2, S7	2	2	S2, S7	2	2
3	S3, S4, S8, S9	4	3	S3, S2a, S8, S7a	4	2
4	S4, S5, S6, S9, S10, S11	6	4	S4, S9	2	2
5	S5, S6, S10, S11, S1	5	5	S5, S6, S10, S11	4	4
..	.....		6	S1	1	1
15	Saturates with 11 states		i	Repeats, R(i) = R(i%5)		

Algorithm

Given a *reducible* flow graph G, we present an *O(E)* algorithm in Sections 13.8.3 and 13.8.4, that identifies the edges corresponding to the transitions in T_E where inserting certain number of *NOP* states will balance the re-convergent paths, including those arising due to loops.

13.8.3 Balancing Re-convergence without Loops

Consider the DAG, $G(V,E^f,r)$ corresponding to a reducible flow graph $G(V,E,r)$ with an entry node *r* and front-edge set E^f. Let *w(e)* denote the weight of the edge, $e=(a,b) \in E^f$. As we shall see later, the weight of the edge *(a,b)* corresponds to one more than the number of NOP states that need to be inserted between nodes *a* and *b*. We define weight for a path $p=<s_1,\ldots,s_k>$, denoted as $w(p) = \Sigma_{1\le i<k} w(e_i)$ where $e_i = (s_i, s_{i+1}) \in E^f$. We now define our problem as follows:

Problem 13.1: For a given DAG $G(V,E^f,r)$ find a weight function, $w:E^f \rightarrow Z$ such that $\forall p_x, p_y \in P(u,v)\ w(p_x)=w(p_y)$, where $u,v \in V$.

Note, if we are able to find a feasible *w*, then the number of NOP states introduced for an edge, $e=(a,b) \in E^f$ will be equal to $w(e)-1$. We say that the set $P(u,v)$ is *balanced* when all paths from *u* to *v* have equal weights, i.e., $\forall p_x, p_y \in P(u,v),\ w(p_x)=w(p_y)$. Let $W(u)$ denote the weight of the paths in the balanced set $P(r,u)$. We define $W(r)=0$.

Lemma 13.1: If $P(r,v)$ is balanced and $P(u,v) \neq \emptyset$, then $P(u,v)$ is balanced.

Proof: We prove by contradiction. Let $p_1(r,u) \in P(r,u)$. Assume $P(u,v)$ is not balanced, i.e., there exists at least two paths $p_1(u,v) \in P(u,v)$ and $p_2(u,v) \in P(u,v)$ such that $w(p_1) \neq w(p_2)$. Let $p_0(r,u) \in P(r,u)$. The weight of the concatenated path $p_0 \oplus p_1$ is $w(p_0)+w(p_1)$ and that of $p_0 \oplus p_2$ is $w(p_0)+w(p_2)$. Since $w(p_1) \neq w(p_2)$, the weight of the concatenated paths are different. However, since $p_0 \oplus p_1,\ p_0 \oplus p_2 \in P(r,v)$ and $P(r,v)$ are balanced, we get a contradiction. Thus, $P(u,v)$ is also balanced.

Using Lemma 13.1, we re-formulate the Problem 13.1 as follows:

Problem 13.2 (Problem 13.1 stated differently): Given a DAG $G(V,E^f,r)$, find a weight function, $w:E^f \rightarrow Z$ and $W:V \rightarrow Z$ such that $P(r,v)$ is balanced, i.e., $w(p_x)=w(p_y)=W(v),\ \forall p_x, p_y \in P(r,v)$

Solution: If $P(r,u)$ is balanced, i.e., $\forall u \in in(v)\ W(u)$ is computed, we can balance $P(r,v)$ *recursively* as follows.

- $\forall_{u \in in(v)}\ w(u,v) = t - W(u)$, where $t = (max_{u \in in(v)} W(u))+1$
- Set $W(v)=t$ as for any path $p(r,v)$ through u will have weight $W(u)+w(u,v)=W(u)+t-W(u)=t$.

We start with an initial set of nodes *S* which are *sink nodes* in $G(V,E^f,r)$. Then, we recursively apply the above steps in the procedure *Balance_path,* as shown in Figure 13.6. Termination is guaranteed as the recursive sub-procedure *Balance_aux* is invoked only once per node. The correctness of the algorithm can be shown easily by an inductive argument.

As discussed earlier, we insert NOP states corresponding to the edge weights obtained after running the procedure *Balance_path.* For edge $e=(u,v)$, we insert $(w(e)-1)$ NOP states. It is easy to see that the algorithm *Balance_path* adds a minimum number of NOP states for balancing paths. However, the inserted NOP states together with NOP states in the original EFSM can generate a *collapsing condition* (discussed in Section 13.8.1). In that case, we collapse the NOP states as discussed earlier. We re-run *Balance_path* as the lengths of the re-converging paths might have changed due to collapsing. Note, one can integrate collapsing with the procedure *Balance_path* to avoid re-running.

```
Synopsis:   Balance path for
            front-edge set G(V,E^f,r)
Input:      G(V,E^f,r)
Output:     w:E→Z, W:V→Z
Procedure:  Balance_path
S = {v | v is a sink node}
W(r) = 0; ∀v∈V,v≠r W(v)=∞;
∀v∈S, Balance_aux(v);
Input:      v
Output:     W(v)
Procedure: Balance_aux
∀u∈in(v) if (W(u)==∞) Balance_aux(u);
W(v) = max u∈in(v) (W(u))+1;
∀u∈in(v) w(u,v)=W(v)-W(u);
```

Fig. 13.6: Pseudo-code of *Balance_path*

13.8.4 Balancing Re-convergence with Loops

Since the flow graph $G(V,E,r)$ is reducible, we know that every loop $L_i=G(V_i,E_i,r_i)$ is a natural loop corresponding to the backedge set $(b_i,r_i)\in E^b$, and has a single entry node r_i. Presence of back-edges in loops and their relative skews can also cause re-convergent paths of different lengths; which in turn, can lead to saturation during control reachability analysis. We say a loop L_i is *saturated* at depth s when $\forall v\in V_i$, $v\in R(t)$ for $t\geq s$. Given balanced $Path(r_i,b_i)$ for each loop L_i, we define the *forward loop length*, C_i for loop L_i as follows:

$$C_i = W(b_i)-W(r_i)$$

where $W:V\rightarrow Z$ is the weight function we obtain for each node in $G(V,E^f,r)$ using the *Balance_path* algorithm (See Figure 13.6).

(13.1)

Observe that the entry node r_i of loop L_i is CSR reachable after $N_i = C_i+w(b_i,r_i)$ steps i.e., $r_i \in R(d_i+n_iN_i)$ where d_i, $n_i \in Z$. We call N_i the *loop period* of L_i. If there is only one loop, it is easy to see that $d_i = W(r_i)$. However, in the presence of multiple loops, we also have to account for the paths from other loops to loop L_i. In particular, if there is a path from entry node r_j of some loop L_j to r_i, then entry r_i also re-appears after N_j. We define *loop clusters LC* as sets of disjoint entry nodes such that for any two clusters LC_x and LC_y, $\forall s \in LC_x$, $\forall t \in LC_y$, $P(s,t)=P(t,s)=\Phi$. Note, a loop in a cluster does not affect the loop in another cluster as far as reachability is concerned. In the following problem statement, we discuss how to prevent loop saturation using suitable transformations.

Problem 13.3: Given a reducible flow graph $G(V,E,r)$ with $E=E^f \cup E^b$, find $w:E^b\rightarrow Z$ and loop period N_i so that loop L_i is not saturated.

Solution: We define a set $to(i)=\{j|\ r_j=r_i \ or\ Path(r_j,r_i)\neq\Phi\}$. Thus, $d_i = W(r_i)+\Sigma_{j\in to(i)}n_jN_j$ where $n_j \in Z$. Define, $D_i = min_{k\in\ to(i)\cup\{i\}}N_k$. It is easy to see that a loop L_i gets saturated at depth $t+D_i$ during CSR if $r_i \in R(t+k)\ \forall k,\ 0\leq k<D_i$. This is captured by the following integer linear equations in terms of s and n_j's for given N_j's, N_i and $W(r_i)$.

$$W(r_i)+\Sigma_{j\in to(i)}n_jN_j + n_iN_i = s$$
$$W(r_i)+\Sigma_{j\in to(i)}n_jN_j + n_iN_i = s+1$$
$$\ldots$$
$$W(r_i)+\Sigma_{j\in to(i)}n_jN_j + n_iN_i = s+D_i-1 \qquad (13.2)$$

To prevent saturation of loop L_i, we need to find N_j's and N_i such that there is no feasible solution to the above equations. One strategy is to choose a weight function $w:E^b\rightarrow Z$ such that the loop lengths match, i.e., $N_i=N_j$ $\forall j\in to(i)$. (It is easy to verify the infeasibility for this solution assuming that each loop has at least two nodes, *i.e.*, $N_i \geq 2$.)

We consider one *loop cluster* at a time. We define *maximum loop period* over all loops in the cluster (i.e. whose entry nodes are in the cluster), $N = (max_i\ C_i) + 1$. We assign a weight to each back-edge (b_i,r_i) as follows:

$$w(b_i,r_i) = N-C_i \qquad (13.3)$$

For each loop L_i in the cluster, the entry node $r_i \in R(W(r_i)+nN)$ where $n \in Z$. Thus, the upper bound on $|R(d)|$ for $G(V,E,r)$ at any depth $d>>1$, $|R(d)| \leq \Sigma_i\ max_t\ |R_i(t)|$, where $R_i(t)$ is the control reachability set (including NOP states) on loop L_i at a depth t. Similarly, the upper bound on $|R^-(d)|$ for

$G(V,E,r)$ at any depth $d>>1$, is $|R^{--}(d)| \leq \Sigma_i\ max_t\ |R_i^{--}(t)|$, where $R_i^{--}(t)$ is the control reachability set (of only non-NOP states) on loop L_i at a depth t.

Example 13.2

We illustrate our algorithm for balancing loops in a flow graph using an example shown in Figure 13.7(a). Let v_i represent the node with index i (shown inside the circle). Note, the flow graph $G(V,E,v_1)$ has three natural loops L_1, L_2, and L_3 corresponding to the back-edges (v_6,v_3), (v_7,v_1), and (v_8,v_3), respectively. The corresponding entry nodes for the loops are v_3, v_1, and v_3 respectively. Note, they all form a *cluster*. The DAG $G(V,E^f,v_1)$ corresponding to the front-edge set, E^f=E-$\{(v_6,v_3), (v_7,v_1), (v_8,v_3)\}$ is shown in Figure 13.7(b). After executing *Balance_path* algorithm, we obtain edge weights, also shown in Figure 13.7(b), that balance all re-convergence paths in E^f. Note that the edge with no weight shown has an implicit weight of 1. Also, shown are the W values of each node. For instance, $W(v_6)=5$ denotes that all the paths in the set $P(v_1,v_6)$ have weights equal to 5. Next, we compute the *forward loop length* of each loop and the weights of the back-edges. The *forward loop length* of loop with back-edge (v_6,v_3) is $W(v_6)$-$W(v_3)$ = 3; similarly, with back-edge (v_7,v_1) is 6, and with back-edge (v_8,v_3) is 5. Thus, value of N, as defined, is 7. The weight of the back-edges (v_6,v_3), (v_7,v_1) and (v_8,v_3) are 4, 1, and 2 respectively as shown in Figure 13.7(c). For each edge with weight w, we insert w-1 nodes corresponding to NOP states, shown as un-shaded circles in the modified flow graph in Figure 13.7(d).

CSR on the original flow graph $G(V,E,v_1)$ in Figure 13.7(a) saturates at depth 6 with 8 nodes. However, CSR on the balanced flow graph in Figure 13.7(d) does not saturate. Instead, the set of reachable nodes $R(d)$ at depth d shows a periodic behavior with period, N=7. If we perform reachability analysis separately on each loop of the modified flow graph in Figure 13.7(d), we obtain $max_t\ |R_1(t)|=2$, $max_t|R_2(t)|=2$, and $max_t|R_3(t)|=2$. Thus, the upper bound on $|R(t)|$ is 6. Similarly, $max_t\ |R^{--}_1(t)|=1$, $max_t|R^{--}_2(t)|=1$ and $max_t|R^{--}_3(t)|=1$ and upper bound on $|R^{--}(t)|$ is 3. In this case, $max_t|R(t)|=4$ and $max_t|R^{--}(t)|=2$. Clearly, the scope of simplification during BMC is significantly improved by balancing the flow graph.

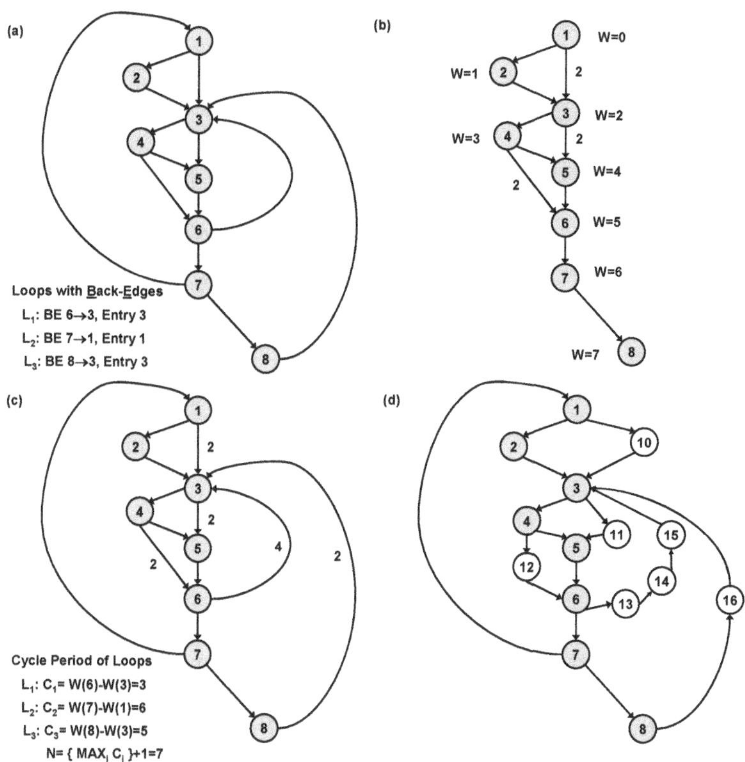

Fig. 13.7: Execution steps of Balancing Re-convergence on an example: a) Reducible Flow graph $G(V,E,v_I)$ where i represents the node v_i, b) DAG $G(V,E^f,v_I)$ with edge weights (=1 if not shown) after executing *Balance_path* procedure, c) weights on the back-edges after balancing loops, d) final balanced flow graph after inserting n-1 NOP states for edge with weight n

13.9 High-level BMC on EFSM

We present the flow of our approach for high-level BMC as shown in Figure 13.8. Given an EFSM Model M and a property P, we perform a series of property preserving transformations (Sections 13.8). After that we perform control state reachability on the transformed model (Section 13.7.1). Using the reachability information, we generate simplification constraints on-the-fly at each unroll depth k (Section 13.7.2). These simplification constraints are used by the expression simplifier (described next in Section 13.9.1) during unrolling to reduce the formula. These constraints are also used to improve the search on the translated problem. We also apply incremental learning technique (described next in Section 13.9.2), i.e., re-use of theory lemmas in high-level BMC framework. We present expression simplification and incremental learning in the following sections.

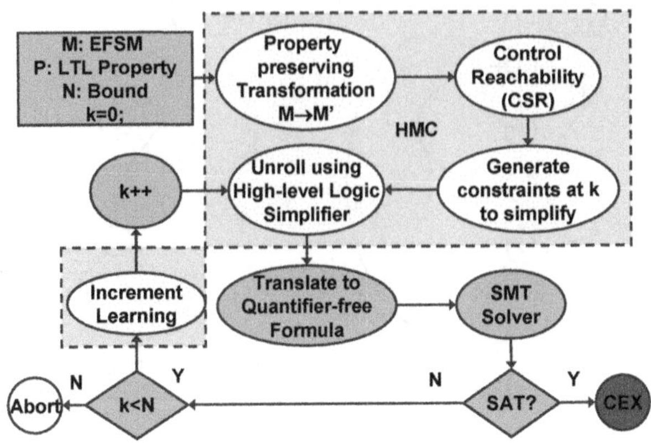

Fig. 13.8: Accelerated High-level BMC (HMC)

13.9.1 Expression Simplifier

High-level expressions in our framework include Boolean expressions *bool-expr* and term expressions *term-expr*. Boolean expressions are used to express Boolean values *true* or *false*, Boolean variables (*bool-var*), propositional connectives ($\vee,\wedge,\neg$) relational operators ($<,>,\geq,\leq==$) between term expressions, and *uninterpreted predicates* (*UP*). Term expressions are used to express integer values (*integer-const*) and real values (*real-const*), integer variables (*integer-var*) and real variables (*real-vars*), linear arithmetic with addition (+) and multiplication (*) with *integet-const* and *real-const*, *uninterpreted funtions* (UF), *if-then-else* (ITE), *read* and *write* to model memories. To model behaviour of a sequential system, we also have a *next* operator to express the next state behavior of each state variable.

Our high-level design description is represented in a semi-canonical form using an expression simplifier. The simplifier re-writes expressions using local and recursive transformations in order to remove structural and multi-level functionally redundant expressions, similar to simplifications proposed for Boolean logic [42, 43] and also for first order logic [123]. Our expression simplifier has a "compose" operator [45], that can be applied to unroll a high-level transition relation and obtain on-the-fly expression simplification; thereby achieving simplification not only within each time frame but also across time frames during unrolling of the transition relation in BMC.

13.9.2 Incremental Learning in High-level BMC

Learning from overlapping instances of propositional formulas [45, 153, 154] as discussed in Chapter 5 are found to be useful in Boolean SAT-based BMC [45, 46, 153]. We use incremental learning of theory lemmas across time-frames, and found this technique to be equally beneficial in the context of high-level BMC.

13.10 Experiments

We have experimented with an HLS tool called Cyber [54], along with our SAT-based BMC tool *VeriSol* [53] and an SMT-based BMC framework (See Figure 13.8) [62]. These tools have state-of-the-art high-level synthesis and verification algorithms, respectively. We conducted our experiments on Red Hat Enterprise Linux 3 with 2.8 GHz Xeon CPU and 4GB RAM. First, in a controlled setup, we evaluated the effectiveness of various HLS parameters in guiding verification. In the second and third experiments, we evaluated the effectiveness of our learning and simplification on industrial designs using proposed EFSM transformations in SAT-based and SMT-based BMC frameworks, respectively.

13.10.1 Controlled Case Study

We investigated the effect of reuse of functional units and registers, modeling of memories, number of control steps, and slicing on the performance of SAT-based BMC approaches. We experimented on a FIFO design described in a proprietary *SystemC*-like language. The FIFO is modeled as an array in memory with address width 7 and data width 7, with a read and write pointer. To make the model interesting, we non-deterministically chose to enqueue the write pointer value or not. For verification purpose, we chose two correctness properties:

- P1: *FIFO gets full eventually* (with FIFO length=24).
- P2: *A state is reachable eventually where the data at the beginning of the queue is 50 less than the current write pointer* (with FIFO length=128).

For both of these properties, a standard static COI reduction is performed to derive the Boolean-level verification model. The results are shown in Table 13.2.

Table 13.2: Impact of HLS parameters on Verification

S	#C	# FU	MEM/REG/ SLICE	P	# MUX	# REG	VER Result (sec, depth, EMM?)
SA	1	4	REG	1	891	1165	243s, D=25
SB	1	4	SLICE	1	93	109	135s, D=25
SC	3	2	REG	2	4398	5748	143s, D=150
SD	3	5	REG	2	4341	5748	52s, D=150
SE	1	5	REG	2	4314	5741	64s, D=52
SF	3	2	MEM: R1,W1	2	177	116	33s, D=150, EMM
SG	3	5	MEM: R1,W1	2	126	116	30s, D=150, EMM
SH	1	5	MEM: R2,W1	2	93	109	10s, D=51, EMM

Columns 1-4 show synthesis results under various synthesis strategies (combination of various available options in HLS) (S) denoted by *SA—SG* in Column 1. Specifically, Column 2 shows the number of control steps (#C), Column 3 shows the number of different and total functional units (#FU) used based on the resources available, Column 4 shows modeling and property specific slicing options — where REG denotes the use of a register file, MEM denotes the use of a memory module with indicated number of read ports (R#) and write ports (W#), and SLICE denotes the use of property-specific slicing performed by HLS to remove the storage elements that are not relevant to the property. (Note, a static COI reduction to derive a Boolean model typically cannot remove the storage elements.) Column 5 shows the property (P) checked by the verification tool. Columns 6 and 7 report the number of muxes and registers, respectively, in the Boolean verification model derived from the synthesized design. Column 8 reports the time taken (in seconds) to find the witness for the properties on the verification model obtained from the synthesized design and the witness length (D); EMM denotes the use of efficient memory modeling in BMC.

Discussion

Comparing the synthesis strategies *SA* and *SB* on property 1 shows that slicing away the irrelevant storage elements by HLS using strategy *SB,* is more effective in reducing the model size compared to strategy *SA*, as indicated by the reduced register and mux counts, This, in turn, improves the performance of SAT-based BMC.

The strategy *SC* on property 2 shows how an increased use of muxes due to resource limitations (*#FU*=2) can degrade the performance of BMC, as compared to strategies *SD* and *SE* with *#FU*=5.

An interesting observation regarding strategies *SD* and *SE* is that although the witness length and the number of muxes are larger using the strategy *SD*, the time taken to find the witness is less as compared to that using the strategy *SE*. In strategy *SD*, we forced the HLS tool to use 3 control steps (in addition to the reset state) while in the strategy *SE* we let the HLS tool minimize the number of control steps. The forced strategy *SD* leads to synthesis of a model where the operations are partitioned across various control steps, and the FUs binding to them are not shared. For such a model, several unrolled BMC instances can be re-solved quickly during the circuit simplification stage or the preprocessing stage of a SAT solver, thereby leading to a better overall performance than strategy *SE*.

In strategy *SF*, unlike strategies *SC—SE*, the HLS tool uses memory modules with one read and write port instead of register files. We see a significant reduction in the verification model size. Using BMC with EMM on this model, we see a significant improvement in performance. After increasing the resources available for HLS to *#FU*=5 as in strategy *SG*, we see a predictable improvement in the performance. Again using strategy *SH* with fewer control steps and increased memory ports, we see further improvements. Unlike strategy *SE*, the improvements using strategy *SH* are due to:

- availability of a larger number of memory access ports eliminates the reuse and hence, reduces number of muxes, and
- reduction in sequential depth (due to fewer control steps) helps in reducing the size of the SAT problem.

13.10.2 Experiments on Industry Software *bc-1.06*

We experimented on a public benchmark *bc-1.06*, a C program for an arbitrary precision calculator with interactive execution of statements. This has a known *array bound access* bug (checked as an error-label reachability property). The bug is that the initialization of an array is done with a bound of another array (most likely, during copy-paste operation). Using our program verification tool *F-soft* [56], we first generated an EFSM model *M* with 36 control states and 24 state variables. The data path elements include 10 *adders*, 106 *if-then-else*, 52 *constant multipliers*, 11 *inequalities*, and 49 *equalities*. The corresponding flow graph has two loops, with 4 and 8 nodes (control states) respectively. We also used statically generated invariants [204] to provide block specific invariants.

We performed controlled experiments to evaluate the role of various accelerators discussed in improving the performance of high-level BMC. We used our *difference logic solver SLICE* [132] in the backend. We modified the solver to support incremental learning across time-frames. We translated conservatively each BMC problem instance into a difference logic problem.

(A precise translation would have been to a UTVPI – Unit Two Variables Per Inequality – problem.) For understanding the effectiveness of our methods, a conservative translation suffices as long as we do not get false negatives; it was not an issue for this example.

We used a 500s time limit for each high-level BMC run. We present the results in Table 13.3. We experimented on three EFSM models *M, M' and M''*. Model *M* is the original model without any proposed transformations. Model *M'* is the model obtained from *M* using the procedure described in Section 13.8.3 (*Balancing re-convergence without loops* and *Collapsing NOP states*). Model *M''* is obtained from *M'* using the procedure described in Section 13.8.4 (*Balancing re-convergence with loops*). Column 1 shows the loop sizes in each of the models for loops L_1 and L_2; the number of control states (including inserted NOP states); the results of control reachability on each of the models i.e., either saturation depth or max loop period *N*; maximum size of the reachable set of control states over all depths $R_{max}=max_d|R(d)|$; and maximum size of the reachable set of *non-NOP* control states over all depths, $R^{-}_{max}=max_d|R^{-}(d)|$. Column 2 presents various learning and simplification strategies denoted as follows:

- A for Expression Simplification (ES),
- B for Incremental Learning (IL) combined with A,
- C for Unreachable Block Constraint (UBC) combined with B,
- D for Reachable Block Constraint (RBC) combined with C,
- E for Forward Reachable Block Constraint (FRBC) combined with D,
- F for Backward Reachable Block Constraint (BRBC) combined with E,
- G for Block Specific Invariants (BSI) combined with F.

Column 3 shows number of calls (*#HS*) made to the high-level solver when the expression simplifier cannot reduce the problem to a tautology. Column 4 shows the depth *d* reached by high-level BMC under a given time limit ('*' denote time-out). Column 5 shows the time taken (in seconds) to find the witness; *TO* denotes that time-out occurred. Column 6 shows whether a witness was found in the given time limit; if so, the witness length is equal to *d*.

Table 13.3: Comparison of high-level BMC accelerators

Model	Strategy	#HS	D	sec	W?
Original M $\|L_1\|= 4$, $\|L_2\|=8$, #ctrl state=36 $R_{max}=36$, $R^{-}_{max}=33$ Saturation at d=13	A: ES	16	17*	TO	N
	B: A+ IL	26	27*	TO	N
	C:B + UBC	41	64*	TO	N
	D:C + RBC	26	49*	TO	N
	E:D+FRBC	26	49*	TO	N
	F:E+BRBC	28	51*	TO	N
	G: F + BSI	28	51*	TO	N
M': M+Balanced Non-Loop paths + collapsed NOP states $\|L_1\|= 4$, $\|L_2\|=6$, #ctrl state=34 $R_{max}=4$, $R^{-}_{max}=3$, Max loop period, N=6	B	28	29*	TO	N
	C	62	143	426	Y
	D	62	143	159	Y
	E	62	143	159	Y
	F	62	143	120	Y
	G	62	143	65	Y
M'':M'+Loop Balanced $\|L_1\|= 6$, $\|L_2\|=6$, #ctrl state=36 $R_{max}=3$, $R^{-}_{max}=2$, N=6	F	32	205	19	Y
	G	32	205	22	Y

Discussion

Note that fewer calls (#HS) made to the SMT solver directly translates into performance improvement, as the expression simplifier structurally solves the remaining *d*-#HS SMT problems more efficiently. We discuss the effect of various learning scheme in improving the structural simplifications. CSR on Model *M* saturates at depth 13 with 36 control states. Although *UBC* allows deeper search with fewer solver calls, the simplification scope is very limited due to a large set *R(d)*. This also prevents other simplification strategies from being useful. As shown in Column 6, none of the strategies is able to find the witness in the given time limit. When we apply the procedure *Balance_path* with the procedure *collapsing of NOP states* on Model *M*, we obtain a model with *M'* with 34 control states with reduced loop size $|L_2|$. CSR on *M'* does not saturate, and has $max_d|R(d)|=4$ and $max_d|R^{-}(d)|=3$. This increases the scope of simplification significantly. As shown in Column 6, all simplification strategies *C—G* are able to find the witness in the given time limit. Except for *FRBC*, all simplification strategies seem useful in reducing the search time; though only *UBC* can reduce the number of calls to the high-level solver as shown in Column 3. *BSI* constraints, added on-the-fly, are also found to be useful. Note, although strategy *B* with only incremental learning does not find the witness, it still helps to search deeper compared to strategy *A*.

By applying our loop balancing procedure on the model *M'*, we obtain a model *M''* with matching loop lengths of 6, and the total number of control states of 36. We added two NOP-states in the back-edge of loop L_1 to get a loop length of 6. CSR on *M''* has $max_d|R(d)|=3$ and $max_d|R^{--}(d)|=2$, further increasing the scope of simplification as indicated by a decreased number of calls to the high-level solver. This is indicated by the reduced solve time (=19s) using strategy *F*, although there is a small performance degradation with strategy *G*. Not surprisingly, the witness length has gone up to 205. Overall, we see progressive and cumulative improvements with various learning techniques and strategies.

Comparison with Boolean-level BMC

To compare with Boolean-level BMC, we used our state-of-the-art Boolean-level BMC framework *VeriSol* [53] on a Boolean-level translation of the model M (with 654 latches, 6K gates) to witness the bug, and used an identical experimental setup as discussed. Note, like in [56], we add high-level information such as *mutual exclusion constraint (MEC)* and *backward reachable block constraints (BRBC)* in the transition relation beforehand. Thus, all these constraints get included in every unrolled BMC instance automatically, unlike the proposed approach here, where only the relevant constraints are added to a BMC instance. The Boolean-level BMC is able to find a witness at depth 143 in 723s. Not surprisingly, the number of instances solved by structural simplification is merely 15, while 128 calls are made to the SAT-solver. Thus, a reduced scope of simplification can greatly affect the performance of BMC, further supporting the case for synthesizing "BMC friendly" models [57, 58].

13.10.3 Experiments on Industry Embedded System Software

To demonstrate the importance of EFSM transformations on verification using SMT-based BMC, we experimented on an industry example of embedded system software written in C (17K lines of code). We considered 6 properties corresponding to assertability of a control condition. We used 500s time limit on each property.

The EFSM model *M* has 259 control states and 149 state (term) variables. The data path elements include 45 *adders*, 987 *if-then-else*, 394 *constant multipliers*, 53 *inequalities*, 501 *equalities* and 36 *un-interpreted functions*. Control state reachability on *M* saturated at depth 84. After transforming *M* after balancing re-convergence of paths, we obtained a model *M'* with 439 control states. We ran BMC for 500s on each of reachability properties P1-P6 on:

- Method I: Model *M* with strategy *B* (i.e., using only structural simplification and incremental learning),
- Method II: Model *M* with strategy *F*, (i.e., with all simplification constraints UBC, RBC, MEC, FRBC, and BRBC), and
- Method III: Transformed Model *M'* with strategy *F*.

We present our results in Table 13.4. Column 1 gives the property checked; Column 2-4 give BMC depth (*d*) reached (* denote depth at time out), time taken (in sec; *TO* denote time-out) and whether witness (W?) was found (*Y/N*) respectively for combination (I). Similarly, Columns 5-7 and 8-10 present information for combinations (II) and (III) respectively. The results clearly show that strategy *F* is more effective on model *M'* compared to *M*, though at increased witness depth.

Table 13.4: Evaluating the effectiveness of EFSM transformation in SMT-based BMC

P	I: Strategy B on M			II: Strategy F on M			III: Strategy F on M'		
	D	sec	W?	D	sec	W?	D	sec	W?
P1	9*	TO	N	38*	TO	N	41	<1	Y
P2	9*	TO	N	41*	TO	N	44	<1	Y
P3	9*	TO	N	43*	TO	N	92	156	Y
P4	9*	TO	N	30	188	Y	94	151	Y
P5	9*	TO	N	21	6	Y	60	4	Y
P6	9*	TO	N	31	164	Y	70	22	Y

13.10.4 Experiments on System-level Model

The EFSM transformations are also helpful in deriving improved Boolean-level models for SAT-based BMC. To demonstrate the importance of EFSM transformations on verification using SAT-based BMC, we experimented on an industry system-level model written in a proprietary SystemC-like input language of our HLS tool. The synthesized EFSM model has 143 control states, and the synthesized gate level model has 415 latches, 49 primary inputs and 22K 2-input gates. We have 143 properties corresponding to reachability of each control state.

On the gate-level model, SAT-based BMC was able to obtain witnesses for 30 properties in 3 hours, reaching a depth of 44 requiring 481 Mb. Control state reachability on the EFSM model saturated at depth 27 with all 143 control states. Since this would severely limit the scope of on-the-fly simplification and learning, we applied property-preserving EFSM transformations. On the transformed EFSM model, CSR does not saturate.

By using BMC on the gate-level model corresponding to the transformed EFSM model with strategy *F*, we were able to reach depth 60 in 3 hours, verifying 30 properties, requiring 489 Mb. Though we were not able to witness any additional properties, BMC on the transformed model is able to reach depth 44 in only 1033s requiring 83Mb. Clearly, our proposed EFSM transformations also improve SAT-based BMC performance significantly. Further, using the simple test of unreachability, we could prove 19 properties unreachable out of 113 (=143-30) properties. Currently, we are investigating other transformations that can be exploited to resolve the remaining 94 (=113-19) properties.

13.11 Summary and Future work

A Design-For-Verification methodology, i.e., exporting designer's intent to verification tools, has been quite effective in improving verification efforts. We discussed a new paradigm Synthesis-For-Verification (SFV) which involves synthesizing "verification-aware" designs that are more suitable for functional verification. Such a paradigm further improves verification efforts by separating the design optimized for performance, area, and power from the design optimized for correctness, thereby reducing the verification burden. SFV methodology should be applied along with DFV methodology to obtain the full benefit of verification efforts. Note, this SFV paradigm requires support only from automated synthesis approaches, e.g. High-level Synthesis (HLS), and can be easily automated. This is in contrast to a DFV methodology, which requires designers' reliable insights.

As part of SFV methodology, we discussed the following guidelines within the existing infrastructure of HLS to generate verification friendly models (specific to SAT-based FV tools) from the system-level design.

- Optimize area under mux-constraint, i.e., reduce sharing of functional units and registers
- Reduce the number of control steps with disjoint partitioning of the system behavior, under mux-constraint
- Choose functional units from a "verification friendly" library
- Reuse operations as much as possible
- Use external memories instead of register arrays
- Perform property-preserving transformations to obtain sliced EFSM models
- Balance re-converging paths in EFSM models

Continuing our push for more scalability, our future works are very much directed along SFV paradigm, as shown in Figure 13.9. We strongly believe that such an integrated system of synthesis with verification is very beneficial in improving the verification efforts.

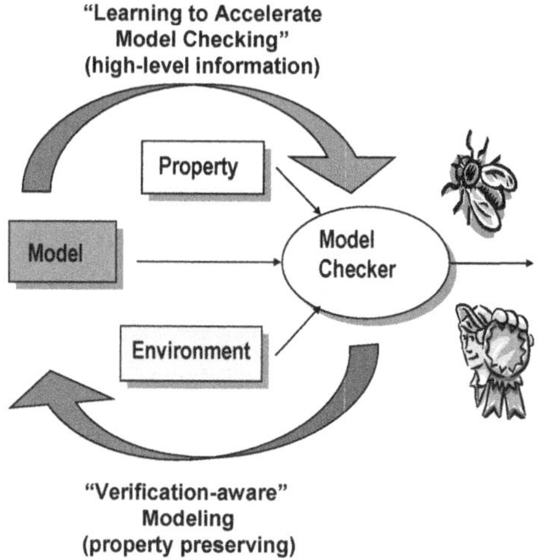

Fig. 13.9: Integrating HLS with verification at high-level

13.12 Notes

The subject material described in this chapter are based on the authors' previous works [57], [62] © ACM 2006, [58] © 2007 IEEE.

References

[1] A. Silburt, A. Evans, G. Vrckovik, M. Diufrensne, and T. Brown, "Functional Verification of ASICs in Silicon Intensive Systems," in *DesignCon98 On-Chip System Design Conference*, 1998.

[2] "EETimes. http://www.eetimes.com."

[3] "CoverMeter: http://www.synopsys.com."

[4] "VN-cover. http://www.transeda.com."

[5] S. Devadas, A. Ghosh, and K. Keutzer, "An Observability-based Code Coverage Metric for Functional Simulation," in *Proceedings of ICCAD*, 1996.

[6] R. Grinwald, E. Harel, M. Orgad, S. Ur, and A. Ziv, "User Defined Coverage: A Tool Supported Methodology for Design Verification," in *Proceedings of DAC*, 1998.

[7] R. Ho and M. Horowitz, "Validation Coverage Analysis for Complex Digital Designs," in *Proceedings of ICCAD*, 1996.

[8] R. C. Ho, C. H. Yang, M. Horowitz, and D. L. Dill, "Architectural Validation for Processors," in *Proceedings of ISCA*, 1995.

[9] Y. Hoskote, D. Moundanos, and J. Abraham, "Automatic Extraction of the Control Flow Machine and Application to Evaluating Coverage of Verification Vectors," in *Proceedings of ICCD*, 1995.

[10] D. Moundanos, J. Abraham, and Y. Hoskote, "A Unified Framework for Design Validation and Manufacturing Test," in *Proceedings of ITC*, 1996.

[11] B. Chen, M. Yamazaki, and M. Fujita, "Bug Identification of a Real Chip Design by Symbolic Model Checking," in *Proceedings of EDAC*, 1994.

[12] E. M. Clarke, O. Grumberg, and D. Peled, *Model Checking*: MIT Press, 1999.

[13] R. S. Boyer and J. S. Moore, *A Computational Logic Handbook.* New York: Academic Press, 1988.

[14] M. Gordon, "A Proof Generating System for Higher-order Logic," in *VLSI Specification, Verification and Synthesis*, 1988.

[15] D. L. Dill, A. J. Drexler, A. Hu, and C. H. Yang, "Protocol Verification as a Hardware Design Aid," in *Proceedings on ICCD*, 1992.

[16] G. J. Holzmann, "The Model Checker SPIN," in *IEEE Transactions on Software Engineering*, 1997.

[17] K. L. McMillan, *Symbolic Model Checking: An Approach to the State Explosion Problem*: Kluwer Academic Publishers, 1993.

[18] S. B. Akers, "Binary Decision Diagrams," in *IEEE Transactions on Computers*, 1978.

[19] R. E. Bryant, "Graph-based Algorithms for Boolean Function Manipulation," in *IEEE Transactions on Computers*, vol. C-35(8), 1986.

[20] J. R. Burch, E. M. Clarke, K. McMillan, and D. L. Dill, "Symbolic Model Checking: 10^{20} States and Beyond," in *Information and Computation*, 1992.

[21] O. Coudert, C. Berthet, and J. C. Madre, "Verification of Sequential Machines based on Symbolic Execution," in *Proceedings of the Workshop on Automatic Verification for Finite State Systems*, 1989.

[22] H. Touati, H. Savoj, B. Lin, R. Brayton, and A. L. Sangiovanni-Vincentelli, "Implicit State Enumneration of Finite State Machines using BDDs," in *Proceedings of ICCAD*, 1990.

[23] R. Ranjan, J. Sanghavi, R. Brayton, and A. L. Sangiovanni-Vincentelli, "High Performance BDD Package Based on Exploiting Memory Hierarchy," in *Proceedings of DAC*, 1996.

[24] A. Aziz, S. Tasiran, and R. Brayton, "BDD Variable Ordering for Interacting Finite State Machines," in *Proceedings of DAC*, 1994.

[25] S. J. Friedman and K. J. Supowit, "Finding the Optimal Variable Ordering for Binary Decision Diagrams," in *IEEE Transactions on Computers*, 1990.

[26] H. Fujii, G. Ootomo, and C. Hori, "Variable Ordering Methods for Ordered Binary Decision Diagrams," in *Proceedings of ICCAD*, 1993.

[27] R. Rudell, "Variable Ordering for Binary Decision Diagrams," in *Proceedings of ICCAD*, 1993.

[28] R. E. Bryant, "On the Complexity of VLSI Implementations and Graph Representations of Boolean Functions with Application to Integer Multiplication," in *IEEE Transactions on Computers*, 1991.

[29] B. Bollig and I. Wegener, "Improving the Variable Ordering of OBDDs is NP-Complete," in *IEEE Transactions on Computers*, 1996.

[30] J. Gergov and C. Meinel, "Efficient Boolean Manipulation with OBDD's can be Extended to FBDD's," in *IEEE Transactions on Computers*, 1994.

[31] S. Minato, "Zero-Suppressed BDDs for Set Manipulation in Combinational Problems," in *Proceedings of DAC*, 1993.

[32] A. Narayan, A. Jain, M. Fujita, and A. L. Sangiovanni-Vincentelli, "Partitioned-ROBDDs - A Compact, Canonical and Efficiently Manipulable Representation for Boolean Functions," in *Proceedings of ICCAD*, 1996.

[33] K. Ravi, K. L. McMillan, T. R. Shiple, and F. Somenzi, "Approximation and decomposition of Decision Diagrams," in *Proceedings of DAC*. San Francisco, 1998, pp. 445-450.

[34] L. Zhang and S. Malik, "The Quest for Efficient Boolean Satisfiability Solvers," in *Proceedings of CAV*, 2002.

[35] M. Davis, G. Logemann, and D. Loveland, "A Machine Program for Theorem Proving," *Communications of the ACM*, vol. 5, pp. 394-397, 1962.

[36] H. Zhang, "SATO: An Efficient Propositional Prover," in *Proceedings of International Conference on Automated Deduction*, vol. 1249, *LNAI*, 1997, pp. 272-275.

[37] J. P. Marques-Silva and K. A. Sakallah, "GRASP: A Search Algorithm for Propositional Satisfiability," in *IEEE Transactions on Computers*, vol. 48, pp. 506-521, 1999.

[38] M. Moskewicz, C. Madigan, Y. Zhao, L. Zhang, and S. Malik, "Chaff: Engineering an Efficient SAT Solver," in *Proceedings of DAC*, 2001.

[39] A. Kuehlmann, M. Ganai, and V. Paruthi, "Circuit-based Boolean Reasoning," in *Proceedings of DAC*, 2001.

[40] E. Goldberg and Y. Novikov, "BerkMin: A Fast and Robust SAT-Solver," in *Proceedings of DATE*, 2002.

[41] M. Ganai, L. Zhang, P. Ashar, and A. Gupta, "Combining Strengths of Circuit-based and CNF-based Algorithms for a High Performance SAT Solver," in *Proceedings of DAC*, 2002.

[42] M. Ganai and A. Kuehlmann, "On-the-Fly Compression of Logical Circuits," in *Proceedings of IWLS*, 2000.

[43] H. Andersen and H. Hulgard, "Boolean Expression Diagram," in *Proceedings of LICS*, 1997.

[44] A. Mishchenko, S. Chatterjee, and R. Brayton, "DAG-Aware AIG Rewriting: A Fresh Look at Combinational Logic Synthesis," in *Proceedings of DAC*, 2006.

[45] M. Ganai and A. Aziz, "Improved SAT-based Bounded Reachability Analysis," in *Proceedings of VLSI Design*, 2002.

[46] M. Ganai, A. Gupta, and P. Ashar, "Beyond Safety: Customized SAT-based Model Checking," in *Proceedings of DAC*, 2005.

[47] M. Ganai, A. Gupta, and P. Ashar, "Efficient SAT-based Unbounded Model Checking Using Circuit Cofactoring," in *Proceedings of ICCAD*, 2004.

[48] A. Gupta, M. Ganai, P. Ashar, and Z. Yang, "Iterative Abstraction using SAT-based BMC with Proof Analysis," in *Proceedings of ICCAD*, 2003.

[49] A. Gupta, M. Ganai, and P. Ashar, "Lazy Constraints and SAT Heurisitcs for Proof-based Abstraction," in *Proceedings of VLSI Design*, 2005.

[50] M. Ganai, A. Gupta, and P. Ashar, "Efficient Modeling of Embedded Memories in Bounded Model Checking," in *Proceedings of CAV*, 2004, pp. 440-452.

[51] M. Ganai, A. Gupta, and P. Ashar, "Verification of Embedded Memory Systems Using Efficient Memory Modeling," in *Proceedings of DATE*, 2005.

[52] M. Ganai and A. Gupta, "Efficient BMC for Multi-Clock Systems with Clocked Specifications," in *Proceedings of ASPDAC*, 2007.

[53] M. Ganai, A. Gupta, and P. Ashar, "DiVer: SAT-Based Model Checking Platform for Verifying Large Scale Systems," in *Proceedings of TACAS*, 2005.

[54] K. Wakabayashi, "Cyber: High Level Synthesis System from Software into ASIC," in *High Level VLSI Synthesis*, R. Camposano and W. Wolf, Eds.: Kluwer Academic Publishers, 1991, pp. 127-151.

[55] "CyberWorkBench. http://www.cyberworkbench.com/english/index.html."

[56] F. Ivancic, J. Yang, M. Ganai, A. Gupta, and P. Ashar, "Efficient SAT-based Bounded Model Checking for Software," in *Proceedings of ISOLA*, 2004.

[57] M. Ganai, A. Mukaiyama, A. Gupta, and K. Wakabayashi, "Another Dimension to High Level Synthesis: Verification," in *Proceedings of Workshop on Designing Correct Circuits*, 2006.

[58] M. Ganai, A. Mukaiyama, A. Gupta, and K. Wakabayashi, "Synthesizing "Verification-aware" Models: Why and How?," in *Proceedings of VLSI Design*, 2007.

[59] R. E. Bryant, S. K. Lahiri, and S. A. Seshia, "Modeling and Verifying Systems using a Logic of Counter Arithmetic with Lambda Expressions and Uninterpreted Functions," in *CAV*, 2002.

[60] A. Armando, J. Mantovani, and L. Platania, "Bounded Model Checking of Software using SMT Solvers instead of SAT solvers," in *Proceedings of SPIN Workshop on Model Checking software*, 2005.

[61] B. Alizadeh and Z. Navabi, "Word-level Symbolic Simulation in Processor Verification," in *Proceedings of IEE*, 2004.

[62] M. Ganai and A. Gupta, "Accelerating High-level Bounded Model Checking," in *Proceedings of ICCAD*, 2006.

[63] M. Ganai, "Ph.D. Thesis: Algorithms for Efficient State Space Search," The University of Texas at Austin, 2001.

[64] P.-H. Ho, T. Shiple, K. Harer, J. Kukula, R. Damiano, V. Bertacco, J. Taylor, and J. Long, "Smart simulation using collaborative formal and simulation engines," in *Proceedings of ICCAD*, 2000, pp. 120-126.

[65] J. Yuan, J. Shen, J. Abraham, and A. Aziz, "On Combining Formal and Informal Verification," in *Proceedings of CAV*, 1997.

[66] A. Biere, A. Cimatti, E. M. Clarke, and Y. Zhu, "Symbolic Model Checking without BDDs," in *Proceedings of TACAS*, vol. 1579, *LNCS*, 1999.

[67] M. Sheeran, S. Singh, and G. Stalmarck, "Checking Safety Properties using Induction and a SAT Solver," in *Proceedings of FMCAD*, 2000.

[68] A. Gupta, Z. Yang, P. Ashar, and A. Gupta, "SAT-based Image Computation with Application in Reachability Analysis," in *Proceedings of FMCAD*, 2000, pp. 354-371.

[69] H.-J. Kang and I.-C. Park, "SAT-based Unbounded Symbolic Model Checking," in *Proceedings of DAC*, 2003, pp. 840-843.

[70] S. Sheng and M. Hsiao, "Efficient Pre-image Computation using a Novel Success-driven Atpg," in *Proceedings of DATE*, 2003.

[71] K. McMillan, "Applying SAT methods in unbounded symbolic model checking," in *Proceedings of CAV*, 2002.

[72] D. Kroening and O. Strichman, "Efficient Computation of Recurrence Diameter," in *Proceedings of VMCAI*, 2003.

[73] A. Gupta, M. Ganai, C. Wang, Z. Yang, and P. Ashar, "Abstraction and BDDs Complement SAT-Based BMC in DiVer," in *Proceedings of CAV*, 2003, pp. 206-209.

[74] D. E. Long, "Model checking, abstraction and compositional verification," Carnegie Mellon University, 1993.

[75] E. M. Clarke, O. Grumberg, S. Jha, Y. Lu, and H. Veith, "Counterexample-guided Abstraction Refinement," in *Proceedings of CAV*, vol. 1855, *LNCS*, 2000, pp. 154-169.

[76] E. M. Clarke, A. Gupta, J. Kukula, and O. Strichman, "SAT based Abstraction-refinement using ILP and Machine Learning Techniques," in *Proceedings of CAV*, 2002.

[77] P. Chauhan, E. M. Clarke, J. Kukula, S. Sapra, H. Veith, and D. Wang, "Automated Abstraction Refinement for Model Checking Large State Spaces using SAT based Conflict Analysis," in *Proceedings of FMCAD*, 2002.

[78] K. McMillan and N. Amla, "Automatic Abstraction without Counterexamples," in *Proceedings of TACAS*, 2003.

[79] A. Aziz, V. Singhal, and R. Brayton, "Verifying Interacting Finite State Machine," Univ. of California, Berkeley, CA 94720 UCB/ERL M93/52, 1993.

[80] R. Schutten and Fitzpatrick, T, "Design for Verification," *http://www.synopsys.com/products/simulation/dfv_wp.html*, 2003.

[81] D. D. Gajksi, N. Dutt, S. Y.-L. Lin, and A. Wu, *High Level Synthesis: Introduction to Chip and System Design*: Kluwer Academic Publishers, 1992.

[82] "SystemC: http://www.SystemC.org."

[83] "Accellera. http://www.accellera.org."

[84] "SystemVerilog. http://www.systemverilog.org."

[85] H. Foster, A. Krolnik, and D. Lacey, *Assertion-based Design*: Kluwer Academic Publishers, 2003.

[86] R. P. Kurshan, *Computer-Aided Verification of Co-ordinating Processes: The Automata-Theoretic Approach*: Princeton University Press, 1994.

[87] A. Pnueli, "The Temporal Logic of Programs," in *Proceedings of IEEE Symposium on Foundation of Computer Science*, 1977.

[88] M. Ben-Ari, Z. Manna, and A. Pnueli, "The Temporal logic of Branching Time," in *Proceedings of POPL*, 1981.

[89] E. M. Clarke and E. A. Emerson, "Design and Synthesis of Synchronization Skeletons using Branching Time Temporal Logic," in *Proceedings of the Workshop on Logics of Programs*, 1982.

[90] B. Alpern and F. B. Schneider, "Defining Liveness," in *Information Processing Letters*, 1985.

[91] "Murphi: http://verify.stanford.edu/dill/murphi.html."

[92] "SPIN Formal Verification: http://spinroot.com/spin/whatisspin.html."

[93] A. Narayan, A. Isles, J. Jain, R. Brayton, and A. L. Sangiovanni-Vincentelli, "Reachability Analysis using Partitioned-ROBDDs," in *International Conference on Computer -Aided Design*, 1997, pp. 388-393.

[94] M. R. Garey and D. S. Johnson, *Computers and Intractability: A Guide to the Theory of NP-Completeness*: W. H. Freeman and Co., 1979.

[95] D. A. Plaisted and S. Greenbaum, "A Structure-preserving Clause Form Translation," *Journal of Symbolic Computation*, vol. 2, 1986.

[96] T. Larabee, "Test Pattern Generation Using Boolean Satisfiability," *IEEE Transactions on Computer-Aided Design*, 1992.

[97] R. Bayardo and R. Schrag, "Using CSP look-back Techniques to Solve Real-world SAT Instances," in *Proceedings of AAAI*, 1997.

[98] J. P. Marques-Silva, "The Impact of Branching Heuristics in Propositional Satisfiability Algorithms," in *Proceedings of Portuguese Conference on Artificial Intelligence*, 1999.

[99] J. N. Hooker and V. Vinay, "Branching Rules for Satisfiability," *Journal of Automated Reasoning*, 1995.

[100] A. Biere, "The Evolution from Limmat to Nanosat," Dept. of Computer Science, ETH Zurich 2004 444, 2004.

[101] C. M. Li, "Integrating Equivalency Reasoning into Davis-Putnam Procedure," in *Proceedings of AAAI*, 2000.

[102] P. Prosser, "Hybrid Algorithms for the Constraint Satisfaction Problem," *Computation Intelligence*, vol. 9, 1993.

[103] L. Zhang, C. Madigan, M. Moskewicz, and S. Malik, "Efficient Conflict Driven Learning in a Boolean Satisfiability Solver," in *Proceedings of the International Conference on Computer-Aided Design*, 2001.

[104] L. Zhang and S. Malik, "Validating SAT Solvers Using an Independent Resolution-Based Checker: Practical Implementations and Other Applications," in *Proceedings of Design, Automation, and Test in Europe*, 2003.

[105] E. Goldberg and Y. Novikov, "Verification of Proofs of Unsatisfiability for CNF Formulas," in *Proceedings of DATE*, vol. olde, 2003.

[106] "SAT Live!. http://www.satlive.org."

[107] J. Pilarski and G. Hu, "SAT with Partial Clauses and Back-Leaps," in *Proceedings of the Design Automation Conference*, 2002, pp. 743-746.

[108] A. Nadel, "Backtrack Seach Algorithms for Propositional Logic Satisfability: Review and Innovations," in *Master Thesis*: Hebrew University of Jerusalem, 2002.

[109] P. Bjesse and K. Claessen, "SAT-based Verification without State Space Traversal," in *Proceedings of FMCAD*, 2000.

[110] A. Cimatti, M. Pistore, M. Roveri, and R. Sebastiani, "Improving the Encoding of LTL Model Checking into SAT," in *Proceedings of VMCAI*, 2002.

[111] T. Latvala, A. Biere, K. Heljanko, and T. Junttila, "Simple Bounded LTL Model Checking," in *Proceedings of FMCAD*, 2004.

[112] K. McMillan, "Interpolation and SAT-based Model Checking," in *Proceedings of CAV*, 2003, pp. 1-13.

[113] "Liveness Manifesto. Beyond Safety International Workshop, 2004."

[114] P. Wolper, M. Y. Vardi, and A. P. Sistla, "Reasoning about Infinite Computation Paths," in *Proceedings of Symposium on FCS*, 1983.

[115] R. Gerth, D. Peled, M. Y. Vardi, and P. Wolper, "Simple on-the-fly Automatic Verification of Linear Temporal Logic," *Protocol Specification, Testing, and Verification*, 1995.

[116] M. Daniele, F. Giunchiglia, and M. Y. Vardi, "Improved Automata Generation for Linear Time Temporal Logic," in *Proceedings of CAV*, 1999.

[117] F. Somenzi and R. Bloem, "Efficient Buchi Automata from LTL formulae," in *Proceedings of CAV*, 2000.

[118] A. Biere, C. Artho, and V. Schuppan, "Liveness Checking as Safety Checking," in *Proceedings of FMICS*, 2002.

[119] M. Awedh and F. Somenzi, "Proving more properties with Bounded Model Checking," in *Proceedings of CAV*, 2004.

[120] C. Eisner, D. Fishman, J. Havlicek, A. McIsaac, and D. V. Campenhout, "The Definition of a Temporal Clock Operator," in *Proceedings of ICLAP*, 2003.

[121] M. K. Iyer, G. Parthasarathy, and K.-T. Cheng, "SATORI An Efficient Sequential SAT Solver for Circuits," in *Proceedings of ICCAD*, 2003.

[122] M. R. Prasad, A. Biere, and A. Gupta, "A survery of recent advances in SAT-based formal verification.," in *STTT 7(2)*, 2005.

[123] J.-C. Filliatre, S. Owre, H. RueB, and N. Shankar, "ICS: Integrated Canonizer and Solver," in *Proceedings of CAV*, 2001.

[124] O. Strichman, "On Solving Presburger and linear inequalities with SAT," in *Proceedings of FMCAD*, Nov 2002.

[125] O. Strichman, S. A. Seshia, and R. E. Bryant, "Deciding Separation Formulas with SAT," in *Proceedings of CAV*, July 2002.

[126] A. Armando, C. Castellini, E. Giunchiglia, M. Idini, and M. Maratea, "TSAT++: An Open Platform for Satisfiability Modulo Theories," in *Proceedings of Pragmatics of Decision Procedures in Automated Resonings (PDPAR'04)*, 2004.

[127] R. E. Bryant, S. K. Lahiri, and S. A. Seshia, "Deciding CLU logic formulas via Boolean and peudo-Boolean encodings," in *Workshop on Constraints in Formal Verification*, 2002.

[128] C. Barrett, D. L. Dill, and J. Levitt, "Validity Checking for Combination of Theories with Equality," in *Proceedings of FMCAD*, 1996.

[129] M. Bozzano, R. Bruttomesso, A. Cimatti, T. Junttila, P. V. Rossum, M. Schulz, and R. Sebastiani, "The MathSAT 3 System," in *Proceedings of CADE*, 2005.

[130] R. Nieuwenhuis and A. Oliveras, "DPLL(T) with Exhaustive Theory Propogation and its Application to Difference Logic," in *CAV*, 2005.

[131] M. Ganai, M. Talupur, and A. Gupta, "SDSAT: Tight Integration of Small Domain Encoding and Lazy Approaches in a Separation Logic Solver," in *Proceedings of TACAS*, 2006.

[132] C. Wang, F. Ivancic, M. Ganai, and A. Gupta, "Deciding Separation Logic Formulae with SAT by Incremental Negative Cycle Elimination," in *Proceedings of Logic for Programming, Artificial Intelligence and Reasoning*, 2005.

[133] L. D. Moura, H. RueB, and M. Sorea, "Lazy Theorem Proving for Bounded Model Checking over Infinite Domains," in *Proceedings of CADE*, 2002.

[134] D. Brand, "Verification of Large Synthesized Designs," in *Proceedings of ICCAD*, 1993.

[135] A. Kuehlmann and F. Krohm, "Equivalence Checking using Cuts and Heaps," in *Proceedings of DAC*, 1997.

[136] A. Kuehlmann, V. Paruthi, F. Krohm, and M. Ganai, "Robust Boolean Reasoning for Equivalence Checking and Functional Property Verification," in *IEEE Transactions on Computer-Aided Design of Integrated Circuits and Systems*, 2002.

[137] A. Kuehlmann, "Dynamic transition relation simplification for bounded property checking," in *Proceedings of ICCAD*, 2004.

[138] S.-W. Jeong, B. Plessier, G. D. Hachtel, and F. Somenzi, "Extended BDD's: Trading off Canonicity for Structure in Verfication Algorithms," in *Proceedings of ICCAD*, 1991.

[139] G. L. Smith, R. J. Bahnsen, and H. Halliwell, "Boolean Comparision of Hardware and Flowcharts," *IBM Journal of Research and Development*, vol. 26, pp. 106-116, 1982.

[140] K. Brace, R. Rudell, and R. E. Bryant, "Efficient Implementation of a BDD Package," in *Proceedings of DAC*, 1990.

[141] A. Kuehlmann, A. Srinivasan, and D. P. Lapotin, "Verity - a Formal Verification Program for Custom CMOS Circuits," *IBM Journal of Research and Development*, vol. 39, pp. 149-165, 1995.

[142] C. L. Berman and L. H. Trevillyan, "Functional Comparison of Logic Designs for VLSI Circuits.," in *Proceedings of ICCAD*, 1989.

[143] M. Schulz, E. Trischler, and T. Sarfert, "SOCRATES: A highly efficient ATPG System," in *IEEE Transactions on Computer-Aided Design of Integrated Circuits and Systems*, vol. 7, pp. 126-137, 1988.

[144] H. Fujiwara and T. Shimono, "On the Acceleration of Test Generation Algorithms," in *IEEE Transactions on Computers*, vol. C-32, pp. 265-272, 1983.

[145] P. Goel, "An Implicit Enumeration Algorithm to Generate Tests for Combinational Circuits," in *IEEE Transactions on Computers*, vol. C-30, pp. 215-222, 1981.

[146] A. Gupta, M. Ganai, C. Wang, Z. Yang, and P. Ashar, "Learning from BDDs in SAT-based Bounded Model Checking," in *Proceedings of DAC*, 2003.

[147] A. Gupta, A. Gupta, Z. Yang, and P. Ashar, "Dynamic Detection and Removal of Inactive Clauses in SAT with Application in Image Computation," in *Proceedings of DAC*, 2001.

[148] M. N. Velev, "Benchmark Suites. http://www.ece.cmu.edu/~mvelev," October 2000.

[149] B. Selman, H. Kautz, and D. McAllestor, "Ten Challenges in Propositional Reasoning and Search," in *Proceedings of IJCAI*, 1997.

[150] J. Warners and H. v. Maaren, "A Two Phase Algorithm for Solving a Class of Hard Satisfiability Problem," in *Operation Research Letters*, 1999.

[151] N. Een and N. Sorrensson, "Translating Pseudo-Boolean Constraints into SAT," *JSAT*, 2006.

[152] M. Ganai, A. Gupta, Z. Yang, and P. Ashar, "Efficient Distributed SAT and Distributed SAT-based Bounded Model Checking," in *Proceedings of CHARME*, 2003.

[153] O. Strichman, "Pruning Techniques for the SAT-based Bounded Model Checking," in *Proceedings of TACAS*, 2001.

[154] J. Whittemore, J. Kim, and K. Sakallah, "SATIRE: A New Incremental Satisfiability Engine," in *Proceedings of DAC*, 2001.

[155] R. E. Bryant, "Symbolic Simulation - Techniques and Applications," in *Proceedings of DAC*, 1990.

[156] C.-J. H. Seger and R. E. Bryant, "Formal Verification by Symbolic Evaluation of Partially-Ordered Trajectories," in *Technical Report TR-93-8, University of British Columbia, Dept. of Computer Science*, 1993.

[157] V. Bertacco, M. Damiani, and S. Quer, "Cycle-based Symbolic Simulation of Gate-level Synchronous Circuits," in *Proceedings of DAC*, 1999.

[158] C. Wilson, D. L. Dill, and R. E. Bryant, "Symbolic Simulation with Approximate Values," in *Proceedings of FMCAD*, 2000.

[159] P. Williams, A. Biere, E. M. Clarke, and A. Gupta, "Combining Decision Diagrams and SAT Procedures for Efficient Symbolic Model Checking," in *Proceedings of CAV*, vol. 1855, *LNCS*, 2000, pp. 124-138.

[160] F. Corno, M. S. Reorda, and G. Squillero, "RT-Level ITC99 Benchmarks and First ATPG Results," in *IEEE Design & Test of Computers*, 2000.

[161] R. Brayton and others, "VIS: A System for Verification and Synthesis," in *Proceedings of CAV*, 1996.

[162] O. Strichman, "Tuning SAT Checkers for Bounded Model Checking," in *Proceedings of International Conference on Computer-Aided Verification*, 2000.

[163] G. Anderson, P. Bjesse, B. Cook, and Z. Hanna, "A Proof Engine Approach to Solving Combinational Design Automation Problems," in *Proceedings of the Design Automation Conference*, 2002, pp. 725-730.

[164] F. Somenzi, "CUDD: University of Colorado Decision Diagram Package, http://vlsi.colorado.edu/~fabio/CUDD/," 1998.

[165] M. Abramovici, M. A. Breuer, and A. D. Friedman, *Digital Systems Testing and Testable Design*: Computer Science Press, 1990.

[166] "The VIS Home Page. http://www-cad.eecs.berkeley.edu/Respep/Research/vis/."

[167] Y. Zhao, "Accelerating Boolean Satisfiability through Application Specific Processing.," Ph.D. Thesis. Princeton, 2001.

[168] A. Hasegawa, H. Matsuoka, and K. Nakanishi, "Clustering Software for Linux-Based HPC," *NEC Research & Development*, vol. vol 44, No. 1, pp. 60-63, 2003.

[169] "VASG: VHDL Analysis and Standardization Group. http://www.vhdl.org/vasg."

[170] B. W. Wah, G.-J. Li, and C. F. Yu, "Multiprocessing of Combinational Search Problems," in *IEEE computer*, pp. 93-108, 1985.

[171] H. Zhang, M. P. Bonacina, and J. Hsiang, "PSATO: a Distributed Propositional Prover and its Application to Quasigroup Problems," in *Journal of Symbolic Computation*, 1996.

[172] C. Powley, C. Fergusion, and R. Korf, "Parallel Heuristic Search: Two Approaches," in *Parallel Algorithms for Machine Intelligence and Vision*, V. Kumar, P. S. Gopalakrishnan, and L. N. Kanal, Eds. New York: Springer-Verlag, 1990.

[173] B. Jurkowiak, C. M. Li, and G. Utard, "Parallelizing Satz Using Dynamic Workload Balancing," in *Workshop on Theory and Applications of Satisfiability Testing*, 2001, pp. 205-211.

[174] M. Boehm and E. Speckenmeyer, "A Fast Parallel SAT-solver - Efficient Workload Balancing," in *Third International Symposium on Artificial Intelligence and Mathematics*. Fort Lauderdale, Florida, 1994.

[175] U. Stern and D. L. Dill, "Parallelizing the Murphi Verifier," in *Proceedings of Computer Aided Verification*, 1997.

[176] T. Heyman, D. Geist, O. Grumberg, and A. Schuster, "Achieving Scalability in Parallel Reachability Analysis of Very Large Circuits," in *Proceedings of Computer-Aided Verification*, vol. 1855: Springer, 2000, pp. 20-35.

[177] M. Ganai, A. Gupta, Z. Yang, and P. Ashar, "Efficient Distributed SAT and SAT-based Distributed Bounded Model Checking," *Journal on Software Tools for Technology Transfer*, 2006.

[178] P. A. Abdulla, P. Bjesse, and N. Een, "Symbolic Reachability Analysis based on SAT-Solvers," in *Proceedings of TACAS*, 2000.

[179] J. R. Burch and D. L. Dill, "Automatic Verification of Pipelined Microprocessor Control," in *Proceedings of CAV*, 1994.

[180] M. N. Velev, R. E. Bryant, and A. Jain, "Efficient Modeling of Memory Arrays in Symbolic Simulation," in *Proceedings of ComputerAided Verification*, O. Grumberg, Ed., 1997, pp. 388-399.

[181] R. E. Bryant, S. German, and M. N. Velev, "Processor Verification Using Efficient Reductions of the Logic of Uninterpreted Functions to Propositional Logic," in *Proceedings of Computer-Aided Verification*, N. Halbwachs and D. Peled, Eds.: Springer-Verlag, 1999, pp. 470-482.

[182] M. N. Velev, "Automatic Abstraction of Memories in the Formal Verification of Superscalar Microprocessors," in *Proceedings of TACAS*, 2001.

[183] G. Semeraro, G. Magklis, R. Balasubramonian, D. H. Albonesi, S. Dwarkadas, and M. L. Scott, "Energy-Efficient Processor Design Using Multi-Clocks with Dynamic Voltage and Frequency Scaling," in *Proceedings of HPCA*, 2002.

[184] P. Bjesse and J. Kukula, "Automatic Generalized Phase Abstraction for Formal Verification," in *Proceedings of ICCAD*, 2005.

[185] J. Baumgartner, T. Heyman, V. Singhal, and A. Aziz, "An Abstraction Algorithm for the Verification of Level-sensitive Latch-based Netlists," in *Proceedings of FMSD*, 2003.

[186] A. Albright and A. Hu, "Register Transformations with Multiple Clock Domains," in *Proceedings of CHARME*, 2000.

[187] J. Baumgartner, A. Tripp, A. Aziz, V. Singhal, and F. Andersen, "An Abstraction Algorithm for the Generalized C-slow Designs," in *Proceedings of CAV*, 2000.

[188] E. M. Clarke, D. Kroening, and K. Yorav, "Specifying and Verifying Systems with Multiple Clocks," in *Proceedings of ICCD*, 2003.

[189] "OpenCores: http://www.opencores.org."

[190] H. Cho, G. D. Hachtel, E. Macii, B. Plessier, and F. Somenzi, "Algorithms for approximate FSM traversal based on state space decomposition," in *IEEE Transactions on Computer-Aided Design of Integrated Circuits and Systems*, vol. 15(12), pp. 1465-1478, 1996.

[191] I.-H. Moon, J.-Y. Jang, G. D. Hachtel, F. Somenzi, C. Pixley, and J. Yuan, "Approximate reachability don't-cares for CTL model checking," in *Proceedings of ICCAD*. San Jose, 1998.

[192] G. Cabodi, S. Nocco, and S. Quer, "Improving SAT-based Bounded Model Checking by BDD-based Approximate Traversal," in *Proceedings of DATE*, 2003.

[193] "Wikipedia: http://en.wikipedia.org."

[194] E. M. Clarke, O. Grumberg, and D. E. Long, "Model Checking and Abstraction," in *Proceedings of Principles of Programming Languages*, 1992.

[195] R. Majumdar, A. Henzinger, R. Jhala, and G. Sutre, "Lazy Abstraction," in *Proceedings of Principles of Programming Languages*, 2002.

[196] N. Amla and K. McMillan, "A Hybrid of Counterexample-based and Proof-based Abstraction," in *Proceedings of FMCAD*, 2004.

[197] A. Gupta, A. A. Bayazit, and Y. Mahajan, "Verification Languages," in *The industrial Information Technology Handbook*, 2005.

[198] J. Jain, A. Narayan, M. Fujita, and A. L. Sangiovanni-Vincentelli, "A Survey of Techniques for Formal Verification of Combinational Circuits," in *Proceedings of ICCD*, 1997.

[199] P. Ashar, S. Bhattacharya, A. Raghunathan, and A. Mukaiyama, "Verification of RTL Generated from Scheduled Behaviour in a High-level Synthesis Flow," in *Proceedings of ICCAD*, 1998.

[200] D. C. Ku and G. D. Micheli, *High Level Synthesis of ASICs Under Timing and Synchronization Constraints*: Kluwer Academic Publishers, 1992.

[201] J. M. Chang and M. Pedram, *Power Optimization and Synthesis at Behavorial and System Levels using Formal Methods*: Kluwer Academic Publishers, 1999.

[202] A. V. Aho, R. Sethi, and J. D. Ullman, *Compilers: Principles, Techniques and Tools*: Addison-Wesley Publishing Company, 1988.

[203] B. Korel, I. Singh, L. Tahat, and B. Vaysburg, "Slicing of State-based Models," in *Proceedings of ICSM*, 2003.

[204] H. Jain, F. Ivancic, A. Gupta, I. Shlyakhter, and C. Wang, "Using Statically Computed Invariants inside the Predicate Abstraction and Refinement Loop," in *Proceedings of CAV*, 2006.

Glossary

AIG AND/INVERTER Graph. Nodes in the graph represent 2-input AND gate and edge attributes represent INVERSION.

AP Atomic Proposition.

ASM Accumulated Sufficient Abstract Model. *Sufficient Model* that does not have witness to the property of length less than or equal to *d*. It is obtained by accumulating latch reasons in the *unsatisfiable core* at all depth up to *d*.

Backleap A step in SAT process that involves *backtracking* to a previous decision level, not necessarily the immediate one.

Backtrack A step in SAT process, involving undoing all the implications and restoring the decision stack to some previous decision level.

BCP Boolean Constraint Propagation. A step in SAT, when *unit propagation rule* is applied on *unit clauses* until no such clauses exist or conflict is detected.

BDD Binary Decision Diagrams. It is a canonical representation for Boolean functions.

Block Control state in an EFSM.

BMC Bounded Model Checking. Checks for bounded length counter-examples to LTL properties.

BRBC Backward Reachable Block Constraint. It is used to specify the constraint, which *blocks* in the immediate previous step are *CSR reachable* to a given *block*, explicitly.

BSI Block Static Invariant. It is used to specify block-specific invariants.

CEGAR Counter-Example Guided Abstraction and Refinement.

CEX Counter Example to a property. CEX trace refers to a sequence of input vectors, when applied to the design,

	results in the violation of the correctness property. See *Witness*.
Circuit Constraint	Gate clauses capturing input-output relation of circuit gates.
Circuit Truths	Property of circuit. Often, refers to non-local relationship between circuit nodes.
ckt_node	An *AIG* node, to represent latches, primary inputs, and circuit gates.
Clause Replication	Applying *circuit truths* such as in *BMC unrolling*.
Clause Reuse	Reuse of *(conflict) learnt clause* in another SAT problem, overlapping with the one where the clause was learnt.
Clock Constraints	External constraints on the input clock signals of the unrolled transition relations. Used during BMC of multi-clock system.
Clock Generator	Circuit to generate *enabling clock signals* for modeling multi-clock system into a single clock synchronous model.
Clock Schedules	*Unrollings* of *BMC* that correspond to some change in *global clock* state due to some input clock tick. See also *Event queue semantics*.
Clock tick	Time instance when the clock is active.
Clocked Property	Formal specification with a clock operator @ on temporal formula. Used to specify properties for a multi-clock system.
CNF	Conjunctive Normal Form. It consists of the logical AND of one or more clauses.
COI	Cone Of Influence. Used to denote transitive fanin cone of a given circuit node.
COI Reduction	Reduction of model by removing logic not in the transitive fanin cone of the signals of interest.
Collapsing Condition	Scenario in STG wherein all outgoing states from a state are NOP state, and none of them are needed for reachability check.
Conflict Clauses	Conflict-driven learnt clauses during *diagnosis* phase of SAT solver.
Conflict-driven Learning	Process of learning clause by applying *resolution rule* on the implication graph during conflict analysis
Conflicting Clause	Clause whose all literals are false.
Contributing state	Any control state in EFSM that can affect a variable of interest.
Contributing transition	Any transition in EFSM that can affect a variable of interest.
CSR	Control State Reachability. Static unrolling of STG, ignoring the conditional guards between the control states, to compute a set of control states or *blocks* reachable statically at every depth.

CSR Reachable	A control state or block is *CSR reachable* at *d* if there is a static transition path to the control state from the initial control state in an EFSM.
CTL	Computational Tree Logic. It is a branching-time logic and used for specifying branching properties on states.
Customization	A verification method that applies dedicated optimization such as partitioning and incremental formulation using the property specific structures.
d-BMC	BMC distributed over a network of workstations.
Deterministic EFSM	EFSM where all outgoing transition from a control state are mutually exclusive.
DFV	Design For Verification methodology. Leverages designer's intent and knowledge to strengthen verification efforts.
DPLL	Davis Putnam Logemann-Loveland. Complete backtrack-based search process for deciding the satisfiability of propositional logic (SAT).
d-SAT	Boolean reasoning (SAT) on a problem distributed over a network of workstations.
Eager Constraint	1-literal clause. DPLL SAT process applies *unit propagation rule* eagerly on such constraint.
EFSM	Extended Finite State Machine, with finite control states and conditional guards on the transition between the control states, and update on datapaths. Datapaths are typically expressed at RTL level. See *Funtional RTL*.
Embedded Memory	Memory in design, with explicit memory interface logic for reading and writing. Data access to memory involves read or write cycles.
EMM	Efficient Memory Modeling. This modeling preserves the data forwarding *memory semantics* without explicitly representing each memory bit.
EMM Constraint	Constraints capturing data forwarding *memory semantics*.
Enabling Clock Signals	Signal that has *true* value when the corresponding clock has the active phase or edge.
Enabling Transition	Transition from one control state of EFSM to another state as a function of control states, primary inputs and datapath values.
Event queue Semantics	Queue of clocks events, each corresponding to a change in *global clock state* due to some clock tick. It is used to identify redundant unrolling in BMC for *multi-clock system*.
Explicit Unrolling	Unrollings where circuit structure at each depth are replicas of the transition relations.
Fanin cone	See *COI*
FIFO	First-In-First-Out
Flip-flops	DFF. Also, known as a register. A state element whose output D value is same as input value Q at the active edge (*posedge* or *negedge*) of the clock input.

Flow Graph	Directed graph with an entry node.
FRBC	Forward Reachable Block Constraint. It is used to specify the constraint, which blocks are *CSR reachable* from a given block in one step, explicitly.
FSM	Finite State Machine.
Functional Hashing	A hashing scheme where the circuit sub-structure is rewritten into a canonical, i.e., functionally equivalent, substructure to allow reusing, hence, reducing number of circuit nodes in the cone of logic.
Functional RTL	EFSM after *allocation* and *scheduling* steps in HLS.
FV	Formal Verification
Gated Clocks	Enable signals to flip-flops and latches that are driven by output of gated logic. It is used to disable the update function of state elements non-periodically.
Global Clock	Clock in synchronous single clock design.
Global Clock State	Product state of the clock generators that drive the input clocks of a multi-clock system. For modeling multi-clock synchronous system, for example, we use tuple $<t^i,c_1^i,c_2^i>$ to denote a global state of two clocks, where at t^i c_1^i or c_2^i changed from previous values.
hESS	hybrid Exclusive Select Signal representation for EMM.
hITE	hybrid If-Then-Else representation for EMM.
HLS	High Level Synthesis. It is also referred to as *Behavioral Synthesis*. It is a process of generating a structural RTL model from a system-level model (SLM).
Hybrid SAT	Uses CNF and circuit representation of the problem. It combines the strengths of CNF-based and circuit-based SAT solvers.
Implicit Unrolling	Unrolling where circuit structure at each depth reuses circuit structure of previous time frame. Note, circuit structure at each depth may not be a structural replica of the transition relation.
Initial Constraint	Constraints corresponding to the initial state of the state elements.
Interface Constraints	Constraints corresponding to the next state circuit node of time frame *i* and current state node of time frame *i*+1.
ITE	If-Then-Else operator. $ITE(s,x,y) \equiv$ if (s) then x else y.
Latches	D-Latch. A state element whose output value Q is same as input value D, as long as enable signal is active; otherwise, Q retains the previous value. Latch enable signals are usually *level-sensitive*.
Lazy Constraint	2-literal clauses whose resolution would give 1-literal clause. Goal is to prevent DPLL SAT solver to apply *unit Propagation rule* eagerly.

LIFO	Last-In-First-Out. Also, known as *stack*.
Liveness Property	Specifies that something good will eventually happen, always.
Localization Reduction	Static abstraction technique. It converts the latches that are farther in the dependency closure of the property signals, into *pseudo-primary inputs*.
Loop Checks	BMC satisfiability problem to check if a state s_k have a transition T to state s_l along the path, where $l < k$. Often, denoted as ${}_lL_k = T(s_k, s_l)$
Loop Saturation	A loop is said to saturate in CSR at depth d, if all the nodes in the loop are *CSR reachable* at any depth greater than or equal to d.
LTL	Linear Temporal Logic. Properties are specified for paths.
MEC	Mutual Exclusive Constraint. It is used to specify the constraint, that only one block is reachable active at any given BMC depth, explicitly.
Memory Semantics	Data read from a memory location is the same as the last data written to that location. It is used for *abstract interpretation* of embedded memory, without modeling each memory bit explicitly.
Multi-clock system	A system with multiple clocks, with arbitrary frequencies and ratios, gated clocks, multiple phases and latches.
Multi-level Hashing	A *functional hashing* scheme where a signature of local substructure, obtained by looking back several levels, is computed dynamically during AIG construction, and then a pre-computed table is used to obtain the unique mapping from the signature to the canonical representation of sub-structure.
Negedge Clock	Negative edge of the clock that causes flip-flops to change state.
NNF	Non-negated Form. LTL in NNF allows negation only on atomic propositions.
NOP	No Operation State. Control state in EFSM with no update transition and single outgoing *enabling transition*.
NOW	Network Of Workstations. Often refers to workstations connected by ubiquitous Ethernet LAN.
On-the-fly Simplification	Reduction of circuit-structure during graph construction.
PAP	Performance, Area, Power (HLS design constraints).
PBA	Proof-Based Abstraction. This method generates bounded depth property-preserving reduced-size abstract model.
PBIA	Proof-based Iterative Abstraction. PBA method applied iteratively, until the model size converges.

PI	Primary Inputs.
PI logic	Gates in the transitive fanout of the primary inputs that are exclusively driven by primary inputs.
Posedge Clock	Positive edge of the clock that causes flip-flops to change state.
PPI	Pseudo-Primary Inputs. Output signals of those registers whose next state logic are removed and the signals behave as primary inputs in the resulting abstract model.
prop_tree_node	Property tree expression with node in the sub-expressions are of types AND, NOT, propositional atom, and X (next LTL operator).
Pseudo-satisfied Clause	Satisfied clause with only one literal true. The true literal is assigned at level higher than the assignment level of the remaining literals. This happens during dynamic addition of clause.
PSL	Property Specification Language. A formal specification language, standardized by Accellera, for specifying temporal logic, sequential regular expression and clocked operators.
RAM	Random Access Memory.
RBC	Reachable Block Constraint. Used to specify the constraints that at least one block is CSR reachable.
Reachability Property	Specifies that along some path, a state is reachable eventually where something good happens.
Reconverging Paths	Two paths in a flow graph are reconverging if the paths are different but the end nodes are same.
Recurrence Length	Repeat length of the *global clock state*, in terms of tick of the global clock, with frequency equal to LCM of all the clocks in multi-clock systems.
Reduced AIG	AIG obtained using functional hashing.
Register File	Array of registers. Read and write to registers typically are done in one clock cycle.
Relevant Clause	Conflict clause that is relevant to current SAT process. Relevancy is typically determined by how useful it was during past conflict analysis.
Repetition Period	Refers to the time period between the repeating *global clock states*. See *Recurrence Length.*
Resolution Rule	$(x + y_1 + \ldots + y_n)\ (-x + y_1 + \ldots + y_n)$ *imply* $(y_1 + \ldots + y_n)$
RTL	Register Transfer Level. Hardware representation using registers and finite datapaths.
Safety Property	Specifies that something bad does not happen, ever.
SAT	The Boolean SATtisfiability problem. It consists of determining a satisfying variable assignment for a Boolean function, or determining that no such assignment exists.

Satisfied Clause Clause with at least one *true* literal at current decision level of SAT.

SFV Synthesis For Verification Paradigm. Goal is to generate verification-friendly models automatically by guiding the use of design entities in HLS, to make model checking tasks simpler.

SINK Control state with no update transition and no outgoing state transition.

SLM System Level Model.

SM Sufficient Model. Model abstracted by converting latches, not involved in *unsatisfiable core* analyzed at depth *d*, into pseudo-primary inputs. Such model does not have witness to the property of length *d*.

SMT Satisfiability-Modulo Theories.

STG State Transition Graph. It is used to represent EFSM where nodes in the graph correspond to control state and edges correspond to the enable transition between control states and datapath updates.

Structural Hashing Technique to identify isomorphic sub-structure during circuit construction, and re-use existing structure to avoid redundancy.

Structural RTL Circuit implementation from functional RTL, using *binding* and *controller synthesis* steps in HLS.

Time Frame A given unfolding of transition relation for a sequential design.

Topology Cognizant Process awareness about topology of networked process, so that the inter-process message transfers occur between them selectively, and are not broadcast, to reduce communication overhead.

Transition Relation Combinational logic of the design, including next-state logic for the latches, as well as output logic for the external constraints, either due to the property, or enforced by the designers.

Two-input (level) Hashing A hashing scheme where isomorphic substructure detection is restricted to the use of commutative property of AND node. In other words, the method looks back one more level up to the immediate two inputs.

UBC Unreachable Block Constraint. Blocks that not *CSR reachable* at a given *d* depth, can be used to simplify the unrolled transition relation by constant folding.

uckt_node AIG node in unrolled transition relation.

UMC Unbounded Model Checking.

Unclocked Property Formal specification without clock operator @. Used to specify properties in single clocked designs.

Unit clause	Clause whose one literal is unassigned and remaining literals are false.
Unit Propagation Rule	A rule in DPLL SAT procedure, wherein, unassigned literal of unit clause is set to true, and the effect is propagated to other clauses, which can become *unit clause* thereafter.
Unrolling	Unfolding of the transition relation of a sequential design.
Unsatisfiable core	A subset of clauses of a given unsatisfiable problem that is sufficient for proving unsatisfiablity.
Unsatisfied Clause	Clause with at least two unassigned literals, and no literal assigned true.
Update Transition	Transition from a datapath state to another, as a function of control state, primary inputs and previous datapath values.
Verilog	A C like Hardware Description Language.
VHDL	Very High Speed Integrated Circuit Hardware Description Language
Witness	A witness trace to a sequence of input vectors, when applied to the design, results in the satisfaction of the given property. *See* CEX.

Index

C

D

E

About the Authors

Dr. Malay K Ganai is a Senior Research Staff Member at NEC Labs America, Princeton, NJ. He received B.Tech in Electrical Engineering from IIT Kanpur, India in 1992; MS degree and Ph.D. degree in Electrical and Computer Engineering from the University of Texas at Austin, USA in 1998 and 2001, respectively. After graduation, from 1992 to 1995, Malay worked at Larsen & Toubro on embedded system design projects. In 1995, he joined Cadence Design Systems, India where he worked until 1997 on various projects involving high-level synthesis and Signal Processing Workbench (SPW). While at UT Austin as a graduate student, he designed a semi-formal verification tool, *SIVA* that combines simulation with symbolic algorithms for efficient state space search. Since 2001, he is working with NEC Labs America in System LSI and Software Verification group on devising various SAT-based scalable model checking algorithms. His current research includes SAT solvers, SMT solvers, SAT-based and SMT-based formal verification methods, semi-formal verification and verification methodology. He has published several papers in the related subjects. He is the principal architect and author of *VeriSol* (formerly *DiVer*), SAT-based Formal Verification Platform, currently available commercially as C-level Property Checker in NEC's CyberWorkBench tool.

Dr. Aarti Gupta is a Department Head (Verification Group) at NEC Labs America, Princeton. She received a B.Tech. degree in Electrical Engineering from IIT Delhi, India; and a Ph.D. degree in Computer Science from Carnegie Mellon University. For the last decade she has worked at NEC Labs, in the areas of hardware and software verification. Her research interests are in formal verification methods, design and systems verification, automatic constraint solvers, electronic design automation, and more recently in software program analysis. She has published numerous papers and book chapters on these topics. She serves/has served on the technical program committees of the IEEE International Conference on Computer Aided Design (ICCAD), the IEEE Design Automation and Test, Europe (DATE), the International Conference on Computer Aided Verification (CAV), the IEEE International Conference on VLSI Design, the IEEE/ACM International Conference on Formal Methods in Computer Aided Design (FMCAD), and the International Conference on Theory and Applications of Satisfiability Testing (SAT). She is also serving as an Associate Editor for Formal Methods in System Design, Springer.